Critical Issues and Bold Visions for Science Education

Bold Visions in Educational Research

VOLUME 63

The titles published in this series are listed at *brill.com/bver*

Critical Issues and Bold Visions for Science Education

The Road Ahead

Edited by

Lynn A. Bryan and Kenneth Tobin

BRILL
SENSE

LEIDEN | BOSTON

All chapters in this book have undergone peer review.

The Library of Congress Cataloging-in-Publication Data is available online at http://catalog.loc.gov

ISSN 1879-4262
ISBN 978-90-04-38964-9 (paperback)
ISBN 978-90-04-38965-6 (hardback)
ISBN 978-90-04-38966-3 (e-book)

This book is printed on acid-free paper and produced in a sustainable manner.

CONTENTS

FIGURES AND TABLES

FIGURES

TABLES

KENNETH TOBIN AND LYNN BRYAN

1. BOLD VISIONS FOR SCIENCE EDUCATION

A Metalogue

ABSTRACT

We participate in a metalogue to generate and expand several critical issues in science education. The issues concern science teacher education, science curriculum and its relationships to sociopolitical issues and connections to everyday life, and the role of the Institutional Review Board (IRB) in design and conduct of research in science education. Each of these central areas of science education is complex, and we identify and elaborate on our experiences as science educators to probe deeper into an expanding vortex of possibilities. Scientism and crypto-positivism emerge as pervasive referents for practices that maintain the status quo. Whereas our metalogue describes new pathways to consider in a quest for a bold new, relevant and sustainable science education, our purpose is to prepare readers for what is to follow in a book with 16 chapters authored by 31 scholars from 11 countries.

Keywords: metalogue, science education reform, science curriculum, science teacher education, human subjects

We first learned about metalogue (Bateson, 1972) from the work of Wolff-Michael Roth, Campbell McRobbie, and Keith Lucas (1998), who used metalogue to show co-authors' complementary and divergent perspectives in co-authored publications. Roth and his colleagues built on an earlier publication by Marie Larochelle, Jacques Désautels, and François Ruel (1995), which also adapted Bateson's work on metalogue for research in science education. Since these pioneering works, science education scholars have used metalogue extensively to represent difference in terms of polyphonia and polysemia (e.g., Roth & Tobin, 2001). Typically, a metalogue is a dialogue between two or more people who have shared experiences, which are the focus of the dialogue. The metalogue is synergistic in that it allows issues being addressed to expand in terms of their scope, and perhaps depth of understanding shared by participants. The purpose is not to converge toward consensus, but to expand, increasing understandings in depth, while maintaining difference in all of the ways differences are felt, understood, and expressed. Roth often maintained that for science education researchers, metalogue could be used to "ratchet up" theory that emerged from research. Often theory was grounded in

 DOI:10.1163/9789004389663_001

hermeneutic-phenomenological inquiry and ratcheting up was to afford theoretical generalizability (Eisenhart, 2009). In the following sections, we use metalogue descriptively, to identify, describe, and expand upon selected issues in science education that we have experienced during decades of participation in science education, and regard as central to transformations that will improve a status quo that has been maintained for far too long. By way of metalogue, we introduce what is to follow in the book and engage in preliminary dialogues about complex issues and associated bold visions we consider essential foundations for much-needed transformation of science education, and possibly survival of life on Earth, as we know it. The issues we address in the following sections can be regarded as appetizers, to whet the appetites of readers – preparing them for what comes in the chapters to follow.

In the critical issues and bold visions offered by the authors of this volume, we can see how unequivocally complex teaching is and the vast knowledge, skills, and dispositions it takes to be able to teach science, for example knowing how to:

- offer the type of support that many learners need to produce the kinds of knowledge that afford individual satisfaction and fulfilment, and the achievement of personal and professional goals;
- frame science instruction in ways that students can build an understanding of the social, political, cultural and economic implications of critical issues and grand challenges;
- address science teaching, learning and research challenges in increasingly diverse contexts that are shaped by migration, multiculturalism and diversity;
- engage in science in meaningful ways, with an awareness of and attention to racialized and marginalized students;
- foster creativity in science to address scientific challenges from classroom-based tasks to the larger societal issues;
- negotiate and survive the high demand for emotion work required on a daily basis in classrooms;
- guide young people in developing a sociocultural critical awareness that illuminates the situation in which they are living and builds different and better options for their futures;
- resist the "performative dominance" of schools today and instead teach with democratic and collaborative pedagogical approaches in the face of a climate of high-stakes testing and encourage students to experience what it is like to encounter new and unexplained scientific phenomena;
- confront unexamined biases about students and consider how one's beliefs shape perceptions about learners and the value of science education for students with disabilities;
- consider and address power relations embedded in all learning environments, including informal learning and out-of-school learning contexts; and

- help students make the connections needed to build integrated knowledge structures, to develop explanations and reasoning for why and how phenomena occur rather than learning bits and pieces of disconnected factual information.

THE PLACE OF EDUCATION IN TEACHER CERTIFICATION

Lynn: Against this backdrop, consider a recent development in the state in which I live and devote my career to preparing science teachers. The Indiana State Assembly recently considered a bill to allow up to 10% of K-12 classroom teachers to be unlicensed. In other words, they need no preparation beyond a college degree to become a teacher (ultimately, the bill died). This would have amounted to more than 6,000 teachers in Indiana schools who would have had no preparation for the realities of the classroom, and a large concentration of these teachers would likely have been high school teachers in the STEM disciplines.

Ken: When I came to the US in 1987 there was a strong movement in many states to require only science credentials to teach science. The movement was backed by the Chamber of Commerce in Florida, where I was then situated. Furthermore, within Florida State University, where I worked, there was a parallel endeavor to move science education to the College of Arts and Sciences and to minimize education coursework. So, what you are experiencing in Indiana seems to be a continuing effort that is most likely grounded in acceptance of many of the tenets of positivism, scientism, and failure to understand and value education as a knowledge system.

I should point out that when I taught in England, in 1970, it was also permissible for graduates with a degree in science to teach high school science, mathematics, and computer science. Furthermore, graduates with a science degree also could become department heads and school level administrators. My point in mentioning this is to emphasize that what you have experienced in Indiana is historically constituted.

Lynn: Yes, but in the mid-eighties, there was a distinct shift in thought about teaching as both an activity and a profession. Until then, teaching was viewed from a technicist/behaviorist perspective. Since that time, we have amassed a significant body of empirical research that has emphasized not only the complexity of knowledge for teaching (e.g., Lee Shulman's canonical publications in 1986 and 1987), but also the complexity of developing that knowledge. The image of teachers as reflective practitioners emerged as a more productive way to view teacher learning and teacher thinking. In many ways, at least in the U.S., the accountability measures that have materialized since the No Child Left Behind Act of 2001 (No Child Left Behind [NCLB], 2002) reflect the complexity of teaching (however, that is not to say that the assessments are designed or applied well).

Thus, the irony is astounding – the governing bodies that have proposed this solution to the current teacher shortage are the same governing bodies that require prospective teachers to demonstrate proficiency through these accountability measures, via multiple (and expensive) written tests and practice-based evidence of a myriad national, state, and local content standards and developmental/pedagogy standards. For prospective science teachers, this amounts to more than three dozen developmental/pedagogy standards alone.

As a science teacher educator and parent, I am concerned about prospective and practicing teachers developing into effective educators who are able to consider, reflect upon and grow in their ability to tackle some relatively provocative issues regarding their students and the teaching and learning of science. How has the regard, respect, and value for the teaching profession fallen to such a low level that large decision making bodies think it would be okay to put someone in a classroom without any preparation?

SCIENCE EDUCATION REFORM AND SCIENCE CURRICULUM

Ken: For at least the last 60 years there have been calls from educators to teach science in ways that connect to the interests and lives of learners, and in a larger sense, to society. Despite the calls, and despite the seemingly exponential expansion of canonical science knowledge, school curricula have a familiar ring. It is not just how science is taught that hasn't changed all that much, it is also what is being taught. So, I begin with some questions about science curricula. If we consider the nearby public schools as a reference point for these questions – in what ways have science curricula been responsive to changing times? To what extent are science curricula related to the lives of learners? How do present-day science curricula connect with some of the major challenges facing the planet, including climate change, mass extinctions, and sustainability? Finally, in what ways do science curricula address Anthropocene?

Lynn: Indeed, in the U.S., nearly every report on the status of K-12 science education, every speech made by a politician who references science education, every new set of science standards that is released, every reform effort over the past 60 years has implored educators to make science more interesting and relevant to the lives of learners, and for good reason. In fact, in the time since I was a physics teacher, there have been numerous waves of reform efforts to do just that, including the inquiry movement spurred by the 1996 *National Science Education Standards* (National Research Council, 1996), to more recent efforts to integrate the STEM disciplines – specifically, integrating engineering design into science and mathematics learning through problem and project-based pedagogies (NGSS Lead States, 2013). But overall, have science curricula been responsive to changing times? Changed to connect with some of the major challenges facing the planet? If we consider the nearby public schools as a reference point for these questions, I would say that at best changes are seen anecdotally, but by and large not much has changed in science classrooms in the decades since I was a student sitting in those small, cold metal seats with an attached laminate desk top.

Ken: I see little value in learning science that is separated from everyday life and the grand challenges associated with Anthropocene, sustaining Earth and the galaxy in which it coexists, mass extinctions, climate change; and sustaining healthy lives. Is it possible to forge a science curriculum that addresses criteria like those I just listed, in the community – in much the way that Amat Arnau describes in his chapter? Can science education draw on science as it is constituted in the local communities and associated local ecosystems? Can the curriculum and many of its teachers be from the community, and in the way Ashley learned about making cheese, can students learn about science as it is interconnected with aspects of everyday life?

Lynn: It is especially important for science teachers and science teacher educators to engage students in examining the nature of science and to not shy away from teaching about controverial issues, particularly in the current sociopolitical climate of growing resistance and hostility to science – e.g., efforts to muzzle communication between government agencies and the public, elimination of national science advisory positions, severe cuts to government-funded research, persistent regulatory rollbacks. In her chapter within this book, Chantal Pouliot (2019) describes the various ways in which she engages in sociopolitical discussions and warns against "epistemological anesthesia":

> Our sociopolitical engagement can aim to tackle citizens' epistemological anesthesia and support individuals and collectives in the process of empowerment. Among other possible actions, we can lead people to conceive of knowledge relationships (Charlot, 1997) that involve experts, corporations and citizens (users, consumers, and producers of lay knowledge) (Bencze & Alsop, 2014; Pierce, 2013). The decision to strive towards this form of critique can be driven by *thinking about the consequences of our inaction* (for example, a continuation of inequality and the trivialization of citizen struggles related to health and environmental issues) as opposed to the consequences of our action. (p. 228 [emphasis added])

I understand and do not want to minimize the fact that there are risks involved in teaching science through controversial issues. I have heard of situations in which teachers have been advised to refer a student to the school counselor if the student becomes too emotional about an issue in class. Not only do teachers feel pressure by administration, parents and/or the community to either ignore or tip-toe around controversial issues, but there are also legal considerations in some cases. Nonetheless, there are consequences of inaction. Chantal's chapter is a poignant reminder that we cannot/should not ignore the sociopolitical sphere that intersects with our research in and teaching of science. We need to allow students opportunities to inquire into the origins of a controversial issue and understand the structures of competing arguments; engage in reasoned, informed debates across the myriad differences within and between societies; develop a critical consciousness in preparation for the responsibilities of active citizenship.

Ken: I acknowledge that some high schools already provide externships that provide workplace learning experiences in numerous professions. We might do more of this in the high school and, I argue, we might also plan activities in the everyday world so that pre-K through PhD students gain science learning opportunities through coparticipation in selected aspects of everyday life. I view a curriculum oriented toward science in social life, as aligned with emancipatory interests, including choice. To accommodate a curriculum within these components, there would necessarily be a need for radical reform of present-day practices in schools.

Lynn: Two of your questions above are particularly salient: "To what extent are science curricula related to the lives of learners? How do present-day science curricula connect with some of the major challenges facing the planet?" We should consider how "curriculum" seems to have taken on a new meaning in the last few decades, during an era of educational standards. I say this not to incite a debate about the concept of standards, but to reflect on what "curriculum" has come to mean in U.S. K-12 classrooms. The standards-based movement of the early 1990s, wedded with the accountability mandates of NCLB (2002), has limited teachers' decision-making power in the classroom. Teachers are obligated to "cover" a litany of topics in their science courses, often without any guidance for contextualizing the teaching and learning of science in a meaningful way. Teachers will tell you that there is little time to teach more than what is listed in the standards, as there is so much to "cover". Deprofessionalization of teaching is one of the biggest casualties of the standards movement.

Often stated in the opening description of most standards documents are repeated claims that "standards are not a curriculum." However, this is not what I have witnessed in classrooms over the last couple of decades. Teachers feel very little, if any, latitude to deviate from the standards or the science textbooks that have been produced to correlate with standards. Under scrutiny of administrators (who themselves are under the scrutiny of the state and community) and others who oversee accountability measures in schools, teachers have increasingly less autonomy to make decisions about curriculum and often have to submit evidence of how they address each standard. I have been in many K-12 science classrooms in which the "standard of the day" was written in the corner of the chalkboard, should a principal or department head walk by and check which standard the teacher is addressing in instruction that day. I recount in a recent book chapter that I co-authored with Lucy Avraamidou (Avraamidou & Bryan, 2018) a brief anecdote that captures the intensity and unintended outcomes of the standards/accountability movement that we are witnessing. In brief, at the end of an academic semester, my teenage daughter came home from school with a study guide for her science final exam. The study guide for this class was literally a word-for-word list of the state science standards. The standards had become the curriculum. So back to you your question about science curricula addressing Anthropocene – if Anthropocene is not in the science standards, it is not addressed in the science classroom.

Ken: The tendency to see Western Modern Science as a superior way of knowing is widespread. What this means in everyday life depends a great deal on context. In his chapter on cheese making, Mitch Bleier makes it clear that there is an enormous knowledge base associated with making cheese that is safe to consume, tasty, and with characteristics that appeal to a group of consumers who consider cheese to be a field in which they are expert. Ashley learned about cheese by being with experts as they practiced their profession. That is, she worked alongside of people who knew and could do. As Ashley learned, her expertise grew, and so did her enthusiasm and positively oriented values concerning the cheese making knowledge domain.

As I think about Ashley's knowledge in an epistemological sense, it appeals as a good example of the crisis of representation (Greene, 1994) – what Ashley knows and can do extend beyond what can be represented. On the one hand, the bottom line about Ashley's knowledge is in the cheese that she produces. This is a simple enough benchmark – but a complex array of the characteristics for that benchmark needs to be assembled in constructing a response to a question like – how good is the cheese that Ashley makes? And, of course, there are many more ways to think about cheese world.

Making cheese involves doing – as well as talking and writing. Although I have never made cheese, my assumption is that it involves the use of all of the senses. Also, values, interests, emotions, intuition, and other examples of affective knowledge are important. In making these points about Bleier's chapter I want to illuminate potential areas for transformation of science education. These relate to what is considered to be science and other curriculum issues; the places in which the curriculum is enacted; arrangement of teachers and students for learning (i.e., learning environments); and how can and should learning be assessed viably?

Lynn: To address the relevance of science education to the lives of learners, we should also consider values and priorities. Who determines what is relevant to science learners? Who sets the agenda? Science education reform recommendations most often are disseminated through legislation, policy reports and publications, and primarily come top-down from outside the public school system. Further, the ideology driving science education reform is often framed in terms of global economic competitiveness and dominance, privilege and power. There is little, if any, consideration given to global citizenship, a lens through which major challenges facing the planet, including climate change, mass extinctions, and sustainability could be addressed in the science classroom. Perhaps opinions, beliefs, and overall discourse about the major challenges facing the planet have become so highly politicized, at least in the U.S., that framing classroom science instruction around global issues is too controversial?

Ken: I am struck by the way that the general public regards science as truth, a superior discourse that is in many ways beyond challenge. Is science taught as a knowledge system that coexists with other knowledge systems that people use (successfully) in their everyday lives?

Lynn: I earned a degree in chemistry and worked as a chemist prior to becoming a physics teacher. Never during these years of instruction and practice had I been introduced to or considered the explicit notion of *science as a culture*. However, science as a culture unequivocally was a part of the implicit instruction of my learning science, learning to become a scientist, and being a scientist and teacher of science. It was not until my graduate studies and a philosophy of science course that I seriously considered the demarcation between science and non-science; the process by which science builds optimally rational knowledge; how and why scientific enterprises have come to don such an elitist reputation; how science knowledge is trusted, contested and changed, and science in relation to a wider social system.

On the surface, science in the K-16 classroom instruction is devoted to developing knowledge of facts, concepts, phenomena that are foundational to science disciplines as well as practices and methods of science. K-16 science also is replete with messages of the authority of science, the hierarchy of scientific knowledge and the special epistemic status of "the" scientific method, as well as who can participate in and access science. The widespread belief that science is truth, a superior discourse that is in many ways beyond challenge underlies the growing hostility and mistrust of science and its position as "a major casualty in the new politics of disruption," at least in the U.S. (Lewis, 2017).

Ken: My preference is for science to be learned with a hermeneutic-phenomenological goal, whereby students learn for the purpose of understanding and applying what they know to their lived experiences. In this way, science could be a way to make sense of lived experiences, including what is in the news, the decisions politicians are making, and technological breakthroughs such as the cloned macaque monkeys born in China. Driverless cars, the quest to colonize Mars, and the implications of the presence of humanlike robots are other potential foci for learning about science in a sociopolitical context in which ethics are central to progress.

I wanted to comment on my recent research that involves oral breathing while speaking. No sooner did I learn that a teacher was potentially a mouth breather, as she taught, I realized that I taught this way too! Furthermore, watching TV anchors, I realize that most of them were mouth breathers as well. It took some time to track down the science – but it is unequivocal. There are important reasons why we should almost always breathe in and out through the nose, including when we are sleeping. I was astonished to read how mouth breathing contributes to structural problems in the face, severe dental issues, and widespread problems in the body's functioning. Nitric oxide (NO), which is produced in the nose, is essential for effective functioning of blood cells and transportation of oxygen, nerve innervation, and killing pathogens (e.g., bacteria, fungi, viruses). If a person breathes in or out of the mouth, the NO reservoir is bypassed. In 1998, the Nobel Prize in medicine was awarded for NO as a signalling molecule in the cardiovascular system. My query is this – why don't science educators teach citizens, birth through death, about the science of breathing and the consequences of mouth breathing? Since I learned about nasal breathing and the benefits of NO, I am wearing tape across my mouth when I work at home and when I sleep – therefore having to breathe through my nose. The tape that I have found most suitable, because it is simple to use, hypoallergenic, and light, is 3M Micropore tape, which can be bought from pharmacies and of course Amazon.com. To me it seems like a priority for all citizens to know that mouth breathing is bad for health and may inhibit learning. Just as an aside, smoking reduces the amount of NO produced in the human body and humming during the outbreath increases the amount of NO 15-fold.

Lynn: Teaching in K-12 classrooms (not only in the U.S., but also in countries around the world) is shackled by standards in ways that were not originally intended. Standards are not contextualized, whether in everyday phenomena or pressing global issues. "The science of breathing" is a metaphor for innumerable ways in which science can be framed in terms of students' lived experience. The purpose of science education has been lost. A bold vision for science education that resonates with many authors in this book is one that will steer us away from a perspective of science as the vehicle for reaching international preeminence in commerce, industry, science, and technological innovation. In their chapter within this book, Lucy Avraamidou, Shakhnoza Kayumova, and Jennifer Adams make an interesting observation from their review of current policy documents on science education reform in Europe such as *Science Education for Responsible Citizenship* (European Commission, 2015); that is, goals related to reducing inequality, promoting social change and social justice are completely absent. They assert that the goals of science education should reflect the relational nature of science. Rather than preparing our students for global competition, science education should be preparing students for functioning in an increasingly globalized world in which different knowledge systems can co-exist and be respected. In his chapter, Masakata Ogawa also writes about the relational nature of science. He establishes that the ultimate goal of science education practice is "to contribute to develop a better relationship between science (scientific enterprise) and society (people living there.)" Ogawa leaves us with the question (among many): "Whose value orientations should be reflected to an image of a better (desired) relationship between science and society or better (desired) coexistence of science and humans.

Ken: It is time we enacted an overarching framework grounded in collaboration and cooperation as a foundation for teaching and learning science. Almost 20 years ago, Wolff-Michael Roth and I realized that coteaching and cogenerative dialogue were collaborative/cooperative oriented activities that could support and enhance learning and teaching in inner-city schools, that many regarded as challenging and even dysfunctional. Our research catalyzed many studies throughout the world, reinforcing the potential of collaboration/cooperation, especially in the contexts of coteaching and cogenerative dialogue. Having said that, the massive amount of research on cooperative learning is highly persuasive – and largely influential.

It seems high time for science educators to address the folly of competition, individualism, and commodification that is so highly valued by policymakers and the public at large.

Lynn: I have seen firsthand the impact of your work with Wolff-Michael regarding coteaching and cogenerative dialogue. At Purdue, we are transitioning to a co-teaching orientation in our preservice teacher education program.

Ken: To what extent is science education a useful form of literacy? For the most part, I regard the focus of science education as mis-oriented, focusing on understanding science for the purpose of becoming a scientist and doing science. A bold vision for science educators to consider would be educating for life–now. The two parts to this vision are to connect the goals of science education to the enhancement of everyday life, and to focus on now, rather than the future. For example, understanding ourselves in terms of health seems like an obvious priority. I regard much of what we have studied in our work on contemplative inquiry and wellness to be highly relevant – including expression of emotion, physiological responses to changes in perceived social and physical safety, and health issues associated with stress, diet, and breathing patterns. Self-study might be a productive means of learning while honing inquiry skills that would serve learners more generally in everyday life.

RESEARCH WITH HUMAN SUBJECTS

Ken: It is now four decades since the National Commission for the Protection of Human Subjects of Biomedical and Behavioral Research wrote the Belmont Report (1978). Notably, the members of the Commission consisted of highly qualified men and women with expertise in medicine, science, race, ethics, law, physiology, and psychology. There were no educators/science educators on the writing team. The basic ethical principles of the report required researchers to show respect for participants, beneficence, and justice. The report laid out principles for informed consent, assessment of risk and benefits, and selection of subjects. Initially, the Report provided a framework to review and approve federally funded research with human subjects. However, during the time from the publication of the Belmont Report to the present, the chief change I have noticed is that most institutions that do research require IRB review for all research proposals – not only those that are federally funded. Accordingly, the IRB, personnel and process, have become gatekeepers for research undertaken in universities, museums, hospitals, and other institutions that do educational research.

I served on the IRB for several years and readily acknowledge its importance. However, most graduate students and faculty I know, roll their eyes and show a range of emotions concerning the IRB process – including disdain, disrespect, and they communicate an impression that gaining approval is a necessary evil. The fact that the IRB process is grounded in crypto-positivism (Kincheloe & Tobin, 2009) and assumes statistically-oriented designs and generalizability is accepted as "the way it is" and most colleagues regard the process as one that needs to be "worked around." A common refrain is: "give them what they want to hear and then do your research ethically." Clearly, this is not an optimal situation. I maintain that it is time for the IRB processes to be reconstituted to legitimize present-day research methodologies and associated methods. For example, research that is emergent and contingent, grounded in frameworks such as hermeneutics, phenomenology, and ethnomethodology would incorporate much different methods than are assumed, and required by most IRBs. Also, the type of generalizability issues would not involve generalizing from the sample to a population, but would likely involve theoretical generalizability (Eisenhart, 2009), and ripple effects (Tobin, 2018).

Importantly, guidelines should be developed for the ethical use of technologies such as, for example, video, photographs, sound recordings, and smart phones. Also, the participant nature of research should be identified as legitimate and roles such as teacher-researcher and student-researcher should be accepted and valued. They should be worked into the guidelines for conducting ethical practices in a context where we never do research on – but instead do research with. Perhaps it is time for discipline-related panels to be established by professional associations to produce guidelines for conducting ethical research with human subjects, for example, in science education – acknowledging a diverse array of designs and associated methodologies. Even though the Belmont Report has served us well, it is time to address the omission, in the 1970s, of educational researchers in writing the initial report. In addition, ethical conduct of present-day researchers should be guided by awareness of what is proposed and being done in a study, including the tools used, emergent goals, and potential outcomes. To be fully cognizant of such criteria it is clear there is need to update and likely revamp substantially sets of guidelines for fields such as science education.

IRB personnel and researchers should accept as legitimate, methodologies such as authentic inquiry and event-oriented inquiry (Tobin, 2017), which I use in most of my research. Such methodologies share common ground with most educational research while having somewhat unique criteria for ethical treatment of participants. Rather than reviewing proposals in a distant and objectified fashion, it is important for IRB members to be peers, understand the type of research being proposed, and work with researchers to maximize the probability that ethical issues do not arise in a study – as designed. Rather than insisting that all proposals adopt tenets of positivism in the design – selected peers might work with researchers to produce a proposal in which all participants are treated ethically.

Lynn: I absolutely agree. What you're describing is a classic case of "mission creep." It is not an IRB's role to critique research design beyond ensuring that the methods proposed for research are ethical and minimize the risk to human participants. IRBs should not prioritize, much less require, any particular methodological paradigm over another. It is not their job/mission to control methodologies. Moreover, qualitative research has been around long enough that you would think a responsible IRB panel would have adequate representation so as to be able to evaluate qualitative methods – grounded in diverse frameworks such as hermeneutics, phenomenology, and ethnomethodology – in terms of ethics and risk to human participants/co-researchers.

Ken: When my doctoral students seek approval for the dissertation research, they are frequently required to submit interview protocols even though the research they do is dialogic in nature (Bakhtin, 1986). I regard it as important for all researchers to document how dialogues are likely to turnout, it is important to acknowledge that they cannot know what might be said by any of the participants – what might be said can be described in a narrative and what will be done if undesirable topics arise can be addressed in the proposal to prepare researchers to act ethically in such circumstances.

Open-ended questions in a questionnaire or heuristic are similar. If space is provided for an open response, it is responsible for researchers to have a plan to ensure that no human subjects are harmed by an inappropriate response. Anonymity and confidentiality are part of respect for persons, and should be part of guarantees included in informed consent documents. Furthermore, when a study is enacted, ethical conduct must be valued and maintained in ways that are consistent with informed consent documentation.

Lynn: Speaking of informed consent documents, voluntary informed consent of research participants is absolutely essential to the safety, protection, and respect of the participants as well as the integrity of the research. However, the process of obtaining written consent cannot simply be transferred from one context to another without taking into consideration sociocultural factors that influence the research. It is incumbent upon IRB panels to consider that signing documents is a cultural practice and therefore carries different meanings, assumptions, and beliefs depending on context. I have encountered participants' reluctance to signing a consent document for fear that it could somehow place them at risk, especially if they have lived under repressive government control. I have encountered participants' distrust of signing a consent document because they do not understand why I would disclose all of the information in the consent form. In essence, I was imposing requirements that are grounded in beliefs, assumptions, and practices that make sense in one culture (i.e., the U.S., academic research), but not in another. IRB panel members should be aware and informed of the cultural and practical issues relevant to the protection of human participants. Further, researchers should be able to modify required consent form legalese and/or request and be granted waivers for alternatives to written consent if the researcher is able to verify that the research participants will be able to give voluntary informed consent.

Ken: In my research, I frequently identify research participants because I know them and want to collaborate to learn from doing research with them. Often there is only one, or at most several participants. All participants are selected serially and contingently, where contingence usually means they are capable of contributing in ways that will enhance the research. The selection is purposeful and never random.

Many members of the IRB expect random selection of a sample that affords outcomes that are generalizable to a population that is well defined. When such IRB members review a proposal that has one participant, purposefully selected, they insist that a larger, randomly selected sample is included in the design. They make sense of selection and generalizability in terms of statistical methods that are usually built on tenets of positivism. To better serve research that is worth doing, I maintain that an IRB structure be established so that appropriately educated peers review proposals and practices to ensure high levels of ethical conduct.

Lynn: These IRB members need to be educated in the philosophy of science, and more specifically, about the ontological, epistemological, and methodological assumptions underlying various research paradigms! It is irresponsible conduct on the part of an institution if the IRB panel is a methodological gate-keeper. It is incontrovertible overreach to "insist that a larger, randomly selected sample is included in the design" when the research design includes a single, purposively-chosen participant. In fact, one might say that the IRB members are exercising coercive power by making a proposal's approval contingent on such a design change – a change that has nothing to do with evaluating the study's ethical treatment of human participants and the risks and benefits involved.

WHERE TO NEXT?

Lynn: I believe that we have compiled a volume of essays authored by leading science education scholars from around the world who offer provocative, bold visions for the future of science education. Through diverse theoretical perspectives, the authors raise critical issues, question axiological commitments, and offer fresh ideas that undoubtedly will enlighten, inspire and even provoke readers' thinking about (and action toward) ways in which science education can be transformed for future generations.

Ken: Science education lives in the shadow of science and pervasive scientism. To some, science education has no distinctive identity, and when it comes to making policy decisions about science education, higher status is often given to scientists, and to a lesser degree other disciplines, such as psychology, and successful business people. In my history of more than 50 years as a science educator, the status of our field has not increased, and in many universities, the same is true of education. Commonly, science education is nested within education, which is not considered to be a high-status college – for example, in comparison to a college of arts and sciences. As the world enters a phase in which science education for all citizens is increasingly important, science educators might seize the moment to lead an endeavor to educate citizens to act as stewards for the Earth, while increasing personal and communal sustainability and quality of life. That is, it is an opportune moment for science educators to expand their roles.

In our metalogue, there is a common foundation to many of the critical issues. As humans, we act in a complex, ever-changing world with simultaneous awareness and unawareness. The hidden (crypto) referents that underpin what Pierre Bourdieu refers to as habitus (i.e., action without awareness), are central to what most individuals accept as common sense. When it comes to science education, crypto-positivism is pervasive in making sense in terms of such tenets as: objectivity, discoverable truth, reductivism, and causal relationships between variables. Similarly, especially in the West, science is regarded as a superior way of knowing and neoliberal principles are taken for granted. Accordingly, schemas such as competition, free markets in which the chief criterion is opportunity to participate, and individualism, are pervasive and, via their impact on accountability and assessment, have an inordinate impact on planned and enacted curricula. Together, referents such as these, drive periodic efforts to transform science education – framing standards, textbooks, assessment of students and teachers, and accountability models and practices. Perhaps solutions lie in bold visions for education, adopting practices that emphasize reflexivity, whereby individuals become aware of what they do and why they do what they do, and critically review, preferably with others, whether changes are desirable.

REFERENCES

Avraamidou, L., & Bryan, L. (2018). Science education reform: Reflecting on the past and raising questions for the future. In L. Bryan & K. Tobin (Eds.), *13 questions: Reframing education's conversation: Science* (pp. 410–418). New York, NY: Peter Lang.

Bakhtin, M. M. (1986). *Speech genres & other late essays* (V. W. McGee, Trans., C. Emerson & M. Holquist, Eds.). Austin, TX: University of Texas Press.

Bateson, G. (1972). *Steps to an ecology of mind: Collected essays in anthropology, psychiatry, evolution, and epistemology*. San Francisco, CA: Chandler Publishing Co.

Belmont Report. (1978). *Ethical principles and guidelines for the protection of human subjects of research*. Washington, DC: The National Commission for the Protection of Human Subjects of Biomedical and Behavioral Research. Retrieved from http://ohsr.od.nih.gov/guidelines/belmont.html

Eisenhart, M. (2009). Generalized notions from qualitative inquiry. In K. Ercikan & W.-M. Roth (Eds.), *Generalizing from educational research* (pp. 51–66). New York, NY: Routledge.

European Commission. (2015). *Science education for responsible citizenship*. Brussels: Directorate-General for Research and Innovation, Science with and for Society. Retrieved from http://ec.europa.eu/research/swafs/pdf/pub_science_education/KI-NA-26-893-EN.pdf

Greene, M. (1994). Postmodernism and the crisis of representation. *English Education, 26*, 206–219.

Kincheloe, J. L., & Tobin, K. (2009). The much exaggerated death of positivism. *Cultural Studies of Science Education, 4*, 513–528. Retrieved from https://doi.org/10.1007/s11422-009-9178-5

Larochelle, M., Désautels, J., & Ruel, F. (1995). Les sciences a l'ecole: portrait d'une fiction. *Recherches Sociographiques, 36*, 527–555.

Lewis, T. (2017, December 14). A year of Trump: Science is a major casualty in the new politics of disruption. *Scientific American*. Retrieved from https://www.scientificamerican.com/article/a-year-of-trump-science-is-a-major-casualty-in-the-new-politics-of-disruption/

National Research Council. (1996). *National science education standards*. Washington, DC: National Academy Press.

NGSS Lead States. (2013). *Next Generation Science Standards: For states, by states*. Washington, DC: The National Academies Press.

No Child Left Behind Act of 2001, P.L. 107–110, 20 U.S.C. § 6319 (2002).

Pouliot, C. (2019). Being a science education researcher and a concerned citizen against epistemological anesthesia. In L. Bryan & K. Tobin (Eds.), *Critical issues and bold visions for science education* (pp. 221–231). Leiden: Koninklijke Brill.

Roth, W.-M., McRobbie, C. J., & Lucas, K. B. (1998). Four dialogues and metalogues about the nature of science. *Research in Science Education, 28*(1), 107–118. Retrieved from https://doi.org/10.1007/BF02461645

Roth, W.-M., & Tobin, K. (2001). *At the elbow of another: Learning to teach by coteaching*. New York, NY: Peter Lang.

Tobin, K. (2017). Researching mindfulness and wellness. In M. Powietrzynska & K. Tobin (Eds.), *Weaving complementary knowledge systems and mindfulness to educate a literate citizenry for sustainable and healthy lives* (pp. 1–18). Rotterdam, The Netherlands: Sense Publishers.

Tobin, K. (2018). The role of mindfulness in harmonizing sustainable lifestyles. *Learning: Research and Practice, 4*, 112–125. Retrieved from https://doi.org/10.1080/23735082.2018.1435039

ABOUT THE AUTHORS

Kenneth Tobin came to the Urban Education doctoral program at the Graduate Center of CUNY in the fall semester of 2003. Prior to that he had positions as tenured full professor at Florida State University (1987 to 1997) and the University of Pennsylvania (1997 to 2003). Also, he held university appointments at the Western Australian Institute of Technology (now Curtin University), Mount Lawley College and Graylands College (now Edith Cowan University). Prior to becoming a university science educator in Australia in 1974, Tobin taught high school physics, chemistry, biology general science, and mathematics for 10 years. He began a program of research in 1973 that continues to the present day – teaching and learning of science and learning to teach science.

Lynn Bryan is a Professor of Science Education and the Director of the Center for Advancing the Teaching and Learning of STEM at Purdue University. She holds a joint appointment in the Department of Curriculum and Instruction and Department of Physics and Astronomy. Dr. Bryan has devoted her career to collaborating with science teachers across the nation to bring innovative practices and cutting-edge science content into the K-12 classrooms. Her current research focuses on science teachers' development of knowledge and skills for teaching physics concepts through technology-enhanced, modeling-based approaches.

GLAUCO SILVA, MARCÍLIA BARCELLOS
AND ELISABETH SOUZA

2. MIND THE GAP BETWEEN SCIENCE, TEACHING AND EDUCATION

Challenges for Enacting Bold and Transformative Visions in Science Education

ABSTRACT

In the last years, Brazilian politicians have been through a crisis that has affected several sections of the population, as well as some important and contradictory changes. The 2013 protests, the presidential electoral race in 2014, and subsequent events that led to the process of the very controversial president's impeachment are important events that contextualize the current moment in Brazil. The polarization between conservative thoughts and progressive social movements have become strong and evident. Consequently, this controversial and delicate moment imposed by politics places Brazilian schools in a dilemma and an associated spread of radical attitudes. On the one hand there is the defence of a 'School without Party', which intends to be technical, apolitical, without ideologies, reduced to the function of transmitting information. On the other hand, there is a history of liberating and transforming education, emphasizing reflection on and through action, as it was figured by Paulo Freire. Although the university frequently relates itself to the second perspective, its practices in the education of teachers do not create conditions for the emergence of a critical and liberating education. Our aim with this chapter is, firstly, to relate the recently political phenomena with the increase of the distance between the school, science content and reality (the gap!). From this vantage point, we are going to present our bold vision based on Freire's perspective of a liberating education, grounded on our concrete experiences. We are going to emphasize the urgent need to rebuild that relation as a way to overcome the technical and apolitical formation of science students and teachers departing from a correlated and dialogical education.

Keywords: Paulo Freire, education with science, critical thinking, transitivity

MIND THE GAP

The warning expression 'Mind the gap' is heard by train and subway passengers in various stations around the world. The voice in the subway alerts for the gap

 DOI:10.1163/9789004389663_002

between the train and the platform. The quite big magnitude of such gaps, and the rush to get on and off the wagons can cause serious accidents.

'Mind the gap,' which is used as an alert, was created in 1968–1969 by the London subway. Its originality consists on the automatization of the message reproduced in the stations through an audio system, since the train drivers and the platform attendants were not succeeding in alerting the passengers. The short message made it easy to assimilate and to visualize the alerts which were also painted on the platform's floor. Besides, since the audio recording and reproduction system was expensive at that time, a short message was needed to facilitate the transmission. 'Mind the gap' was recorded in the voice of famous English actors which gave a tune of charm and importance to the message.

This idea of 'Mind the gap' was so successful that it has become the trademark of London's subway. It is possible to find souvenirs with the London subway logo and instead of the name of the station you see the words mind the gap in the center. Another curiosity regarding 'Mind the gap' happened more recently, in 2013. The Embankment Station restored the original sound 'Mind the gap' recorded by the British actor Oswald Laurance in order to allow his widow to still hear his voice (British Broadcasting Corporation, 2013).

On the other hand, in big cities, such as London, New York, Rio de Janeiro or São Paulo, 'Mind the gap' can serve as warning for other gaps and distances, which, just like the distance between the train and the platform, may not be noticed. This metaphor can indicate the distances between people, including social differences. The pace of life in big urban centers is so frenetic that it shapes the rhythm peoples' lives such that they stop noticing each other, and, therefore, end up losing a sense of solidarity.

In this sense, the contemporary world, marked by deep social inequalities, causes a transitivity of people from the poorest places with lack of job opportunities, well-being and social security, to places which they hope to experience better prospects – such as the big urban centers. The agglomeration of people from different locations (sometimes even from different countries) on those big urban centers is a consequence almost immediate of the transit of people, promoting as consequence the increase of the diversity. People from different places start sharing same spaces. People who are distant from their beloved people (family, friends, fellow countrymen) also start living distant from those people who are near (workmates, neighbors). The 'Mind the gap' needs, thus, to go beyond the subway stations to reduce the distances between people, respecting their cultural, social and ethnical differences.

However, our intention is not to approach the complexity of the social problems on big cities, firstly because there would not be space in this chapter, secondly because we could get into the trap of naivety as we talk about social relations. On the other hand, in order to notice the gap, not only that one between the train and the platform, it is necessary to make a minimum analysis of our reality. So, we support the idea that it is not possible to notice the gap between people, between people and their realities, if we do not 'put in relation of organicity our historical-cultural contexture. Relation of organicity that puts us immersed in our reality and that we can emerge with critical consciousness.

> The relation of organicity which we refer to implies in an increasingly critical position of man in the face of his context so that he can interfere on it. (…) On the same way, the organicity of the educational process implies in its integration with the conditions of time and space that it applies so as it can alter or modify those conditions. Without integration the process is inorganic, overlapping and inoperative. (Freire, 2003, p. 11)

From the social gap we move to another one, mentioned in the title of this chapter, the one between science, teaching and education. We want to point out in the text the lack of the relation of organicity in science education. The lack of relation of organicity which should connect the science contents the school and its own reality. Relation of organicity between teacher and students, who should connect themselves using the dialogue as a pedagogical and democratic principle. In our metaphor, we propose that the gap is the lack of relation of organicity between science, teaching and education.

Generally, science teaching at basic education schools means the emphasis merely on the science content: concepts, laws, formula, mathematics and experiments. In the science teachers education the content scenery is also the most common. Paulo Freire (2003), whose text was originally published in 1959, clarifies the lack of organicity in the teacher education programs, criticizing the gap between the universities, colleges or other formation centers and elementary and high schools. Even the pedagogical courses are criticized by him:

> The subjects of pedagogical formation, worried that would be with the educational acting, with the experience of the educational process, lose themselves, on the contrary, great part on digressions almost always academic. Worse, they lose themselves, not rarely, on the sonority of the word. Once more our reunion with our matrices where the strongly 'wordly' tendency of our culture was born, wrongly called, sometimes, 'theoretical'. (Freire, 2003, p. 99)

In the this text, on one hand we are going to point out the gap, i.e., the lack of organicity, and on the other hand we will show some concrete attempts of overcoming the gap. Thus in the first part we are going to approach the Brazilian historical, political and cultural contexture, arguing that the lack of organicity in (science teaching) education is built historically in Brazil, especially due to our "democratic inexperience" (Freire, 2003), in which predominated an authoritarianism in Brazilian politics and education. In the second we are going to present Freire's (2003) concepts of intransitivity, naïve transitivity, and critical transitivity. In the third part we present some concrete attempts to overcome the lack of organicity in the science education in the context of a project which has been carrying out for two years together with Mauá High School, located in district of Mauá, in the state of Rio de Janeiro.

We decided to overturn the text organization! Our bold vision to the science education will be, in fact, our conclusion. Throughout the text we are going to bring up the aspects which lead us to our boldness. Certainly these years working with Mauá High School, in association with the hardworking in the production of this chapter during the last months, enabled us to confidently come up with our bold vision of science education. Bold because it grounded on Freire's concepts. Bold because it overturns the dominant logic of science education without dialogue, ideology nor culture. Bold because it brings to stage the student in dialogue with their social contexture which tries to fill the gap.

Paulo Freire Said, "Mind the gap!"

In the book, *Brazilian Education and Actuality*, written in 1959, Freire supports that "the Brazilian educational problem (…) is one of those that must be seen organically. They must be seen from our actuality point of view" (Freire, 2003, p. 10). In order words Paulo Freire discusses the gap between the Brazilian education and reality. For that discussion he proposes the fundamental thesis of the antinomy of Brazilian actuality, at that time, that is, "from one side the democratic inexperience, formed and developed by typical traces of our colonization, and from the other, the emersion of the people on the national public life, caused by the country's industrialization" (p. 26). In this part of the text we are going to better understand the historical-social and political contexture of Brazil, bringing up the elements that would be the same fundamental antinomy of new Brazilian actuality.

THE GAP: HISTORICAL-SOCIAL AND POLITICAL CONTEXTURE OF BRAZIL

> Never in Brazil will we have lived better conditions for a revision on our education, in terms of authenticity, as today, despite all the contradictions which characterize our actuality. One of the tasks, in fact of education that we need, will really be to decrease and even overcome the antinomies of today. (Freire, 2003, p. 57)

This quotation describes the educational and political context of 1959 however this can undoubtedly be applied to the present time. When reporting the favourable conditions Paulo Freire refers to the democratic context and the changes happened in the 1950s after the 'Estado Novo' (Vargas Era) period. Brazil was living a democratic experience, despite all the political after the former president Getúlio Vargas's death.

In the cultural sphere, Brazil was trying to find its identity freeing itself from imported models. In the education field there was also a desire to build a national identity from a process which allowed the Brazilians to understand their historical trajectory and rebuilt new perspectives from that comprehension. For Freire (2003) only a critical education could make man and women the subjects of their own history and not objects manipulated by extra forces

> The euphoric climate that characterize the period from the 50s until the middle of the 60s was fertilised by a series of events (for example the New Cinema,

> passing by Bossa Nova, 1958 World Cup Championship etc.), adding the popularity of Juscelino Kubstcheck [President at that time], associated with elements of populism and nationalism. Those elements produced a necessary equilibrium to the production of the ideological atmosphere favourable to the beginning of the development start. (Xavier, 1994, p. 209)

At the end of the 1950s Brazil started to go through social changes resulting from the historical moment that it lived and supported by the search for a proper identity set up on the national development. Among the several fights, the democratization of the education became a wish from the population and it raised a great discussion on the expansion of school opportunities. In this context, the school begins to be noticed as a place for everyone, and not only for a more favoured class. The inclusion of other social classes in the school brought a new student profile due to the diversity of subjects from different classes who started to take part in the schools.

The educational institution that emerged in the 1960s context was marked by that change on the student profile. The school became, then, a place for everyone with the perspective of a less encyclopaedic education but more connected to the Brazilian reality. It searched beyond the academic formation, a political one. However, that objective could not be accomplished because of the Military Dictatorship that was installed in 1964. The wish for an education system with national characteristics that would represent the desires of the young students was replaced by a technicist, authoritative and little reflexive education. The educational legislation launched in the early 1970s consolidated the relation school and labour market. This relation limited the education of youngsters, those who were able to enter in the secondary education, to a technical one, aimed only at the labour market, and without a human and social formation.

> The worry about reforming the primary and secondary education manifested the intention of creating supposing equal conditions to the great mass at a moment of extreme inequality and repression and at the same time ensure the legitimacy of the actions of the State through a discourse which would justify the lack of resources to primary areas (health, providence, education). But if we examine attentively, there is not any reference to the citizenship exercise. The vision of education is technicist and utilitarian, preparing man to the labour market. (Xavier, 1994, p. 250)

In this sense, that suitable moment reported by Freire in 1959 was contained by the installation of the authoritarian regime and of state exception. Such containment costed a delay on the development of the democracy and educational processes, due to the utilitarian characteristic of education. We could only recover from that after the 1990s, and to be more exact, after the promulgation of the Education Law (Lei de Diretrizes e Bases da Educação Nacional – LDB), in 1996. After 30 years of the process known as the Brazilian re-democratization to the reestablishment of a relation of organicity in the education, we should ask ourselves about the new Brazilian actuality.

Over the last 20 years we have been through a gradual changing process regarding the aims of the basic education school. The LDB, on the twentieth of December, 1996, opened possibilities to the school so that it could break with its reproductive characteristic and become a place of formation to citizenship. Thus the LDB provided a favourable moment to build our identity from an authentic education which focused on the reflexion of the Brazilian problems and on the search for solutions departing from democratic practices.

The legal documents that began to circulate after the LDB echoed the discourse of curriculum flexibility, content contextualization, linked to the human and technical education and mainly education to citizenship, that is, to act in the society. However, in practice, school in general kept their encyclopaedic and propaedeutic roots. Little we planted nothing we harvested!

One of the most effective changes we could point out is the democratization of the access, especially to higher education, but this access does not correspond to any improvement on teaching, mainly on high school, but to specific politics to the entrance of young people in the universities. Thus, we did not seize the moment and continued, according to Freire, to keep the same antinomia from 1959.

According to Romão (2003), what Freire names antinomia is, in fact, the dialectical tension established between the poles from the same reality, which is the populist propose: give the people political voice, without giving them voice. The authentic democracy will not be built to the people, without the people, it will only be possible when built with the people (Freire, 2003). In accordance with Freire (2003), for the consolidation and establishment of the democratic processes it is needed an educational process which gets people to leave their naïve consciousness, characterized by a magical vision of the reality, and to build a critical consciousness, capable of analysing their context and effectively take part in it. And in this sense we lost once more the favourable moment for transformation!

It is worth pointing out that in the 1959s context the rupture with the possibilities of progress on the society and, consequently, on the most critical educational processes occurred due to the Military Dictatorship installation in 1964, and the antagonistic fights between conservative and progressive led to the exception regime in the country. On one side the conservative power searched for the consolidation of the capitalism not worried with the common good. The discourse used was the one on defence of propriety and family, which could be seen in the Family Marches with God for Freedom, happened between March and April in 1964. Those marches claimed the resignation of President João Goulart, considering him a communist because he stimulated core reforms that contradicted the interests of the upper middle class, since he supported the distribution of goods and land.

Not very different from 1959 picture, currently in Brazil we live the same fight between progressives and conservatives, also characterized by protests that, in a certain way, culminated in the controversial impeachment of Dilma Roussef, in 2016. She was the president who had been democratically elected by direct vote in 2014. However, in different scenery from the 1950s, when the manifestations happened in a short period of time, the current protests that led to the impeachment started in 2013. Those protests agenda were against the price increase in the bus fare and the 2014 World Cup, but not necessarily the president's impeachment. Actually, different groups with different agendas were protesting nationwide!

The movement for impeachment arises in 2015 with great power, associated with the fight against corruption. Protests restarted in March 2015 at the beginning of Dilma Roussef's second term of office, and mildly extended to her definite impeachment, in August 2016.

The 2013 protests were trigged by the price increase in the bus fare in some capitals and big cities. Headed by the Free Fare Movement (Movimento Passe Livre – MPL), these demonstrations claimed to be nonpartisan and were marked not only by the big number of people that they could mobilize, but also by episodes of violence and depredation of the public patrimony. Amplifying the list of claims, other movements started to question the various problems.

> The lists of claims were not reduced to the price of the bus fare, even though this is a relevant political strategy. This movement refers more widely to the citizen's rights concerning urban mobility in general, which should be considered as a fundamental right as education, health, etc. Thus, it was historically proposed the demercantilization of the public transport, founding on an ideology of systemic transformation, as other had in the past or have in the present. (Scherer-Warren, 2014, p. 418)

The most evident characteristic of the 2013 manifestations was the rupture with the traditional powers of protest, since there was not a clear leadership, nor a unique list of claims. The movement was multiple and for this reason covered several chains, but without a clear proposal. Another point to be taken into account regarding this movement was the importance of the social networks to its consolidation and confirmation of its nonpartisan origin.

Differently, the demonstrations in 2015 and 2016 had a more specific focus: the fight against corruption, which, in fact, meant the fight against a political party, the Worker's Party (Partido dos Trabalhadores [PT]). Our interest on this passage is to highlight the nonpartisan discourse that took shape in 2013 by the Free Fare Movement, and gave strength to the School without Party Movement. Thus, we are not going to extend the discussion on the 2015s and 2016s manifestations. However we are going to refer to them when we feel necessary.

The nonpartisan discourse echoed with the 2013 protests got the School without Party Movement to gain shape, which was born in 2003 and proposed a nonpartisan posture in the schools. This movement created the Programme School without Party. The Programme School without Party is a draft bill which makes compulsory to display of a chart on every classroom in elementary and secondary schools with teachers' obligations:

1. The teacher will not take advantage of the loyal audience of the students to promote their own interests, opinions, conceptions or ideological, religious, moral, political and party preferences.
2. The teacher will not favour, not prejudice and not embarrass the students for their political, ideological, moral or religious convictions or lack of them.
3. The teacher will not make political-party advertisement in classroom nor initiate their students to participate in manifestations, political acts or demonstrations
4. When dealing with political, social-cultural or economic issues, the teacher will present to the students, in a fair way – that is, with the same deepness and seriousness – the main versions, theories, opinions and perspectives regarding the issues.
5. The teacher will respect the parents' rights to their children receive the moral education according to their own convictions.
6. The teacher will not allow that the rights previously listed be violated by the students or others in the classroom.

The justification for the program is to finish the 'ideological Left indoctrination in the school'. According to the draft bill's author, the students are subjected to teacher's indoctrination. The bill that provides the implementation of the program School without Party is in the Chamber of Deputies for appreciation. When proposing this idea, the author defines the students as blank sheets, subjected to the teachers' marks. For him the school must be nonpartisan and should not discuss ideologies!

The bill, on its Article 2, defends that the 'Education will attend the following principles: political, ideological and religious estate neutrality.' Besides that, the bill provides punishment with imprisonment of teachers who commit ideological harassment, which means a clear curtailment of teachers' autonomy. However we believe that the main motivation for this program is to transform the school into a place of truth transmitting and thought controlling. On the program website, there is a quotation from Max Weber about the relation student-teacher:

> In the lecture-room we stand opposite our audience, and it has to remain silent. I deem it irresponsible to exploit the circumstance that for the sake of their career the students have to attend a teacher's course while there is nobody present to oppose him with criticism. The task of the teacher is to serve the students with his knowledge and scientific experience and not to imprint upon them his personal political views. (Weber, 1946)

It is known that in a specific moment of the history of school and education, the teacher used to be the center of attention in the classroom. He was considered the holder of the truth, and just he could transmit the contents. The students were seen as blank sheets. But, it is also known that the movement of New School from de 19th century broke with the teaching dogmatism and gave the students, since then, a more effective participation in the knowledge construction process. If we think about current students' profiles, who were born and grew up in a globalized and interconnected world, it is quite naïve to imagine they are susceptible to the teacher's discourse, to use Freire's expression. Well, students today can access to a variety of information, even in many situations over class.

So, if you add to the young formation the access to information, how much would the school represent on that sum? Very little or nothing, once it is wrecked, it is not up dated and do not fulfil the projects of young people. The courses of teacher formation seem to follow the same way, since they do not track changes in the society on the same pace they happen, and almost never succeed in incorporating them in the curriculums or teaching practices.

In 1967, Freire in his book 'Education as practice of Freedom' warned that the Brazilian education always needed to be controlled not to give possibilities of thinking to the masses. The program School without Party which is being discussed in the congress as a draft bill is, therefore, the consolidation of the antinomy of an odd democracy to the people without the people's participation as expressed by Freire. It is the consolidation of the people only as manipulating mass because it is taken from them the relation of organicity with its time and space. It is the consolidation of the gap in the education without party, ideology dialogue and so ignorant.

> Never the author has thought naïvely that the defense and practice of that type of education, which respected the man and their ontological vocation to be subject, could be accepted by those powers, who were interested in the alienation of the man and the Brazilian society. In the maintenance of that alienation. That is why the need to regiment themselves using all the weapons against any attempt of enlightening the consciousness always seen as strong threatens to their privileges. It is true that, by doing so, yesterday, today and tomorrow, here, there or anywhere, those powers always distorts the reality and insist on appearing as defenders of Man, his dignity, freedom, pointing to the efforts of real liberty as 'dangerous subversion', 'massification', 'brainwashing – all of this as products of demons, enemies of man and Christian occidental civilization. In fact, they massificate as far as they domesticate and evilly control the most naïve layers of the society. As far as they leave in each man the shadow of oppression that crushes them. *Expel those shadows through awareness is a fundamental duty of a real liberty Education, and for being so, respectful of man as a person.* (Freire, 1967, p. 36, emphasis added)

Our failure to meet the students' expectations is evident when they look for means to solve some of the problems caused by the lack of organicity in the school with its time and place by themselves. One of the clearest ways of the students' discontent protests was the occupations of the public high schools. In the month of October, 2016, more than 1000 public schools were occupied in several regions in Brazil, as a clear demonstration that our school is not working, but the young people want it to work organically.

The students occupied the internal spaces of the schools and they themselves organized schedules with activities and relevant points they would like to discuss, as an expression of boldness. Our students dare to challenge the educational inertia established and put themselves on the search for a quality education through the occupations. Those occupations, unlike the manifestations in 2013, had a clear claim list: improvements on the quality of the education and to what they call dismantling of the public school trough the approval of some constitutional changes. The occupations of the public schools are the bold actions of the Brazilian students!

The current presidency published a Provisional Measure – PM (Medida Provisória – MP) that reforms the secondary school. This reform was edited without any discussion with education professionals, students and associations linked to the education development. The PM 746/2016 excludes from the curriculums the subject-matters of Arts, Physical Education, Philosophy and Sociology, and emphases the technical education. Besides PM 746/2016 opens possibilities to someone else to teach a (technical) content regardless its undergraduate background as it already happens in São Paulo.

Such reformulations that intend to transform the high schools into education for the labour market are not new. The educational reforms of the New State had the same aim, as well as the one occurred in 1971, under the Law 5692/71. We found in Paulo Freire some aspects to reflect on that.

> It was what recently in Baviera, a German educator, friend of mine, said he had heard from the militants of 'left', when commenting my proposals: Paulo Freire has no sense anymore. 'The education we need has nothing to do with dreams, utopias and awareness. It has to do with technical, scientific and professional formation.' And by formation they understood training. This is exactly what has always interested the dominant classes: the depolitization of education. In fact, however, the education needs as technical, scientific and professional formation as dream and utopia by the shadow of this mango tree. (Freire, 2015, p. 49)

The school could be a place more concerned about citizenship education, as integrated to the human formation as to the technical one. Yet, it seems we are losing this favourable moment to let our students build their worldviews grounded on constantly dialogue with the historical-cultural reality and on an organic way.

Up to this point we presented some of the main points to understand the idea of the gap in Brazilian education from elements of the Brazilian actuality: the one from 1959, by Freire and the current one from 2016. We tried to present the texture of the context, the Brazilian historical and educational contexture in order to understand the lack of organicity, which is the gap.

THE TRANSITIVITIES: POSSIBILITIES AND CHOICES

After presenting this contexture it is possible to identify that we are at a crossroads concerning our democracy project. In our view, our society today is cracked as Freire saw it in 1967.

The political turbulences that we have been through in the past four years (2013–2017) have taken us to the radicalization of positons in the progressive and conservative fields. Although what we call radicalization is more closely to the sectarian defined by Freire (1967):

> To be radical does not imply self-flagellation. Radicals cannot passively accept a situation in which the excessive power of a few leads to the dehumanization of all. Unfortunately, the Brazilian people, elite and masses alike, were generally unprepared to evaluate the transition critically; and so, tossed about by the force of the contending contradictions, they began to fall into sectarian positions instead of opting for radical solutions. Sectarians are predominantly emotional and uncritical. It is arrogant, antidialogical and thus anticommunicative. It is a reactionary stance, whether on the part of a rightist (whom I consider a "born" sectarian) *or* a leftist. The sectarian creates nothing because he cannot love. Disrespecting the choices of others, he tries to impose his own choice on everyone else. Herein lies the inclination of the sectarian to activism: action without the vigilance of reflection; herein his taste for sloganizing, which generally remains at the level of myth and half-truths and attributes absolute value to the purely relative. (p. 9)

In the educational field such radicalization occurs through the contraposition between Freire's proposal, which here we characterize it bold, and the emerging proposals such as the School Without Party, or as the reform of the secondary school, in the political field, imposed by the current President through the PM 746/2016. Thus, although Freire's ideas are not new, they are still modern and bold. And maybe so bold that little left of the paper!

As some critics point out, Freire's ideas little arrived at the Brazilian schools. Freire is much more quotation on paper covers than reference inside the school. There is some superficiality when it comes to Freire's epistemology, treated in many cases as a Freire's method of teaching. For this reason, it is bold to keep on the same fight for educational transformation which was urgent 60 years ago.

It is bold to stand up for those ideas in a transitivity moment as the one we live today. From what we can see in this turbulent historical time, the moment is of setbacks in the education field, moreover, the official (governmental) context is of setbacks. It is a setback which is happening even before the liberating aspirations from the LDB could become praxis.

Assuming Freire's antinomy idea of the Brazilian current society, built on one hand by our democratic inexperience, and on the other hand by the emerging of the people on the national political life, we can see clearly two ways to follow.

The first is to deepen the educational structure which still restricts us to our democratic inexperience (inexperience characterized mainly by historical aspects since that the colonial Brazil never allowed the Brazilian people, slaves, immigrants or workers, to live decisions, ideas or wills originated from the people). This is the way of the authoritarian, content-based school that does not lead to reflexion or criticism. This is the school which restricts us until today, despite the efforts applied since the democratization, to a naïve transitive consciousness.

Naïve transitive consciousness that, as define by Freire (2003), is the consciousness which arises from the man as a historical subject who relates with the others and with the world, but does not do it in a critical way, and, therefore, does not do it authentically democratic. The urban man cannot be on the intransitive state anymore, which is characteristic from the country man, who lives guided by their biological instincts, but do not exist in the historical and cultural consciousness of being a subject, relating in a consciousness and collective way with the world around him. The urban man cannot avoid the transitivity as the verb that needs a complement so does the man who needs a complement for his actions. The transitivity which provides the relation of organicity with the other and with the world.

Just as Freire (2003) alerted us, only the urbanization and industrialization would not be enough to lead the Brazilian to overcome their democratic inexperience. The overcoming of the intransitivity would happen almost automatically through the Brazilian urbanization and industrialization – as long as it was an inside out movement, looking for national solutions despite of imported from developing models. From this process emerges the naïve intransitivity, from the people who participate on the political decisions, but can only do that in a massive way, once the proposed development is not discussed with the subjects. It is imposed from o the outside in. In Freire's (2003) words:

> *Naïve transitivity,* (…) is characterized by an over-simplification of problems; (…).By the transference of responsibility and authority, instead of simply delegating it. By a lack of interest in investigation, accompanied by an accentuated taste for fanciful explanations; by fragility of argument; by a strongly emotional style; by the doubt of everything that is new; by the practice of polemics rather than dialogue. By a tendency to conformism. (p. 34)

Would this naïve intransitivity be the characteristic of our democratic inexperience? If to the Brazilian people, throughout their history, it was denied all or almost all way of democratic experience, of voice to act as the subject of their history, how could it be possible to overcome this democratic inexperience and move towards an authentic democracy? Democracy as a way of life, which goes beyond a form of government since it demands the genuine participation of the people. The democracy as a just a form of government is an empty one, because it is not made with people able to exercise it. It is the result of a people who had an education to alienation and domestication. A school for the man object to the democracy for the naïve transitive man. Democracy as a way of life is the one from a critical and consciousness people who do not allow them to be manipulating mass acting only as characters of a democracy which in fact is "disturbed" by the people.

Freire (2003) already said in 1959 that the education would be the way to overcome that democratic inexperience towards the construction of a democracy as a way of life. It was necessary to give people voice, bring to the man his vocation to subject and not to object. It was necessary to educate with the people and not the people, it was needed to break up with the authoritarian education, it was necessary to found the education as the practice for freedom (Freire, 1967).

The idea of an education that was a genuine democratic experience was severely repressed and its practices interrupted by the Military Dictatorship from 1964 on. The re-democratization opened the way to very slow changes which materialized in the field of official public politics only in 1996 with the LDB. Even though, the educational practices that were consolidated in the dictatorship period still resist in the practical context. And before we had the chance to make more advances in overcoming that authoritarian educational model, we are facing a turnaround in the official governmental area which opens space for seatbacks.

It seems that in 2017 we still did not overcome our democratic inexperience. The people who rise in the political life in 2013, 2015 and 2016 do not look as if they have overcome the fundamental antinomy designed by Freire (2003). We do not believe that those people have arisen their critical transitive consciousness.

The critical transitive consciousness, that would be necessary to found a Brazilian democracy as a way of life, and which the possibility of construction would happen through education, potential tool to overcome our democratic inexperience is characterized, different from the ingenuous transitivity,

> (…) by depth in the interpretation of problems; by the substitution of causal principles for magical explanations; by the testing of one's "findings" and by openness to revision; by the attempt to avoid distortion when perceiving problems (…); by refusing to transfer responsibility; by rejecting passive positions; by soundness of argumentation; by the practice of dialogue rather than polemics; by receptivity to the new for reasons beyond mere novelty and by the good sense not to reject the old just because it is old – by accepting what is valid in both old and new. (Freire, 2003, p. 34)

In this sense, we return to that image of the crosswords that we are at. Our democratic experience led us to a situation similar to the one lived in the mid 1964 in Brazil. The sectarian of positions, result of the dynamism of a transit phase where we seem to be puts us in front possible advances and resources.

We defend the idea that it is necessary to leave the way which will only reinforce the naïve transitivity. Not only the School Without Party, but also the Provisional Measure that imposes a reform on the secondary education reinforces the authoritarian character from which our education could never be free of.

Our idea of boldness goes into the direction of defending liberating education, because it is based on dialogue which is, quintessentially, a horizontal relation, not authoritarian. The dialogue communicates since it establishes a reliable relation based on humility and hope (Freire, 1967). We believe on an education that:

> (…) to be education, it should be brave, enabling the people to reflect on themselves, their time, their responsibilities, and their role in the new cultural

> climate of a transition era. Education which allowed them to reflect on their own power of reflection and that had its own tools. The resulting development of this power, by clarifying their potentialities, would mean an increased capacity for choosing. (Freire, 1967, p. 57)

Consequently, science education should also be intrinsically connected to the concept of citizen formation. According to Delizoicov and Angotti (1990) the full exercise of citizenship is only possible with a minimum of basic science education, so as to offer instruments which enable a better comprehension of the society. Thus every individual should receive a minimal education in natural sciences to their cultural formation, once the scientific knowledge is part of culture built by the man.

Frequently the markedly universal character of the knowledge from the natural sciences ends up associating with forms of unproductive education, which content and methodology are characterized especially by a cultural invasion (Freire, 1977).

A science education in a way that can improve the conditions of life in a community and the education of a critical consciousness of the subjects involved implies on dialogical relations. Based on an education for 'problematization', Delizoicov (1982) proposes an approach to science education which departs from the abstraction from the phenomena of nature that influence directly on the social and economic aspects of a certain reality. By discussing this reality and using the scientific knowledge to better understand it, we can establish contents which have direct connections to the problems of the community.

The beginning of changing on teachers view about science education would make transformations that could achieve the students and the communities involved in their practices and relations with the physical, natural and social world, respecting the cultural realities and local costumes.

THE BOLDNESS WE LEARN IN PRACTICE!

The boldness we are presenting in practice refers to our experience developing a project at Mauá High School (a public state school), which will be presented on this part of the text.

The project 'Citizenship and Environmental Education: rediscovering the potentialities at High School Mauá' is financed by Fundação de Amparo à Pesquisa do Estado do Rio de Janeiro (Research Support Foundation of the State of Rio de Janeiro – FAPERJ), through the Public Notice 36/2014 for improvements in the public schools of the state of Rio de Janeiro. The main aim of this project is to develop action plans, with the students and teachers, which promote concrete improvements in the school through environmental education and the exercise of citizenship (Silva, Barcellos, & Souza, 2016).

High School Mauá is located at Anil Beach, district of Mauá, in the city of Magé – RJ. The school works on the three shifts for Elementary and High School students. The school is small-sized, containing eight classrooms, a cafeteria, a kitchen, an auditorium with capacity for 30 people, a library, a teacher's room, a reception, a principal's room , a court and wide outdoors area.

The district of Mauá, located at the back of Guanabara Bay, was named after Irineu Evangelista de Souza, Baron of Mauá, a Brazilian banker and businessman. In the past it was the main entrance to the city of Rio de Janeiro, where the vessels from Portugal used to arrive. Another important historical landmark is the South American first railway built by the Baron of Mauá. The railway linked the coast to the interior of the country, passing by the city of Petrópolis-RJ, and became the route of the gold. Despite the historical importance, the district of Mauá is affected by environmental pollution from the wastes of the Bay, and also from the community who still lives in a poor urban infrastructure and with lack of basic sanitation. Then, given its social context, the school presents potential for discussions regarding the environment, once that before the beginning of this project, there already been some efforts such as selective waste collection and worries about the water quality.

Currently the group is composed by the three of us, professors at CEFET/RJ, who were responsible writing and submitting the project, and by students and teachers at the School, who received scholarship from FAPERJ. The project was submitted in September and approved in December 2014. However, the scholarships allowance for the students and teachers began in May 2015, and the budget to the project was released in October 2015. Thus, we are working on this project since August 2014, when we started to visit the schools. It is important to highlight the support we received from the School's Principal.

One of the most important activities in the Project is the meetings with the group that happen fortnightly, which started in the middle of May 2015. In order to have a register, the meetings are recorded in audio, and the conversations, issues and solutions, as well as conflicts and dilemmas are later transcribed, as a way to collect data and as an instrument of reflexion to the group.

Into Freire's perspective, the meetings are characterized by the equal rights to talk and give opinions, making possible to insert the generative theme and resulting actions. This way, at the very beginning of the meetings two themes arose: the origin of the water in the school, proposed by the teachers; and the necessity to know what the other students thought about the school, proposed by the students in the group. The group decided for the second proposal, since before proposing to study the water it had to become a problem (Freire, 1989) to the students. So, as a process of awareness it was necessary to know what the students thought, that is, what were their opinions and perceptions about the school where they studied.

The search to know what the students thought about the problems that were there mobilized the team in the following way: (a) elaboration of a questionnaire to be applied to most students, asking about the problems in the school, the methodologies used by the teachers and their own motivation to study; (b) tabulation and organization of the data in graphics; (c) presentation of the results to all the students in the school by the scholarship holders, in October 2015; (d) implementation of a Curricular Integration Workshop, with participation of the students, the scholarship holders and the teachers, in May 2016; (e) production by the students of a documentary introducing Mauá in a different perspective, finished and presented in May 2017.

Minding the Gap: The Seminar as a Critical Transitivity Generator

The seminar took place on October 6th 2015, with several sessions so that all the students could participate. The scholarship holder students of the project were responsible for presenting the data and promoting the dialogue with the other students. In order to explain the reason of that presentation, one of the scholarship holders made a rap describing the aims of the project and the questionnaire they had answered. The rap was sung in all the sessions. After the rap the students took turns showing the results of the questionnaire.

Soon after the presentation of the rap, Esther and Robert, scholarship holders, started each session asking their school students partners what they thought about the environment. This procedure was adopted because the project initially had the perspective of environmental education, as already described previously. As expected, the students participating in the Seminar answered nature, erosion, deforestation etc.

Based on their partners' answers, Esther and Robert guided the Seminar intending to broaden their classmates' notion of environment. To do so, other elements that define environment were given as examples by Esther and Robert. They kept on that discussion until they could draw the conclusion that the school is part of the environment, as well as their houses and the relation among people.

The question '*do you think that the education at school allow students' critical sense?*' was the one that grabbed more attention from everyone. During the questionnaire application it was reported that the students did not know what critical sense meant, however most of them answered yes to that question. During the Seminar that question returned and was widely discussed. The arguments presented by scholarship holders were that it would not be possible to change the reality without critically noticing it. At that moment, the problem of the water and environment were put into discussion again, looking for a new critical perspective.

Following we are going to exemplify the discussion promoted by Esther and Robert, presenting their speeches and some other's students who expressed their opinions. We do not intend to make a deep analysis of the speeches here, but to show why we think that the Seminar worked, in some way, as a generator of critical transitivity for those students.

1. Esther: Here [pointing to the graphic on the question about critical sense] most answered yes, but did most know what critical sense was? [students answer 'no'] Why did they answer yes then? [silence] Make a guess?
2. Students: lack of interest…
3. Esther: Lack of interest, and what else? Let's see if you can come to the point [students speak little]
4. Esther: They didn't put their opinion… didn't have critical sense [someone says they were manipulated]

5. Esther and Robert: Were manipulated…
6. Esther: Exactly, this one got to the point; they were manipulated, because here there wasn't critical sense to answer (…) it was lack of critical sense to answer the questionnaire … as a whole.
7. Esther: In the social surroundings, what can contribute with you later… the critical sense?
8. Student1: Have you own opinion isn't it?
9. Esther: Have your own opinion, exactly! … What else?
10. Student 2: Not allow yourself to be manipulated […]
11. Esther: I'm gonna put a situation for you: if we do not have critical sense today to answer this research, will we have in bigger facts as voting… as other rights we have? [silence]…No!
12. Robert: Look, pay attention! I'm gonna ask you three questions: who thinks the school should change raise your hands, please? […] ok, you can lower down… now… eh… Who wants to change? […] ok! Among the ones who want to change, who does anything … to change?
13. Robert: Ok, who's already thought about doing something to change, at least thought? [someone answers negatively]
14. Robert: And why didn't you do?
15. Student 3: Because nobody did anything […] don't you remember I was suspended…
16. Student 4: Nobody ever took us seriously since the first grade… before I had this attitude, you know, to try to change something… there were other questions on paper…
17. Student 3: Robert, I was suspended […] because during the recess we can't do anything… so I went there [Principal's] ask why.
18. Robert: Let me ask you something: it was you and how many more?
19. Student 3: Just me! …
20. Robert: Exactly…
21. Student 3: I was suspended!
22. Student 5: There is no point in only one person want to change… want to make some change, if no one will help that person
23. Esther: It was the case she talked about [referring to student 3] they move, but nobody let them do anything… (…) there is where the critical sense comes, those people who are not interested, they don't have the power to criticize, to contest where they live, they are people who accept anything, you see… I think we don't want to be like that (…) but I'm here to tell you that it is possible, like it or not, you have to have determination… because it will help you ahead and not only now… And I say that because I know it is possible! (…)

The speeches related above show a part of one of the Seminar's sessions. As we said, the aim was that all the students at the School had access to the results of the research and the questionnaires they had answered. We selected the moment when the students were discussing about critical sense and their actions because this represents well the students' empowerment process. It is the transitivity in which students start to transit from the "simplicity in interpreting problems" (Freire, 2003, p. 33) to the increasing of the "capacity of noticing the problems" (Freire, 2003, p. 31).

The question Esther asked her partners on the 7th shift indicates her perception of the problems and a tendency to the organic thought, as defined previously. She questions her partners about their capacity to use their critical sense in what she calls social surroundings. On the 11th shift, Esther makes her main conclusion when she inquires her partners pointing out that the lack of critical sense to answer the research will be noticed on crucial moments the students' political lives, as voting and claiming for their rights.

On the other hand, Robert also dialogues with his partners when he proposes three questions expressed on the 12th and 13th shifts, initiating a series of speeches, especially Student 4 on the 16th shift and Student 3 on the 17th shift. Those students' speeches signalize a certain critical consciousness because they can notice the problems in the school. However, they express passivity in face of reprisal from the school's direction in the past. According to Freire (2003) this movement of repression in the school is because "we insist, almost automatically, on the past formulas, which characterized our passivity or our democratic inexperience" (p. 97).

The dialogue established between Robert and Esther and their schoolmates reinforces the characteristics of collaboration, in special the question of the critical sense that became a communicative tool among the students. The structure of the Seminar contains some important elements to push up students the movement, the transit, that is, from naïve transitivity to the criticism. For example, the scholarship holder Robert who created the rap communicates with his schoolmates through a kind of artistic-cultural language proper from the community where the school is located. On the other hand, the arguments brought up by Esther signalled to a democratic leadership whose critical sense is the content of the communication and transformation.

To sum up, the Seminar was an important event in the Project because brought us important elements to the relation between content, school and social reality, pointing, though, to a relation of organicity. At the beginning of the Seminar, Esther asked her partners what they thought about environment and she finishes pointing out the importance of critical sense in the social surroundings. We can affirm that there is a passage, a transit, from a view of the environment, related only to the nature, to a wider vision including social surrounding. Esther goes from the environment to the social surrounding and tries to indicate this movement to her schoolmates.

Minding the Gap: Curricular Integration Workshop as a Critical Transitivity Generator

> When I got home Saturday I started to reflect on this (…) I started to think about this: how to transform a class that be at the same time attractive which do not have to sell a capitalist idea, of consumerism, or that is relate to pass

> in an exam? How can I make my subject attractive for its own? Then comes the big boom! Correlation! By always trying to correlate my subject with others, isn't it? I'm giving an example of a class today that I felt the students were really motivated. I'm working with electrochemistry, when you get that subject and transform it only in content, as oxidation, reduction, Daniell cell etc. [everything gets difficult] in the students' mind, isn't it? When I try to build, they can't build anything. Then I got today's lesson and deconstructed everything. (…) at the first moment I talked about what's going on, pollution, the use if batteries (…) I don't know what happened (…), we had a bit of philosophy in the lesson. Then came the historical concept. I … felt different looks in class. *Besides, to make the lesson less content-based is harder work than to make content-based one*. Seriously! Because I was preparing the lesson and forgot the book! (…) I don't want to take a battery and the student thinks it's the most interesting thing. Everyone has batteries at home! (…) I simply threw the theme and let the class guide it. (…) then when I got into the pollution part, everything mixed up in their minds, and then each one gave their opinion on what they liked most. (…) And I came to a beautiful conclusion about our profession (…). Our profession is much better than any other because the engineer receives the plan with what he is has to do in that plant. What he must build is already programmed. Our profession builds itself. You are not the same teacher you were yesterday and won't be the same within 10 days (…). (Chemistry teacher, May 23rd, 2016)

We started this part of the text with some words of the chemistry teacher at Mauá High School because they represent for us a rich moment of critical transitivity. The teacher seems to notice the lack of organicity between the chemistry content in the curriculum and teaching material and the school's and student's reality. The context of that speech is a meeting which happened a week after a Curricular Integration Workshop, as a part of the project that we are developing with the school as will be explained following.

The Curricular Integration Workshop was an important part of the project. It was an option of the team as a way of integrating the teachers to the project, once that the Seminar could be closer to the students. Initially subject matter of the workshop would be about interdisciplinary projects involving questions relating to Environmental Education and Curriculum. However, during the planning phase we noticed that the main element was the lack of curricular integration, and not necessarily the lack of interdisciplinary projects. We assumed the position that "man does not live authentically if he does not find himself integrated with his reality. Critically integrated to it" (Freire, 2003, p. 11), organically integrated to it. Education is a process through which the integration happens, and the school is its privileged place. Thus, the Workshop was grounded on that perspective.

Students and teachers who were scholarship holders from the Project organized the promotion of the Workshop in the school. At the end we had three more teachers and eight other students, of which five continued to work with us. The first part of the Workshop took place at the campus of CEFET/RJ in Petropólis, and lasted one day. The other phases happened at the school and it culminated with the production of a documentary by the students.

In the first part, we discussed some results of the questionnaires, the same ones that were presented to the students during the Seminar. We also treated some issues related to the Curriculum. However, the most important moment of this first part was the exhibition of a documentary about the relation School, mathematic teaching and students' reality which made the students feel the necessity to promote the rescue of the culture of Mauá. As was told, despite all its historical importance in the development of Brazil, Mauá is forgotten today. Thus, the idea of the students producing a documentary about Mauá came up.

Throughout the whole process, the students organized themselves to produce the documentary about the place they live in, Mauá. The group of students was responsible for the complete process, since the initial idea, script to the edition of the video. They chose the sequence of images and which real life characters from Mauá would appear in the video. They gave the documentary the title 'Mauá, for a different perspective'. The video lasts 19 minutes and is ready. The official exhibition in the school was in May 2017.

The chemistry teacher talk occurred in the second part of the Workshop, in the school, during a meeting with newcomers in the Project and the teachers who already were in it. The chemistry teacher's perception that a content-based lesson is easier to be prepared than a no contend-based one shows us the first step to a critical transitivity and a relation of content organicity with the students' and teachers' social and cultural contexture.

Finally, the same way as the Seminar, this whole process enables the students to make an organic relation between content, school and reality. There is no content without School; there is no school without reality. This way, the documentary produced by the students and the chemistry teacher's perception are the bond with the relation of organicity.

IS THE GAP MINDED? IS THE MIND OPEN?

> Existing is a dynamic concept. It implies an eternal dialoguetion [dialogue] of man with man, of man with the circumstances. Of man with his Creator. No man can be admitted apart of dialogue. And there is no authentic dialogue without a minimum of transitive consciousness. And this man's dialoguetion [dialogue] about suggestions even with suggestions is what makes him historical. (…). The man regardless his phase, it is always an open being. Ontologically open. (Freire, 2003, p. 35)

All this trajectory of the Project, together with our basis on Freire's work, leads us to systematize our contexture as a bold perspective.

Did we overcome the gap along this chapter? Somehow we feel we are on that way. We built with the students an educational experience that goes to the direction of the organicity defined by Freire. Our educational experience was democratic because the students were the ones who guided the Project. From their voice we followed a learning path, providing learning for all of us. The chemistry teacher's speech also revealed the need for an empowerment of the students in the chemistry lesson.

Those students' and teachers' words along this path, which we built together, give important evidences of the overcoming of the naïve transitivity towards the critical transitivity. The students' empowerment through the process is notable, since the idea of producing the video, the format and the content came from them.

As already mentioned along the chapter, we considered bold to work in the sense of building an education to the real democracy, from the organicity relation. It is also bold to work from Freire's perspective in a moment in which educational projects (reactionaries) are been explicitly put in the field of political dispute.

In this sense it would be bold to recognize that our bold praxis took us into a project for science education that does not teach science properly. The field of science education already presupposes a content. The Science. Although our initial perspective had been on environmental education that is deeply linked to science education, our Freirian boldness took us to other routes.

We notice that it would not be possible to have a critical environmental education, organic, before those subjects could historically locate themselves. We recognized on the interest to produce the documentary students and teachers looking for their ontological vocation to be historical subjects. "Ontologically open" (Freire, 2003, p. 35).

When they were producing the documentary they looked for their origins, interviewed their parents, grandparents, typical characters in the city, trying to understand how the city became what it is today. We could notice that giving value to those narratives, through the documentary, a narrative built from the point of view of the people, would be our possibility of building an authentic dialogue.

And so, on this first moment we put aside our toolbox of science education, of environmental education, to look for the historical roots of our subjects. On this sense it was bold to overturn the logic of the contents in the school space and treat the environmental education on its wider sense, back to the relations between subjects and between the world, not only between subjects and their biological environment.

We are convicted that the on next steps of the Project, on the problematization of the reality, the science contents will be necessary and will be present. This way, we intend to build a science teaching that is organic, and perhaps more an *education with science* than teaching itself. The chemistry teacher's speech shows this aspect, correlating. An education *with* science.

Then our bold vision is the science education *without* science content which shall transform itself into an education *with* science, to a critical transitivity. Is your mind open?

REFERENCES

British Broadcasting Corporation. (2013, March 9). *Mind the gap tube announcement returns after wife's plea*. Retrieved from http://www.bbc.com/news/uk-england-london-21719848

Delizoicov, D. (1982). *Concepção problematizadora do ensino de ciências na educação formal* (Dissertação de mestrado). IFUSP/FEUSP, São Paulo.

Delizoicov, D., & Angotti, J. A. (1990). *Física*. São Paulo: Cortez.

Freire, P. (1967). *Educação como prática de liberdade* [Education: the practice of freedon]. Rio de Janiero: Paz e Terra.

Freire, P. (1977). *Extensão ou Comunicação?* Rio de Janeiro: Paz e Terra.

Freire, P. (1989). *Pedagogy of the oppressed* (31st Printing). New York, NY: The Continuum Publishing Company.

Freire, P. (2003). *Educação e Atualidade Brasileira* [Brazilian Education and Actuality] (3rd Printing). São Paulo: Cortez; Instituto Paulo Freire.

Freire, P. (2015). *À sombra desta mangueira*. Rio de Janeiro: Paz e Terra.

Romão, J. E. (2003). Paulo Freire e o pacto populista. In P. Freire (Ed.), *Educação e Atualidade Brasileira* [Brazilian Education and Actuality] (3rd Printing, pp. XIII–XLVIII). São Paulo: Cortez; Instituto Paulo Freire.

Scherer-Warren, I. (2013). Manifestações de rua no Brasil 2013: encontros e desencontros na política. *Caderno CRH, Salvador, 27*(71), 417–429.

Silva, G. S. F., Barcellos, M. E., & Souza, E. G. (2016). Mauá project: Citizenship and environment educational as pathway to critical thinking and student's empowerment. In C. K. Looi, J. L. Polman, U. Cress, & P. Reimann (Eds.), *Transforming learning, empowering learners: The International Conference of the Learning Sciences* (Vol. 2). Singapore: International Society of the Learning Sciences.

Weber, M. (1946). *Science as a vocation*. Retreived from http://anthropos-lab.net/wp/wp-content/uploads/2011/12/Weber-Science-as-a-Vocation.pdf

Xavier, M. E. S. P., Ribeiro, M. L. S., & Noronha, O. M. (1994). *História da educação: a escola no Brasil*. São Paulo: FTD.

ABOUT THE AUTHORS

Glauco Silva, PhD, has been working since 2010 as physics teacher educator at Federal Center of Technology and Education of Rio de Janeiro (CEFET/RJ), campus Petrópolis where he is associate professor. Graduated on Science Education (2013) at University of São Paulo, Glauco was high school physics teacher in Petrópolis (2000–2001), Juiz de Fora (2002–2004) and São Paulo (2007–2009). In 2012 he was a visiting scholar of the Urban Education Program (now Learning Sciences) at the Graduate Center of CUNY, taking part in the Brooklyn College research project on emotional climate and co-teaching. He held at CEFET/RJ the chair position of the Physics Teacher Education Department (2013–2015). Currently Glauco is responsible for the Institutional Program of Scholarships for Teaching Initiation (PIBID) and he is part in the Mauá Project team, partially presented in this chapter.

Marcília Barcellos, PhD, has been working since 2008 as physics teacher educator at Federal Center of Technology and Education of Rio de Janeiro (CEFET/RJ), campus Petrópolis where she is associate professor. Graduated on Science Education (2013) at University of São Paulo, she taught physics for several years in public and private high schools. Marcília held at CEFET/RJ the chair position of the Physics Teacher Education Department (2009–2011) and coordinated the Institutional Program of Scholarships for Teaching Initiation (PIBID) for five years (2010–2015). She is part in the Mauá Project team, partially presented in this chapter.

Elisabeth Souza, PhD, has been working since 2014 as teacher educator at Federal Center of Technology and Education of Rio de Janeiro (CEFET/RJ), campus Petrópolis where she is assistant professor. Graduated on Linguistic Studies (2012) at Federal University of Minas Gerais, she worked in both elementary schools (1996–2013) in Barbacena and the State University of Minas Gerais (2003–2014). Currently Elisabeth is responsible for the Mauá Project team.

S. LIZETTE RAMOS DE ROBLES AND
ALEJANDRO J. GALLARD MARTÍNEZ

3. TEACHING SCIENCE IN TIMES OF CRISIS

How Can Science Education Help Young People Construct Different Futures?

ABSTRACT

In this chapter, we reflect on the role that science education can, or should play, in helping young people construct options for a future in a society in crisis. We analyzed the current situation of young people concerning education, employment, health and emotions – demographic data concerning young people in México and the USA, with only the occasional global broad stroke. We used the sociocultural perspectives of positionality, and spaces to situate our data within multiple social contexts. From these perspectives, we question the influence and consequences that neoliberalism has caused in the social development of young people. We note how development policies which focus primarily on global economic growth have led to greater inequalities and social injustices, increased levels of stress, despair, a diminishing sense of place, and fewer illusions for the future amongst young people. Not only must young people deal with a sense of place, but also the host of emotional issues we underscored earlier as well as an understanding that their futures are dependent on their competitiveness in a socially inequitable and unjust society connected to the global economy. Finally, we offer the reader of this chapter a series of reflections in which we recognize that education has become more of a service for the development of the global economy than the development of people. Consequently, education has lost its humanistic goal of developing a free and critical thinking society. We invite and challenge educators to strengthen forms of teaching, learning, and research focused on youth by creating forms of literacy that are useful in their lives – not only literacy to fulfill an enterprise like science. Young people need to be guided in developing a sociocultural critical awareness that illuminates the situation in which they are living and builds different and better options for their futures.

Keywords: education, science education, neoliberalism, young people

 | DOI:10.1163/9789004389663_003

HISTORY AND EVOLUTION OF OUR THINKING

In planning and outlining this chapter we kept returning to two questions:

> How is the function of education – and science education – defined by today's policy makers?
>
> When the function of education is actualized, how are young people affected?

These two questions created an intellectual restlessness as we prepared to and eventually wrote this chapter. Our *inquietude* created a need to dig deeper and gain insight into the sociocultural contextual mitigating factors (CMFs) (Gallard et al., 2018) that seem to influence and perpetuate education generally and science teaching specifically, as well as learning models that are not inclusive of all students. Specifically, our attention was drawn towards trying to understand the paradox between thematic issues such as social injustice and the reproduction of social injustice in science classrooms, and how young people are affected (Barton, 2003). It is puzzling to us that even though in the science education community there is an awareness of both societal inequities and injustices, they continue to occur in science classrooms (Tobin et al., 2017). It is our contention that regardless of the existence of awareness of social injustice, this recognition is negated by the powers of Western science and neoliberalism (Ellison & Allen, 2016) – both of which are powerful macro genic CMFs. CMFs are an infinite set of socio-historical-political contexts which are fluid, dynamic, and intertwine with communities, families, gender, identity, ethnicity, socioeconomic status (SES) of individuals, education, and other factors (Gallard et al., forthcoming). Another macro genic CMF influencing education and the teaching and learning of science is the function of education as defined in economic terms – that is, neoliberalism and the development of future workforces. Today, we live in a world filled with emotional student casualties in which students realize just how replaceable they are in the workforce (Giroux, 2001).

SOME OF OUR FRAMEWORKS: A WAY TO PROBLEMATIZE OUR REALITY

For readers of this chapter, it is important to note that that we have confined the demographics of young people to México and the USA, with only the occasional global broad stroke. Also, it is important that the reader understand the lenses we used to frame our ideas. The two most prominent lenses are positionality (Mayer & Tetreault, 1994), and spaces (Bourdieu & Wacquant, 1992). To a lesser extent, we also use standpoint theory (Wylie, 2003), because it is our belief that all forms of science knowledge are subsumed under one function: neoliberalism's insistence on the advancement and preservation of western science over all other forms of knowing (Tobin, 2006b) through homogenization of people or students (Harding, 2003). The aforementioned influences framed our thinking, leading us to various questions about the role of teaching and learning science when foregrounded against the emotions of young people today. Our purpose in offering questions is to encourage the reader to entertain and interrogate the idea that education and the teaching and learning of science – in the lives of many young people today and tomorrow – rather than being a form of emancipation are in fact delimiting in the social places in which they are actualized. We also remind the reader that even though the questions revolve around science education, the teaching and learning of science is contextually situated within the system of education. We ask the following:

> Should the function of Science Education and its associated curriculum be dedicated to science content or be inclusive of other political and sociocultural issues?
>
> Should the function of Science Education include reducing the impact of the financial and emotional crises in students?
>
> Should the function of Science Education build and offer different options for future students?
>
> What are some of the causes of the reproduction of social injustices in science classrooms – and how are they manifested?

We understand and suggest that the above questions imply a positioning of education in general, and science education in particular. The importance of recognizing socially significant issues in science education emanates from identifying and understanding that within social places there exist CMFs which contextually situate people – including young people. Understanding the existence of CMFs allows us to situate the currently prevailing teaching and learning practices which young people encounter during the learning of science as being influenced by a host of inter- and intra-connected cultural, economic and political macro-, meso-, and microgenic factors. Additionally, using a CMF perspective allows us to raise critical questions about the social construction of knowledge in general – and science knowledge in particular – and surface how current neoliberal schemes affect young people. These and a host of other preoccupations regarding the sociocultural influences on science classrooms – as well as how they are actualized – have been sources of serious reflections between the authors of this chapter.

EDUCATION AND NEOLIBERALISM: A SOCIOCULTURAL VIEW

Understanding the function of education as defined by education policymakers and other power brokers today is critical in developing a *sens pratique* (Bourdieu, 1976) of the social fields within which education is situated. For example, if the function of education is to produce more and better workers than we remind ourselves of the following observation: “On the whole, members of racial minorities and low-income groups are less likely to do well in school, and they are also less likely to do well in the job market” (Carnoy & Levin, 1985, p. 1). Extending Martin Carnoy’s and Henry Levin’s observation, we suggest that the focus should not only be on the development of a future workforce, but rather an understanding that there is a greater emphasis on low- and not highly-skilled or professional wage earners. Realizing this we began to wonder if success and failure in education generally – and in science education specifically – speaks to the reproduction of labor relations (Bowles & Gintis, 1976) instead of scientific literacy. If so, this would imply a bifurcation of students in that for some, the function of education is to develop a low-skilled labor force, while for others, it is to develop a highly skilled labor force. Our belief is that it is not only a question of a low or highly skilled labor force, but also recognition of the power and influence of neoliberalism which has situated labor relations and sociocultural categories such as gender and the workforce. For example, regarding women, "The United States sees a drop in its ranking due to a decrease of its Economic Participation and Opportunity score" (World Economic Forum, 2016, p. 21), whereas "Mexico sees a decrease in the female share of professional and technical workers and remains among the lowest performing countries in the region on this indicator" (World Economic Forum, 2016, p. 20). In plain language, both countries have a worsening economic gender gap despite the educational opportunities that are available in both countries. If it is not opportunity, at least at the educational level for women, then what else is driving this mediocre performance amongst educated women?

Our standpoint is that neoliberalism creates and frames sociocultural places in which students are positioned to fulfill economic goals and objectives under the guise of admonishments such as: *you must get a good education if you want to be economically successful in life*. Indeed, this may be true for many people, but we suggest that many more are disenfranchised by the power structures which stand in the way of their dreams of getting ahead. Looking at Blacks and Hispanics, they have more than double the poverty rate than that of non-Hispanic Whites (Proctor, Semega, & Kollar, 2016). Our belief is that being well-educated in general – and successfully negotiating science courses in particular – is tempered by spatial enactments such as the power of neoliberalism positioning subjects through macro genic CMFs such as race, ethnicity and gender, situating them within the hegemony of cultural and social capital (Bourdieu, 1986). In addition, we point to the status of the pay gap between men and women in the workforce. The World Economic Forum (WEF) estimates that it could take 170 years to close the pay gap between men and women (World Economic Forum, 2016). Connecting the WEF 2016 report to STEM pipelines may be a bit of a stretch for some but we use the data in the spirit of Carnoy and Levin, to underscore that the dearth of women and other underrepresented people in the sciences mimics their academic performances in K-12 and tertiary settings. Accordingly, even when the underrepresented *make it*, there are still large hurdles of inequity and a lack of social justice they must overcome. It is not the STEM pipeline which is the problem, but rather the causes of underrepresentation existing in society within which the STEM pipeline is situated.

Education has a centrally critical role, this being the development of critical thinking necessary for young people's constructions of different futures. The development of critical thinking in today's world, according to William Sewell (1999), is necessary to develop agency in individuals in such a way that with available and necessary materials and human resources, young people could modify the social conditions of the specific social fields in which they participate. Within the social spaces populated by neoliberalism, power is not the only issue. There are critical tensions to be considered. Examples are poverty and a good education, gender and the science and mathematics fields, and race/ethnicity and agentic power. These aspects must be considered in understanding the social spaces which young people occupy. Considering these aspects is critical in order to define and/or construct elements that allow for change. Stuart Aitken (2014) reminds us about power and change within social spaces:

> Spaces pose form the question of how change is negotiated. Understanding the embeddedness of power relations in space helps us to see how change plays out in particular locales, but considering power alone is insufficient to understand the dynamic relations between people and places because, as a habituated aspect of our lives, spaces are often envisaged as contexts through which events play out rather than the other way around or both simultaneously. (p. 10)

If we recognize the importance of education as a tool for society in developing critical awareness and active participation of young people, as well as for the integral development of all their capacities, then México is a country greatly indebted to its youth. According to the Organización para la Cooperación y el Desarrollo Económicos (OCDE), México has one of the lowest proportions of students between 15 and 19 years of age enrolled in formal education as compared to other member countries (see Figure 3.1). In addition, México has a high school dropout rate of nearly 50%, one of the highest in Latin America (OCDE, 2015). By comparison in 2014, the USA high school dropout rate was only 6.5% (National Center for Education Statistics, 2016). However, this figure is greatly misleading, because of the 6.5% who did not get a high school diploma 7.4% were Blacks and 10.6% were Hispanic – representing a disproportionate number of Blacks and Hispanics, which also are disproportionately represented among those in poverty (Proctor, Semega, & Kollar, 2016). School desertion amongst the neediest is not only isolated to México and the USA; sadly, it is a global issue. According to the United Nations Educational, Scientific and Cultural Organization (UNESCO, 2015), the figures are rather stark. In 2011, 57 million children were not enrolled in school. Those out of school are also from the poorest, most remote areas of their countries, and thus most vulnerable to sexual and reproductive health problems. At the secondary level, 64 million adolescents were out of school in 2011, of which 21 million lived in sub-Saharan Africa (UNESCO, 2011).

School dropout is a complex phenomenon which, according to the OECD, is related to multiple factors ranging from personal and family context, academic performance and educational policies, labor market and socio-economic conditions. For example, school dropout rates amongst young pregnant women have many economic and social consequences that we should consider. The fertility rate of Mexican women between 12 and 19 years of age is 37 births per thousand women (Encuesta Nacional de Salud y Nutrición, 2012). Pregnancy rates of young women in México constitute the second largest cause of school drop-out (World Economic Forum, 2016). In México, this translates into an annual average of 280,000 young women who drop out of school prior to receiving a high school degree. Dropout rates for young women are closely linked to poverty levels (Secretaría de Educación Pública [SEP], 2011). Compounding the issue of female dropouts through the Global Gender Gap Report, the World Economic Forum reports that in the specific case of México only 48% of women in the economically active population are working as compared to 83% of men. Similarly, 31% of young women in México are unemployed or uneducated in contrast to 9% of young men (World Economic Forum, 2016). Some readers may think that México is a male-dominated society and as such try to explain away the differences. However, to do so would be to fall victim to some of the false doxas used to portray México. We could vigorously debate this, but that is best reserved for another paper. For the purposes of this chapter we want to underscore that dropout rates amongst young women have severe social, and economical consequences. And while this also applies to young males, it occurs to a much lesser degree.

The complexity of the dropout situation is reinforced by a labor market in which there are few or no employment opportunities for highly professionalized activities. For example, the Asociación Nacional de Universidades e Instituciones de Educación Superior (ANUIES), reported that jobs with highly professionalized activities fell from 69 to 61 percentage points, while those of medium to no labor skills increased from 19 to 24 points from 12 to 16, respectively (ANUIES, 2012). We have a global economy that increases its demand for a low-skilled labor force while decreasing the need and opportunities for prepared professionals such as engineers, scientists, and educators. Young people are victims of the labor market emphasis in at least two ways: economically and emotionally.

DEMOGRAPHICS OF YOUNG PEOPLE

Paying attention to young people is important for practical reasons which go beyond the idea that they are in our schools and science classrooms. There are approximately 1.8 billion young people representing almost 25% of the world's population (Office of the Secretary-General's Envoy on Youth, 2015). The sheer number of young people in the world has not gone unnoticed in the last five decades, as evidenced by initiatives of international organizations such as the United Nations (UN). The UN General Assembly in 1965 passed one such initiative: Resolution 2037. Resolution 2037 was developed to promote in young people the ideals of peace, mutual respect, and understanding across world-wide communities. Subsequently, the UN's Economic and Social Council emphasized three basic themes: youth participation, development, and peace. Both international initiatives led to the development of several international youth policies which resulted in making 1985 the International Year of Youth. The agentic implications for young people in these initiatives appears paramount:

> It is imperative that young people from all over the world take an active part in all relevant levels of decision-making, as it affects their lives today and has repercussions for their future. In addition to the intellectual contribution and the capacity to mobilize support for youth, young people have a way of analyzing things that need to be considered. (United Nations-Agenda 21, 1992, p. 1)

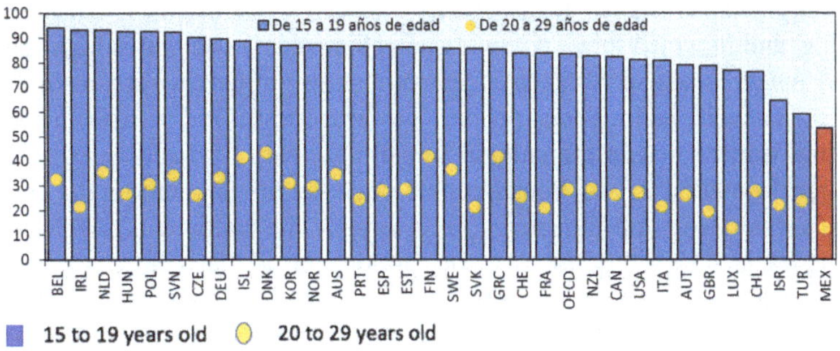

Figure 3.1. 2012 School enrollment rate of 15 and 29-year-olds
Resource: Gupta et al. (2014), OCDE Publishing

Each social place and the associated social, economical, political, environmental, and even religious components may favor or hinder the integral development of young people. Understanding that there are contextually situated dynamics helps us to problematize their lived experiences on a global scale. Almost 85% of young people in the world live in developing countries, 60% neither study nor work, and 28% survive on less than $2.00 a day (Gupta et al., 2014). Young people in industrialized countries make up a smaller proportion of the total population because of lower birth rates and higher life expectancy (UN, 2010). In response to this inequitable world situation, priority lines of action were established by the UN through the World Program of Action for Youth framework (UN, 2010). Fifteen priority areas were established as the framework for improving the welfare and moral of young people:

1. Education
2. Jobs
3. Hunger and poverty
4. Health
5. Environment
6. Drug abuse
7. Juvenile delinquency
8. Recreational activities

9. Girls and young women
10. Full and effective participation of young people in society and decision-making.
11. Globalization
12. Information and communication technology
13. HIV/AIDS
14. Armed conflict
15. Intergenerational questions

These 15 areas of concern underscore the significant disadvantage and bleakness of future prospects for the world's young people – especially those from so-called developing lands. However, regardless of where they are from, the world they are inheriting, and in which they are contextually situated, is one that is highlighted by poverty, hunger, war, and violence underscoring how limited they are in contributing to their own development. Young people's powerlessness is sad because the potential for young people to enact the cultural capital necessary to facilitate and allow for social transformation exists. Young people possess the imagination, the creativity, the ideals, the talent, and the energy necessary for the continuous development of the societies in which they inhabit. The implication is that young people can be agents of change as long as there exist sociocultural conditions which allow them to channel and use their talents. If impeded, young people will be reduced to being responsible for reproducing established hierarchies of power and oppression. However, regardless of the desire for increasing agency among young people, peace and their development, these goals are at odds with the premises of neoliberalism: competition and individuality.

MENTAL HEALTH ISSUES

According to the United Nations International Children's Emergency Fund (UNICEF, 2011), 20% of young people worldwide have mental or behavioral problems. Depression is the largest contributor to global morbidity among 15- to 19-year-olds, and suicide is one of the three leading causes of death among people aged 15–35. Suicide is the second leading cause of death in the age group of 15–29 years (World Health Organization, 2016). In México, there has been a dramatic increase in suicides (Mexican Foundation for Health [MFH], 2016). The Mexican National Institute of Statistics and Geography (INEGI), reported in 2013 that 5,909 suicides were recorded, 41% of which occurred in young people 15–29 years. This translates into a rate of 7.5 suicides per 100 thousand young people in this age group. Disaggregated by gender, suicide rates among young people aged 15–29 are 12 per 100,000 men and 3.2 per 100,000 women. In the United States suicide was the second-ranked cause of death among young people (Jiang, Mitran, Miniño, & Ni, 2015).

In the last 25 years, suicides have increased by a tremendous 270%, currently representing the third leading cause of death in young people. The MFH associates suicide with depression, frustration, and drug use leading to dependency. The realities of our young people associated with suicides demands that we ask ourselves:

> What happens in a society where every day there is an increase of young people deciding that it is better to be dead than alive?
>
> Did the institutions responsible for the wellbeing of its citizens rid themselves of the responsibility to meet the needs of the new generations?

YOUNG PEOPLE IN THE WORLD: A REFLECTION OF A NEOLIBERAL SOCIETY IN CRISIS

On many occasions, as teachers both of science and of science educators, we have become cognizant about the indifference of young people towards school – including would-be science teachers and their participation in science methods classes. Our revelation that learning has been reduced to *tell us what to do* is not new – nor does this attitude magically appear in classrooms. As such, we believe that it is important to underscore that attitudes among young people are rooted in the influences of sociocultural phenomena. In other words, the attitudes or indifferences amongst young people have social and economic roots that are embedded within the social places they occupy. There are many positioning factors within social places of which one of these can be traced to neoliberalism and the movement's push for accountability. Perhaps indifferences amongst young people is synonymous with their awareness as to how the culture of education has been situated by outside forces? Yes, we are suggesting that young people are very cognizant that the system of education has become a cornerstone for the neoliberal accountability movement. *Just tell us the answer* is an attitude which underscores their understanding that it is better to be test-ready and credentialed rather than pedagogically prepared. We hold that young people have caught onto what neoliberalism demands today: accountability, with future opportunities determined by good test scores and appropriate credentials.

Additionally, we recognize that there exist marked differences between today's young people and other generations because their forms of coexistence are not congruent. It seems that every day the generational gaps both in, and between school, and daily life are widening. A generational imbalance makes us wonder about the *whys* and *how's* of social problems in which young people are immersed and their relationship to science education. Cathy Sandeen (2008) paints the following picture: "Millennials are graduating from college and are entering graduate school or the workforce. They tend to be very career oriented and expect rapid advancement and perks. They are very accustomed to being in the spotlight, receiving recognition for practically every achievement growing up no matter how trivial" (p. 18). How then do millennials react to labor markets in which the emphasis is on a low- and not a high-skilled labor force? Perhaps part of the answer lies in their indifference to education and science education.

> Cultural difference must not be understood as the free play of polarities and pluralities in homogeneous empty time of the national community … Cultural difference, as a form of intervention, participates in a logic of supplementary subversion similar to the strategies of the minority discourse. The question of cultural difference faces us with a disposition of knowledge or a distribution of practices that exist beside each other, *abseits* designating a form of social contradiction or antagonism that has to be negotiated rather than sublated. (Bhabha, 2002, p. 232)

Bhabha's notion of cultural difference serves as a framework for problematizing the situation of young people in the world in a complex and multi-causal way, underscoring the conditions of each country and region while simultaneously preserving the singularity and cultural differences of each. Regardless of the country, Kenneth Tobin (2006a) reminds us that educational spaces are social fields of cultural enactment. Accordingly, within each country and region, educational spaces as cultural enactments position young people. Adding to this is an understanding that young people live and enact in a society which has pivoted from a local to a global perspective. Positioning young people away from regional to global connections suggests a loss of a sense of place which adds to the emotional baggage they must deal with. Not only must young people deal with a sense of place, but also the host of emotional issues we underscored earlier as well as an understanding that their futures are dependent on the competitiveness in a socially inequitable and unjust society connected to the global economy. Even we have developed a sense of desperation and indifference that many young people must feel when they attend school. We additionally wonder if school desertion is also another form of agentic action by young people.

EMOTIONAL CRISIS AS A RESULT OF A SICK SOCIETY

Whether it be education generally or science education specifically, we are mindful that we cannot disregard that sociocultural-influencing aspects contextualize the social places in which education takes place. Our inclusiveness of the social, environmental, psychological, economic, and cultural factors which position science education enables us to develop more robust and complex contributions to the integral development of our young people. Accordingly, we believe that it is necessary to problematize the situation and role of young people in our society within at least the following positional influences.

Emotions are very complex psychophysiological reactions experienced by humans and are multi-causal in nature. As an example, a quick review of the global social situations helps us to realize that what prevails today is shocking news: terrorist attacks, unemployment, rising levels of poverty, suicides, pollution, climate change denial, and climate change itself. We are not pessimists, and do not believe that it is a question of others being pessimistic, but when one is bombarded constantly with negative news and good news is rare, we understand that it is easy for young people to conjecture that their future is bleak. According to Mario Sandoval (2012), each context can either make possible, favor, hinder or prevent young people from living and fulfilling their youth. Living one's youth is not solely an issue of age, but is inclusive of the socio-environmental conditions necessary to fulfill their lives and prepare adequately to enter the adult world with the necessary and sufficient capital for integration into adulthood.

We do not argue a causal relationship, but do believe that the information age and technology are mitigating factors, which together paint a very bleak picture of society lending itself to the development of hopelessness amongst the young. Recently, studies have been published noting a strong correlation between social media use and depression in the youth, with those using three or more social media outlets holding statistically higher rates of depression than those who minimize or refrain from social media use (University of Pittsburgh Schools of the Health Sciences, 2016, December 20).

Additionally, we suggest that technology contributes to this emotional crisis by depersonalizing the occupation of spaces by promoting emotional ties via cyber or virtual connections rather than connecting in-person with other humans. Giroux (2009) suggests another connection – and that is the connection to neoliberalism:

> The havoc wreaked by neoliberal economic policies can be seen in the hard currency of human suffering such policies have imposed on children, readily evident is some astounding statistics that suggest a profound moral and political contradiction at the heart of one of the richest democracies in the world. For example, the child poverty rose in 2004 to 17.6 percent pushing upward the number of poor children to 12.9 million. Moreover, children make up a disproportionate share of the poor in the United States in that they are 26 percent of the total population, but constitute 39 percent of the poor. (p. 3)

NEOLIBERALISM AS A SOCIAL FIELD: ECONOMIC CRISIS AND EDUCATION

"During the last four decades, expanded globalization has been accompanied by increased interests in cultural identity. Indeed, social theorists see the latter as an expectable response to the increased homogenization and universalization of cultural forms created by the globalization of the economies, politics, and of culture itself" (Harding, 2003, p. 49). We suggest that the teaching and learning of science is being held hostage. This is accomplished by those who define the function of education through not only a neoliberal lens, but also through the actualization of this lens. By focusing on the extremes of the global and the individual – as opposed to students being contextually situated in spaces of local communities working hand-in-hand for the betterment of their localized spaces – the teaching and learning of science becomes oppressive rather than liberating. "We're talking about an ideology marked by the selling off of public goods to private interests; the attack on social provisions; the rise of the corporate state organized around privatization, free trade, and deregulation; the celebration of self-interests over social needs; the celebration of profit-making as the essence of democracy coupled with the utterly reductionist notion that consumption is the only applicable form of citizenship" (Note, 2015, p. 449).

The economic crisis surrounding the function of education in general and in particular the teaching and learning of science has been dominated by neoliberal forces and their agenda for many decades. As an economic movement, neoliberalism places control over goods in markets as opposed to the public (Giroux, 2009). Neoliberalism and its tenets are macro-, meso- and microgenic CMFs positioning education, science education, and students to meet the needs of a global economic market. Specifically, neoliberal forces view the function of education as a commodity that benefits the private sector primarily and the common good secondly. The development of education policy such as No Child Left Behind, the Charter School movement in the USA, or the new teacher reform movement in México directed by the SEP stand as examples. All three of these movements – as well as many other global education movements – are framed by an emphasis on the world's development of human capital through standardized testing accountability, and minimizing curriculum for greater efficiency (Sahlberg, 2010). Giroux (2009) states, "What is new about neoliberalism in the last three decades is that it has become normalized-even celebrated by the dominant media-and now serves as a powerful pedagogical force that shapes our lives, memories, and daily experiences, while attempting to erase everything critical and emancipatory about history, justice, solidarity freedom, and the meaning of democracy" (p. 8). As a result, many countries like México have steadily reduced their national budgets for health, the environment, education, and research while increasing budgets for the military, private sector, legislative, and judicial powers (Navarro, 2016). Globally, education has fallen victim to the ideals of neoliberalism in a variety of ways, but the most common seems to be preparing students for the economy of jobs (Lipman, 2011).

We use the word neoliberalism as if it were a commonly known term. We know that it is not. Over the years as we have attended conferences and listened to presentations, read articles and books on education in general and science education in particular. We are struck by the scarcity of the use of this term in scholarship. But it is not just a lack of using the term neoliberalism that draws our attention, but more so the implication of scholars not understanding how neoliberalism has positioned education, science education, and research on the teaching and learning of science and science teachers. Ignorance of, failure to acknowledge, or even a desire to deliberately not recognize how neoliberalism has positioned the function of education and science education has consequences not only for education reform efforts but also for transformative efforts. A good example is the public's insistence on charter schools – especially by disenfranchised communities – because these schools will provide a better education. "The power of the neoliberal agenda is visible in the hard and partly successful work that the Right has done in convincing many people, including some Black and Latino/Latina activist groups, that neoliberal policies offer a more realistic hope for the future of their children than, say, existing state-supported schools" (Apple, 2015, p. 173).

A lack of a clear picture of the connection between schools and the future workforce under the rubric of neoliberalism is missing from an insistence on seeing charter schools as the answer to the woes of the disenfranchised. "In the neoliberal risk society, young people have to 'chase credentials' (Jackson & Bisset, 2005, p. 196) to gain security in future education or workplaces. Failure to achieve is deemed one's own fault, and "human beings are made accountable for their predicaments" (Wilson, 2007, p. 97). Another point we suggest is that getting credentials is only one side of the story. The other side of the story is this: Even though students have obtained the correct credentials, do they have the correct and necessary cultural and social capital (Bourdieu, 1986) to take advantage of future economic opportunities? Perhaps once students realize that they do not have the correct cultural and social capital, they understand that "Life comprises events over which we have considerable, partial, or little to no, control. Lives move forward depending upon how events play out and opportunities open in concert with our reactions to them. The distance between the event and our daily lives suggests a quirky spatial politics" (Aitken, 2014, p. ix).

The social dynamics underpinning neoliberalism have managed to implement an order in which institutions are made to guarantee the reproduction of established social order. Victims of establishing a hegemonic social order under the rubric of neoliberalism include the educational spaces in which schools are situated. Roth (2003) reminds us that "hegemony is a struggle of garnering the existing social order that attributes different social positions to those whose knowledge gives them control over societal processes" (p. 258). As such, under the neoliberal *flag* there has been an intensification in the cultural standardization and organization of feelings and sentiments (Haskell, 1985) in such a way that we can recognize a society educated through "hegemony as a pedagogical process" (Giroux, 1992, p. 186).

For decades, education and other policies have focused on economic development all while ignoring the wellbeing of young people, leading to not only frustration but also a reduction of future options. What role did we play in this deterioration of the health and welfare, as well as this creation of illusory futures? What are our moral, ethical and responsibility mandates to try and counter a focus on economic development? Is there still time to change our future? Do we have any options? Is education in general and science education in particular a way to improve our future? What kind of education do young people need in a society in crisis? Perhaps we should take Roth's (2003) message to heart in that we need to enact critical science education as reflexive practice, deconstructing our own role in the reproduction of an inequitable society by reproducing conservative teachers, administrators, policy-makers and politicians. At what point and time can we make it possible for transforming society through the development of citizens with critical thinking? We believe that as long as the relationship between the economy and education is dominated by neoliberalism, transformation will be possible, but will be looked upon as being impractical.

CONSTRUCTING OPTIONS FOR THE FUTURE OF YOUNG PEOPLE: WHAT KIND OF SCIENCE EDUCATION IS NEEDED?

Some researchers in the field of science education have addressed and analyzed the global crisis of humanity. Tobin (2015), for example, recognizes the great challenges that humanity must solve in order to ensure, among other things, wellbeing and sustainability. The implication of Tobin's words is that priorities for science education must change to address new and emerging lifestyles and sustainability. In other words, we need to strengthen forms of teaching, learning, and research focused on youth by creating forms of literacy that are useful in their lives and not a literacy to fulfill an enterprise like science. Bourdieu (1976) hypothesizes that, if one usually observes a very close correlation between scientifically constructed objective possibilities, opportunities to access this or that, and subjective expectations, motivations and needs, it is not because agents consciously adjust their aspirations to an accurate assessment of their chances of success, but because they have been inculcated in some way preordained to their demands. This observation leads us to wonder to what extent have social institutions, such as education amongst many, have had an implicit and duplicative role in developing individuals with dispositions that have led us to this global crisis amongst young people? Additionally, if we use neoliberalism as a filter, why should the underrepresented and disenfranchised care about scientific literacy when the majority of those who produce this knowledge are represented by enfranchised white males?

In our view, being aware of what is happening in our society implies recognizing that the *crisis* of young people is neither an abstract nor a blurred societal problem. The crisis with young people has many faces and is manifested in peculiar ways in each social group. Prevalent among young people throughout the world are crises associated with health, emotions, education, and employment – problems such as these must be addressed. If we ignore these crises we run the risk of increasing the acuteness of the current situation regarding young people. Thus, we believe that understanding social practices and their origins allows us to problematize reality and understand both why what happens and how our social organizations are reflected in our ways of life. What happens to us is not a coincidence, which means we cannot leave our future in the hands of fate nor in those sociocultural forces that position young people and society. Once we problematize our reality within a sociocultural perspective, we see the possibility of developing in individuals a critical awareness that allows them to re-know themselves as individuals and as social beings in order to imagine and build their future options. Social places such as education generally – and scientific education specifically – can be places that allow for and encourage the construction of individual options for the future to be relevant socially and individually. However, we cannot ignore that education has been positioned as a means to further entrench neoliberalism as a hegemony of oppression for a large number of young people.

Education can be a very powerful tool for the economic and social transformation of people. Accordingly, we propose an education that can develop in individuals a critical awareness of their social context and development of their potential role in the transformation of a society concerned with improving the health and welfare of all, and not just economic aspects. We are in agreement with Xicoténcatl Martínez (2016): "If education is a reflection of a society and if we improve education, then, the ways to achieve this cannot be violence or isolation because both reveal a system whose parts cancel dialogic understanding and construction for transformation" (p. 7). Specifically, we advocate for an education in which the current problems of our society are addressed to include issues that directly affect our students. We have noticed that in science teaching, the emphasis seems to consist of content that is not associated with current global crises. As such, we believe that if science teaching and learning is to be transformed, it is important to situate content with current political, economic, social, and cultural events.

The type of education that we propose is a response to a world in crisis. It is a way for our young people to inherit a better world to live in. It is an education that allows each young person to develop the best of themselves. It is a type of education that promotes critical awareness of sociocultural inequities and social injustices and offers agentic tools to change society on an individual basis. It is an education framed within a true sense of creating not illusive future opportunities but one that promotes the development of cognitive and emotional wellbeing through an attainable future. It is an education that allows all to participate in the development of the next decision-makers, governors, and teachers. That guarantees sustainable development. It is an education that is not based on current practices – which is to say, controlled by neoliberalism, emphasizing the construction of individuals lacking the sense of place necessary to inhabit their world in harmonious ways.

Power should not exist in the hands of a few. Education should be transformed to create the tools necessary for students to have the power to take agentic action against discriminatory practices that produce inequity. We are advocating an education committed to the dignity and freedom of all human beings and the ecosystems that support life.

REFERENCES

Aitken, S. C. (2014). *The ethnopoetics of space and transformation: Young people's engagement, activism and aesthetics*. Farnham: Ashgate.

Apple, M. (2015). Understanding and interrupting hegemonic projects in education: Learning from Stuart Hall. *Discourse: Studies in the Cultural Politics of Education, 36*(2), 171–184. Retrieved from https://doi.org/10.1080/01596306.2015.1013245

Asociación Nacional de Universidades e Instituciones de Educación Superior. (2012). *Mercado laboral de profesionistas en México: Diagnóstico 2000–2009 y prospectiva (2012–2015)*. México, D.F., México: ANUIES.

Bhabha, H. (1994). *The location of culture*. New York, NY: Routledge.

Bourdieu, P. (1976). Le sens pratique. *Actes de la Recherche en Sciences Sociales, 2*(1), 43–86.

Bourdieu, P. (1986). The forms of capital. In J. Richardson (Ed.), *Handbook of theory and research for the sociology of education* (pp. 241–258). New York, NY: Greenwood.

Bourdieu, P., & Wacquant, L. (1992). *An invitation to reflexive sociology*. Chicago, IL: University of Chicago Press.

Bowles, S., & Gintis, H. (1976). *Schooling in capitalist America*. New York, NY: Basic Books.

Calabrese Barton, A. (2003). *Teaching science for social justice*. New York, NY: Teachers College Press.

Carnoy, M., & Levin, H. M. (1985). *Schooling and work in the democratic state*. Stanford, CA: Stanford University Press.

Ellison, S., & Allen, B. (2016). Disruptive innovation, labor markets, and Big Valley STEM school: Network analysis in STEM education. *Cultural Studies of Science Education, 13*(1), 267–298. Retrieved from https://doi.org/10.1007/s11422-016-9786-9

Encuesta Nacional de Salud y Nutrición. (2012). *Encuesta nacional de salud y nutrición. Resultados*. México, D.F., Mexico: Instituto Nacional de Salud Pública-Secretaría de Salud.

Gallard, M. A. J., Pitts, W., Brkich, K., & Ramos de Robles, S. L. (2018). How does one recognize Contextual Mitigating Factors (CMFs) as a basis to understand and arrive at better approaches to research designs? *Cultural Studies of Science Education*. Retrieved from https://doi.org/10.1007/s11422-018-9872-2

Gallard, M. A. J., Pitts, W., Ramos de Robles, S. L., Flores Bustos, B., Claeys, L., & Brkich, K. (forthcoming). *Discerning contextual complexities in STEM career pathways – insights from successful Latinas*.

Giroux, H. A. (1992). *Border crossings: Cultural workers and the politics of education*. New York, NY: Routledge.

Giroux, H. A. (2001). *Theory and resistance in education: Towards a pedagogy for the opposition*. Westport, CT: Greenwood Publishing Group.

Giroux, H. A. (2009). *Youth in a suspect society: Democracy or disposability?* New York, NY: Palgrave Macmillan. doi:10.105719780230100565

Gupta, M., Engelman, R., Levy J., Luchsinger, G., Merrick, T., & Rosen, J. E. (2014). *The power of 1.8 billion adolescents, youth and the transformation of the future: State of World Population 2014*. New York, NY: UNFPA. Retrieved from http://www.unfpa.org/sites/default/files/pub-pdf/EN-SWOP14-Report_FINAL-web.pdf

Harding, S. (2003). A world of sciences. In R. Figueroa & S. Harding (Eds.), *Science and other cultures: Issues in philosophies of science and technology* (pp. 49–69). New York, NY: Routledge.

Haskell, T. (1985). Capitalism and the origins of the humanitarian sensibility. *American Historical Review, 90*(3), 339–361.

Jackson, C., & Bisset, M. (2005). Gender and school choice: Factors influencing parents when choosing single-sex or co-educational independent schools for their children. *Cambridge Journal of Education, 35*(2), 195–211. Retrieved from https://search.crossref.org/?q=doi.org%2F10.1080%2F03057640500146856#

Jiang, C., Mitran, A., Miniño, A., & Ni, H. (2015). Racial and gender disparities in suicide among young adults aged 18–24: United States, 2009–2013. *National Center for Health Statistics*. Retrieved from http://www.cdc.gov/nchs/data/hestat/suicide/racial_and_gender_2009_2013.pdf

Lipman, P. (2011). Neoliberal education restructuring: Dangers and opportunities of the present crisis. *Monthly Review, 6*(3) 114–127. Retrieved from https://search.crossref.org/?q=10.14452%2FMR-063-03-2011-07_13#

Martínez, X. (2016). La libertad comprometida o el diálogo para educar. *Innovación Educativa, 16*(71), 7–10.

Mayer, F., & Tetreault, M. (1994). *Feminist classrooms*. New York, NY: Basic Books.

Mexica Foundation for Health. (2016). *Salud en la prensa digital*. Retrieved from http://funsalud.org.mx/portal/wp-content/uploads/2016/05/Salud-en-la-Prensa-Digital-31-de-mayo-de-2016.pdf

National Center for Education Statistics. (2015). *2016, digest of education statistics 2015, table 219.70. Percentage of high school dropouts among persons 16 to 24 years old (status dropout rate), by sex and race/ethnicity: Selected years, 1960 through 2014*. Retrieved from https://nces.ed.gov/programs/digest/d15/tables/dt15_219.70.asp

Navarro, J. (2016). Bipolaridad esquizofrénica de la SEP. *Editorial Revista Educarnos*. Retrieved from http://revistaeducarnos.com/2016/09/20/bipolaridad-esquizofrenica-de-la-sep/

Note, A. (2015). Henry Giroux on the rise of neoliberalism. *Humanity & Society, 39*(4), 449–455. Retrieved from https://search.crossref.org/?q=10.1177%2F0160597615604985#

Office of the Secretary-General's Envoy on Youth. (2015). *10 things you didn't know about the world's population.* Retrieved from www.un.org/youthenvoy/2015/04/10-things-didnt-know-worlds-population/

Organización para la Cooperación y el Desarrollo Económicos. (2015). *Serie: "Mejores Políticas". México. Políticas prioritarias para fomentar las habilidades y conocimientos de los mexicanos para la productividad y la innovación.* México, D.F., México: OCDE.

Proctor, B. D., Semega, J. L., & Kollar, M. A. (2016). *Reports. U.S. Census Bureau, current population reports, P60-256(RV), income and poverty in the United States: 2015.* Washington, DC: U.S. Government Printing Office.

Roth, W.-M. (2003). On science (education), learning, and hegemony. *Journal of Curriculum Studies, 35*(2), 247–262. Retrieved from https://doi.org/10.1080/00220270210151791

Sahlberg, P. (2010). Rethinking accountability in a knowledge society. *Journal of Educational Change, 11*(1), 45–61. Retrieved from https://doi.org/10.1007/s10833-008-9098-2

Sandeen, C. (2008). Boomers, xers, and millennials: Who are they and what do they really want from continuing higher education? *Continuing Higher Education Review, 72*, 11–31.

Sandoval, M. (2012). La desconfianza de los jóvenes: sustrato de malestar social. *Última década, 20*(36), 43–70.

Secretaría de Educación Pública. (2011). *Encuesta nacional de deserción de la educación media superior 2011.* México: SEP-SEMS-COPEEMS. Retrieved from http://www.sems.gob.mx/work/models/sems/Resource/10787/1/images/Anexo_6Reporte_de_la_ENDEMS.pdf

Sewell, W. (1999). The concept(s) of culture. In V. E. Bonnell & L. Hunt (Eds.), *Beyond the cultural turn: New directions in the study of society and culture* (pp. 35–61). Berkeley, CA: University of California Press.

Tobin, K. (2006a). Aligning the cultures of teaching and learning science in urban high schools. *Cultural Studies of Science Education, 1*, 219–252. Retrieved from https://doi.org/10.1007/s11422-005-9008-3

Tobin, K. (2006b). Global reproduction and transformation of science education. *Cultural Studies of Science Education, 6*, 127–142. Retrieved from https://doi.org/10.1007/s11422-010-9293-3

Tobin, K. (2015). Connecting science education to a world in crisis. *Asia-Pacific Science Education, 1*(2), 1–21. Retrieved from https://doi.org/10.1186/s41029-015-0003-z

Tobin, K., Alexakos, K., Malyukova, A., & Gangji, A.-K. H. (2017). Jin Shin Jyutsu and ameliorating emotion, enhancing mindfulness, and sustaining productive learning environments. In A. Bellocchi, K. Otrel-Cass, & C. Quigley (Eds.), *Beyond cognition in science education* (pp. 221–247). Dordrecht: Springer. Retrieved from https://doi.org/10.1007/978-3-319-43353-0_12

UNESCO. (2011). *Education for all global monitoring report 2011.* Paris: UNESCO.

United Nations. (1992). *Agenda 21.* New York, NY: United Nations.

United Nations. (2010). *Programa de acción mundial para los jóvenes.* New York, NY: United Nations.

United Nations. (2016). *Transforming the world: Parity in science.* Retrieved from http://www.un.org/en/events/women-and-girls-in-science-day/

United Nations Educational, Scientific and Cultural Organization. (2015). *Education for all 2000–2015: achievements and challenges. The 2015 EFA Global Monitoring Report.* New York, NY: UNESCO.

United Nations International Children's Emergency Fund. (2011). *Adolescence an age of opportunity.* New York, NY: United Nations.

University of Pittsburgh Schools of the Health Sciences. (2016, December 20). Multi-social millennials more likely depressed than social(media)ly conservative peers. *ScienceDaily.* Retrieved April 5, 2017, from www.sciencedaily.com/releases/2016/12/161220175543.htm

Wilson, B. M. (2007). Social justice and neoliberal discourse. *Southeastern Geographer, 47*(1), 97–100. Retrieved from https://doi.org/10.1353/sgo.2007.0016

World Economic Forum. (2016). *The global gender gap report 2016.* Retrieved from http://www3.weforum.org/docs/GGGR16/WEF_Global_Gender_Gap_Report_2016.pdf

World Health Organization. (2016). *Mental health: Suicide data.* Geneva: World Health Organization. Retrieved from http://www.who.int/mental_health/prevention/suicide/suicideprevent/en/

Wylie, A. (2003). Why standpoint matters. In S. Harding & R. Figueroa (Eds.), *Science and other cultures: Issues in philosophies of science and technology* (pp. 26–48). New York, NY: Routledge.

ABOUT THE AUTHORS

S. Lizette Ramos de Robles is a professor of Biology and Environmental Health Sciences in the Department of Environmental Sciences at the Universidad de Guadalajara, Mexico. Her research focuses on sociological and cultural aspects of discourse in the teaching and learning of science, especially in multilingual contexts. To understand scientific literacy more deeply, she uses sociocultural and sociolinguistic perspectives. In addition, her research also focuses on socio-environmental aspects related to Environmental Health, specifically climate change literacy.

Alejandro J. Gallard Martínez is a professor and Goizueta distinguished Chair in the Department of Teaching and Learning, and the Director of the Georgia Center for Educational Renewal. Alejandro's interests include researching societal complexities promulgated by contextual mitigating factors (CMFs) that contribute to students' lack of success (or success) in general and in the STEM fields. His frameworks include global perspectives on differences, otherness, polyphony of voices and meaning making that reflects categories used to situate people in social life. His current research efforts include working with an international team to understand what are the characteristics of Latinas who are successful in the STEM fields.

RALPH LEVINSON

4. I KNOW WHAT I WANT TO TEACH BUT HOW CAN I KNOW WHAT THEY ARE GOING TO LEARN?

Creative Science Teaching: An Uncertain, Emancipatory and Perturbing Endeavour

ABSTRACT

Through reflections on my own history of learning and observing significant teaching episodes of others in teaching science I argue for a transformative programme of science teaching which is knowledgeable and non-presumptive. In doing so I draw on Hannah Arendt's depiction of *praxis* as central to the collaborative enterprise and openness of learning, and Emanuel Levinas' conceptualisation of the Other in creating an ontology of nature which is non-dominant. In neoliberal education systems driven by an instrumental and corporatist STEM agenda such possibilities might seem remote but placing social justice at the very core of pedagogy, resistance to short-term instrumental outcomes can be successfully achieved through 'collaborative resonance'.

Keywords: STEM, neoliberalism, transformative pedagogy, creativity, knowledgeability, non-presumptiveness, praxis, the Other, collaborative resonance

THE CRITICAL ISSUE

Those advocating science for social justice pedagogy within school curricula face political, economic and educational challenges in the face of STEM, (Science, Technology, Engineering, Mathematics) education. This acronym – almost an ideology so quickly and so deeply has it pervaded educational discourse and become synonymous with 'science' – has crept up stealthily on the world of science education, accompanied by terms such as 'innovation', 'competitiveness', 'entrepreneurship', 'enterprise' and 'excellence' (The Royal Society, 2009). It points to a move in science and technology education to liaise more closely with business and private enterprise than with the public and state sectors. STEM education proponents make particular claims for human capital to support technological innovation in a consumerist society, as evidence, for example, by encouraging STEM careers in science lessons (Hutchinson, 2012) and is itself a science-society formulation; hence, it melds very well into calls to link the study of science more closely to social concerns (Gough, 2015).

 DOI:10.1163/9789004389663_004

Implicit, too, in the term STEM is inter-disciplinarity. U.S. schools and UK schools need to emphasize a STEM epistemology to justify its social and economic existence and the ways in which science can support technology and engineering, although little work on pedagogic and curricular integration has been done in this direction. But the motivations of STEM are economic and corporatist. They are also political because the question arises as to whose benefits such changes are directed (Owen et al., 2009).

Extracts from the foreword to The Royal Society's (2009) report *Hidden Wealth* exemplify these purposes:

> How has science contributed to this growth of wealth and enhanced quality of life via services?
>
> we set out to answer a simple question: where has science – in the widest sense – already contributed well to fostering innovation in the services sector and where and how might new policies enhance the situation?
>
> We anticipate services delivered much more cheaply, to better quality and personalised to millions of individuals where that is desired. While much of this will be provided by the private sector, government can enhance its own services hugely by cloning the best of private sector developments to maximise value for taxpayers' money and to strengthen democracy…
>
> Ever better collaboration between STEM practitioners and social scientists and those in the humanities will be essential if the services are to be acceptable and fit for purpose. Changes in our educational system would also make a material contribution to such success. (p. v)

STEM has now become a clarion call in the push for national economic competitiveness. Drives towards more STEM education for the labour market have become a cornerstone of EU policy (Caprile et al., 2015). Wealthy economies which fail to reach the heights of the Programme for International Student Assessment (PISA) scores of the powerful East Asian economies resort to introspection and 'standards speak', demands for more content-laden delivery of the curriculum and improved assessment scores. The UK opposition spokesperson on education pointed to "… the importance that these [Pacific rim] high-performing education systems place on the quality and status of the teaching profession as the central lever for driving up standards" (Adams, 2013). The UK education minister at the time, Michael Gove, forefronted his Parliamentary statement in response to PISA 2013 scores with a drive towards social justice, dressed in the language of increased marketisation of the education sector and a call for greater accountability and performance. *The Telegraph*, a centre-right newspaper reported that government reforms to the curriculum will "scrap 'vague', non-scientific topics such as caring for animals and societal context" (Dominiczak, 2013).

Despite the recruitment drive for the STEM corporate market, national attitudes towards science among young people stand in inverse relationship to national GDPs

(Rose, 1998). Scandinavia and Japan score the lowest for young people's positive attitudes towards science (Schreiner & Sjøberg, 2007) while Uganda top the list. The writings among socialists in 1930s Britain (Bernal, 1969) seeking to overturn the excesses of capitalism through increased knowledge of, and engagement in, science stand in a curiously inverse relationship to contemporary policies: one links increased understanding of science to liberation from the shackles of capitalism, the other focuses on rigour to compete in the global market.

The neoliberal assault on school education over the last twenty years has affected teaching and the curriculum as well as school organisation, with a school culture focused on short-term outcomes driven by examination results (Hargreaves & Shirley, 2009), pedagogic conformism and performativity generated through new technologies of control, so that teachers become "fabricated", changing their identity to represent the performative culture of the organisation for appraisal (Ball, 2003).

While globalisation and neoliberalism, and the concomitant drive towards cultural homogeneity, are the dominant economic factors prevalent in STEM, there are local and historical differences which constrain what is possible, and the perspectives through which teachers, school policy makers and students come to these issues: in the UK, for example, the influence of science for the worker since the industrial revolution (Layton, 1973), the relationship of academic subjects to vocational subjects (quite different from the situation in Germany) (Osborne & Dillon, 2008), and the influence of a liberal education distinguish the challenges facing science education in the UK compared with other countries. What might be appropriate pedagogical shifts in response to STEM in one country might not apply to another despite globalisation.

In addressing what I see as the subjugation of science teaching to the consumer market, I want to turn the narrative to an emphasis on epistemology and creative science teaching before coming back to some cautious suggestions to transform the present situation.

What Science Might Mean to a Young Person

When I was 16, I had an epiphany – in a secular sense. One afternoon when I was not focusing on anything in particular it struck me, as if lightning had flashed out of an unlikely sky, that the ideas of atomic chemistry were awe-inspiring. Here was a world of atoms, protons, electrons, spin, orbitals which explained so much; non-perceptible, real yet entirely the product of human thought and social practise. I felt just at that moment how ineffably joyful – and possible – the world is, a sensibility that has never quite left me. Until that point, I had been using my school notes and textbooks to solve chemistry problems but had never imagined, let alone understood, what existential processes were at stake. DNA, continental drift, neurobiology, nuclear physics now became both real and mind-boggling.

Over the years, there appeared to me to be a link between this experience and a commonality among creation stories whatever their sources – Judaeo-Christian, Maori, Hindu – the shift from darkness to light (or in pedagogic terms, illumination), formlessness to structure, in other words, an ontological transformation. Although I regard myself as completely secular I feel the Old Testament depiction of Creation as the shift from 'tohu vavohu' (formlessness) to the illumination of being (God said, 'Let there be light') is more meaningful for the scientific sense of creativity than the ideas of innovation, novelty and value (Kind & Kind, 2007) or through cognitive strategies (Mayer, 1989).

The world was different after my epiphany: it was patterned. It had become describable and hence explicable. This presupposed certain characteristics that gradually became explicit. In the words of Richard Rorty (1989), "The world does not speak. Only we do" (p. 6). The conceptualisation of Nature as explicable through the mediation of human consciousness, therefore socio-cultural – as opposed to supernatural intervention as in creation myths – seems to me to be such a crucial point in justifying science on the curriculum that it is surprising that it is at best implicit – I see a distinction here between the experience I describe which is integral to what it is to be human from concept change theory (Strike & Posner, 1982). The latter seeks to explain how young people's concepts might change without necessarily taking into account the personal meanings associated with such changes. Its immediacy was brought home to me recently through reading a poem of Wallace Stevens, *The Idea of Order in Key West*, which speaks to the nature of reality (Kermode & Richardson, 1997). In this poem, Stevens is relating to a friend of his feelings on hearing a woman singing by the seashore. The woman's voice conveys the sound of the sea which brings to the fore the mediation of human interactions in our perceptions of the world.

> The song and water were not medleyed sound
> Even if what she sang was what she heard,
> Since what she sang was uttered word by word.
> It may be that in all her phrases stirred
> The grinding water and the gasping wind;
> But it was she and not the sea we heard. (pp. 105–106)

My reading of the poem is that the rhythm, patterning and mythical resonances of the sound of the sea do not reach our consciousness directly but are mediated through the voice and consciousnesses of others. Meaning, saturated with cultural and emotional resonance, emerges through human discourse in its most general sense. Yet that meaning, the form and structures, which seem so compellingly real – the Periodic Table, structure of DNA, habitat inter-relationships, mechanisms of energy transfer – is inherently unstable because humans continue to argue and interpret Nature and upset the paradigms to which we become so habituated (Kuhn, 1970). The uncertainty between 'what is' – the canon of school science – and 'what might be' is at the core of human possibility.

The Role of Pedagogy

So how might teaching support these kinds of transformations even in the context of contemporary science education?

Underpinning my view of the environment necessary for creative teaching – challenging the hegemony of STEM – and hence transformative understanding, arose from an incident about ten years ago. It was in a debriefing meeting I had in my role as a teacher educator with a pre-service teacher, Tom, a thoughtful and knowledgeable science graduate.

When I came to observe Tom teach a Year 10 group (14–15-year-olds), I was met before the lesson by his school mentor, Judy, who told me that she had serious concerns about his progress and that "in all likelihood teaching isn't for him." Since his planning for lessons was unusually meticulous as I had seen through the materials he prepared and a narrative he practised at the beginning of each week, I was taken aback. There was little time before the lesson so we agreed to talk about his progress after it. The episode I saw him teach was based on a question: "How can we know what's inside the Earth?" He began with a poem about an earthquake and had a number of models set around the room, which the students had to interrogate in terms of their plausibility as explanatory structures. As they did this, he encouraged the students to articulate two things: first was the difference between their understanding of the earth's structure and inner composition and what the models were portraying; and secondly anything they could say about the validity of the models, i.e. what they represented. He had mistimed the lesson (not uncommon among beginning teachers). It ended rather earlier than he had planned but not before students were expressing their surprise at the differences between their own representations of the Earth's structure and the models, as well as curiosity as to how models come to be constructed. I overheard a comment from one student to his friend as he left the lesson: "It makes you wonder, man. I thought it was all solid rock down there." At least this student's transformed understandings of this aspect of Nature seemed to reflect my own epiphany many years before.

After the lesson, Judy, who had supported many of our student teachers and with whom I had a good professional relationship, asked me if I could now see the problem. I was baffled. After I had a moment to recover, I asked what her concerns were. "The lesson outcomes and objectives. Where were they? And what about the starter, the middle and the plenary?" At this time in England, there was a National Strategy advising that all lessons should have, or rather conform to, a three-part structure.

Judy was a mentor whose judgement I had always trusted, but I realised that if she was undergoing this process of fabrication we were in trouble. When I debriefed with Tom, I asked him why he hadn't put the learning objectives to the class at the beginning of the lesson and why they weren't on his plan. Lesson outcomes and objectives go on the planning sheet and I took it as common-sensical that these were part of the tools of lesson planning. After a moment's thought, peering at me ingenuously, he responded: "I know what I am going to teach. But how can I know what they are going to learn?" It took me some time to understand the power of this observation. It was not simply about learning a few propositions such as "the Earth is made up of different layers" but what *meaning* students made of them. In many ways, I was as fabricated through working to imposed teacher standards as Judy, the teaching mentor.

What Tom was doing was encouraging the students to think about what is known about the Earth's structure, how we come to know this given the provisionality of scientific knowledge, and what it means. He had a deep knowledge of the topic, hence his care in selecting appropriate models and his ability to gently present students with counter-factuals when they made assertions. But he always left the door open for them to come back with further justifications for their own conclusions, nor would he impose the accepted right answer. The students, first implicitly, then explicitly, were coming to appreciate that Nature as represented wasn't 'given': reason, affect, inference and knowledge came to temper the 'rough ground' (Dunne, 1993).

When I was reading Hannah Arendt's *The Human Condition,* inspired by Joseph Dunne's book *Back to the Rough Ground*, I began to see the pedagogic importance of what Tom had taught me and what had taken me some time to assimilate. *The Human Condition* was written after the Second World War and among its many messages is a reflection on the nature of tyranny. Arendt distinguishes between two forms of reasoning: *techne* and *phronesis*.

Techne is associated with purposive reasoning encompassing activities (called *poiesis*), which have formulated ends, where technical proficiency leads to pre-defined outcomes. *Poiesis* is not necessarily a mindless process; it involves a mastery of certain techniques. It is craft with a rationality underpinning it. This has some relation to what is meant in contemporary language by technicism or technicist, the rationale is a means to a specified end. It is to operationalise, to concretise something that is given, e.g. a learning outcome. So, a teacher who prescribes a learning objective and directs their teaching to meeting that particular given objective is in a sense technicist, and in standard terms a very good teacher. The idea is prescribed from somewhere *beyond the person* and their job is to implement it fully. The choice of what is prescribed is not theirs; hence *techne* was something looked on askance by the free citizen in the Greece of Aristotle.

Praxis, as distinct from *poiesis*, is a realisation of one's self, one's inner worth, a mediation of the social and individual, and indicates less control over a means-end relationship over which you had complete sovereignty, a more personal and experiential knowledge. Associated with *praxis* are contingency, uncertainty and heterogeneity, quite distinct from the more circumscribed practise of *poiesis*. This practical knowledge or wisdom is *phronesis*, and although non-technical has an underpinning rationality. Arendt expands the notion of *praxis* in relation to 'action' as opposed to 'making'. To take action is to put one's ideas to the judgments of others where one ceases to be the actor – it is to acknowledge uncertainty and change and accept the limits of control. Human agency, which underpins *praxis,* is deemed to incorporate distinctiveness and equality. Distinctiveness presupposes speech and action because these relay the agent's need to reveal who they are and make themselves understood. Equality is the condition for mutual understanding, hence recognition for the past and an ability to co-operate in planning for the future.

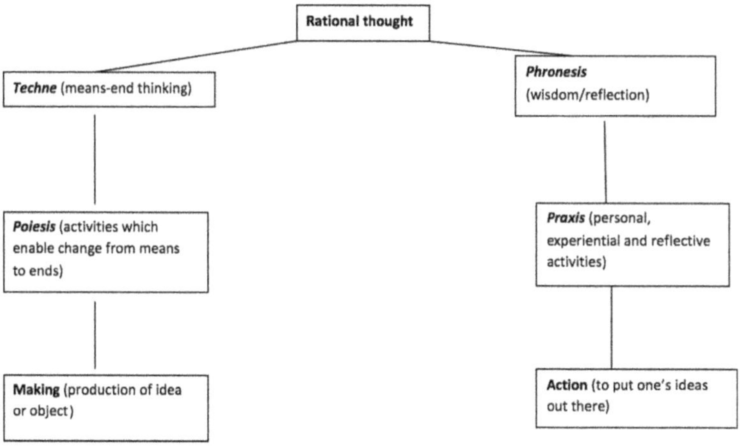

Figure 4.1. Making and action

The rationale behind modern theories of *praxis*, derived from both Hegel and Marx, is the realisation of consciousness through action

> Embedded within a tradition of communally shared understandings and values, that remain vitally connected to peoples' life-experience, that finds expression in their ordinary linguistic usage, and that, rather than being a means through which they achieve outcomes separate from themselves, is a kind of enactment through which they constitute themselves as persons in a historical community. It is through *praxis* that a person comes to have an individual identity, but at the same time it always transpires within an intersubjective medium… The moral subject, the subject of *praxis*, is inconceivable in abstraction from communicative relations with others. (McCarthy, 1978, p. 35)

So the relationship between knowledge and action is turned the other way around, that is in a Deweyan sense, knowledge is accrued through collaborative inquiry in acting upon the world (Tobin, 2014). Action becomes an existential choice which becomes more challenging in a world saturated with discourses promoting a uniformity of consumption.

This is the first distinction I want to illustrate using this example, that Tom understood something about the ways young people might conceptualise reality and that he could not reify any meaning when confronted with something which challenged their prior representations. So Tom could have specified certain concepts for students to explain, such as the Earth's structure is made up of x layers. Instead his approach was to lay out the problem as best he could, indicating the strength and tentativeness of what was and what was not known, and allow students to interpret this knowledge in their multiplicity of ways.

Nature, Science and Pedagogy

The post-Enlightenment relationship of the Self to the Other is one that has human subjectivity (the Self) imposing order and meaning on Nature (the Other). The project of modern science is intrinsically connected with power and domination:

> From its Baconian inception, modern science has been about both knowledge and power, above all the power to control and dominate nature, including human nature. Nowhere perhaps has this Faustian pact been made so explicit as in the programme that has shaped molecular biology since its origins. (Rose, 1998, p. 273)

For the philosopher Emmanuel Levinas (1993), ethics begins with the realisation of the Other, that the Self is constituted by and responsible for the Other by opening oneself to the Other non-reciprocally, (the Other has no ethical responsibility towards me, i.e. the Self), consequently the possibility of social justice. In Western philosophy the Other is appropriated as part of the same and loses its alterity. Ann Chinnery (2000), drawing on Levinas, exposes the problem of neglect of the Other even when people act altruistically towards other human beings in the name of common humanity. This advocacy of sameness, of commonality – "they are after all, one of us" – can also impel people to act barbarously towards human beings because they do not see them as the same. In other words, one's moral obligations are perceived to extend only to those for whom there is a perceived sameness. For Levinas, there is no pre-ethical subject: the lesson for education is moral responsibility, for the other lies outside familial or even species relationship, and the Self becomes realisable through the acceptance of the Other.

This refusal to impose one's subjectivity, the acceptance of the Other, the notion that we always have a responsibility for the Other, has consequences for pedagogy. An everyday example is the difference between the question "Would you like to come into my home?" and "Welcome to my house". The first asserts a subjectivity; it puts the onus of the guest to accept my invitation. The second is an opening up without any pre-conditions. The teacher must welcome difference in terms of the voice of the students and the Otherness of the natural world, both biotic and abiotic. All actions taken on the world must take into account the needs of others, the sharing of the world (Blades, 2006). To impose learning objectives is not to see the otherness of the student, it is to stamp one's own subjectivity on to another. To see students as the Other is to open oneself up to them, where teachers become learners and students as a consequence become teachers as well as learners. This is not to imply a free-for-all, however. As in the case with Tom, the teacher steers a course between inducting the students towards a problem (Le Dit) and the "Saying" (Le Dire). David Blades (2006) refers to the "Saying" as "an invocation to the rupture of my being" (p. 658). Hence, the Other – organisms, their interactions, the students growing knowledge of these interactions – is not a phenomenological construct but is allowed to rupture the experiential world of the subject.

In contemplating the relation of the presence of the Other to Nature, Blades (2006) draws on the notion of a pond. (Levinas did not include non-humans in his call to the Other but others, e.g. Llewellyn (1991) and Blades (2006) have expanded on his philosophy towards the non-human world). A pond has no definable boundaries: the water from the pond overlaps surrounding mud and vegetation, oxygen from the atmosphere slowly dissolves in the water as surface water molecules evaporate into the atmosphere and condense back into liquid water. The call of the frogs from the pond has diminished over the years. But there is a difference between nature in flux – through Natural Selection species come and go – and the responsibility to the Other. One could imagine Nature without humans which then brings starkly forward human responsibility. What would the pond be like without human interference? Might local human-made pollutants be responsible for changing the pond or might it be the relentless change of Nature? We have a responsibility to respect the latter and not to act or, if necessary, to act. Imagination and sensitivity to the Other mediates the choices. Teaching like Tom's episode bring the same ethics to their students as they do in their feelings of responsibility towards Nature, *non-presumptiveness*: a refusal to dominate which accompanies a liberation of the scientific imagination.

The Dawning of Uncertainty

Until now I have dwelt on the point that creative teaching – that which promotes ontological and with it epistemic transformations – sits uneasily with the ways in which neoliberal education systems are run. I want to add a further dimension through the illustration of another teaching episode to further this characterisation of creative teaching and the circumstances in which it might flourish.

John was another student teacher who had completed the course a year or so before Tom. Unlike Tom's school, John's practise was in a school where he had some freedom to make mistakes, and was encouraged to take risks. The lesson I saw him teach was to 11-year-olds about graphing the rise in temperature of water against time, which on the surface looked as bland a lesson as one could possibly imagine. John had shown the students how to read thermometers correctly, to take measurements and to plot a graph. The students were then asked to draw a temperature time graph, heating the water and taking the temperature every 30 seconds until they were confident there was no further rise in temperature. After they had done this he drew their attention to the fact that there were two different types of graph drawn by the students (Figures 4.2 and 4.3). In Figure 4.2, students had drawn a best-fit graph based on their previous knowledge of what happened when water boiled, whereas the others had joined the points together. Without any obvious direction towards a

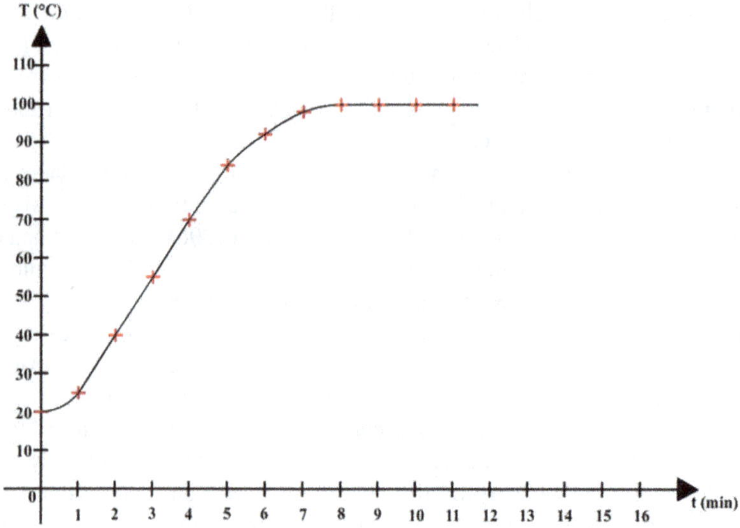

Figure 4.2. Temperature-time graph (best fit)

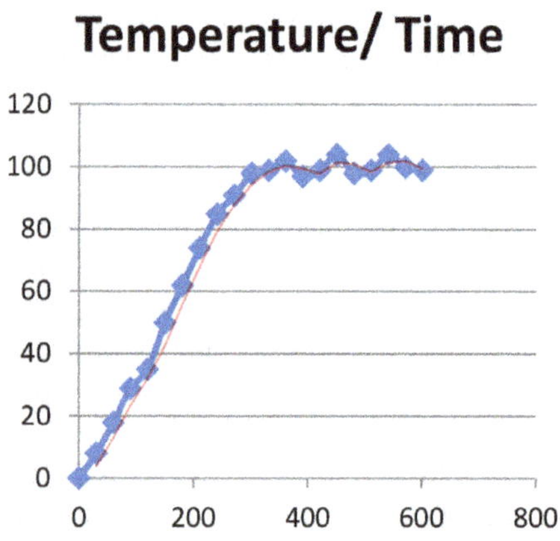

Figure 4.3. Temperature-time graph: Joining up the points

"right" answer John asked the students to talk about the graphs they had drawn. The students who had joined up the data points regardless asked the other students why they hadn't joined up the points. At this juncture the students said they just knew 'that's how the graph should go'. This then led to a discussion of the reliability of data, that all measurements of temperature were prone to error: human error, the constraints of the measuring instrument, and the problem of outliers. How could you know whether a measurement was the right one? So John put the question back to them: how could *they* know the boiling point was 100C? Gradually the students began to realise that no measurements could give certain knowledge and that how you drew the graph was theoretically constrained, drawing on previous knowledge. There was a slight perturbation because they had always accepted as a mantra that water boiled at 100C. Now they could see that those who had not made a best-fit graph had not relied on accepted knowledge. So how do people know that points on a graph formed a pre-determined pattern? What's more John then drew their attention to biological measurements, e.g. of heart rate, which didn't have a best fit. All their certainties were beginning to collapse but at the same time they became conscious that it was human beings like themselves who could infer patterns from data. Rather than a feeling of helplessness there was more a sense of emancipation, that the description of Nature was always provisional, that Nature didn't tell them just how it was, there is always interpretation.

Like Tom, John had refused to impose an answer. Providing a forum for discussion and expressing doubt had allowed the students to accept the problem of certainties, that uncertainty held promise, and that they were intimately connected to shaping their own understanding of Nature. John had no control of where the discussion would go although he had sufficient knowledge to make the grounds for the discussion coherent. It was a risk but one based on respecting what pupils would bring to the discussion, and a respect for the probabilistic account that respect for nature always gave.

Towards a Transformation

One response to the STEM hegemony is to emphasize the links between science and education and activism (Bencze & Carter, 2011; Hodson, 2010), an approach to which I am sympathetic. I have written about how closely the scientific principles of the production of materials for a consumer society are interwoven with the economic ruthlessness which provides cheap goods for the market at the expense of those, unseen and marginalised, who source the raw materials for processing (Levinson, 2014). Activist approaches, however, risk buttressing the rationale for STEM because they are the other side of the instrumental coin. Corporate capitalism is adroit at playing this game because it appropriates resistance to its own advantage (Bauman, 2000). The image of Che Guevara has become commodified into street chic. Coca-Cola, using green iconography, advertise green recyclable bottles to show how much they care for the environment (Balch, 2011), dreamed up, no doubt, by their global head of sustainable packaging (a post in Coca-Cola – it would be impossible to make it up).

I am cautious about aims that counter STEM by stipulating social justice because they risk reproducing the same instrumentality on which STEM thrives. Justice and openness to Nature and the Other have to be core to science pedagogy, not only an aspiration. This is not easy to attain when working in an environment so hostile to such practise. Teachers, whether pre-service or experienced, cannot ignore examination teaching, refuse to teach the set curriculum, or carry out day-to-day duties. They cannot simply opt out of school politics; indeed, it would militate against the meaning of justice and acceptance of the other to do so.

One criticism of acceptance of the Other presupposes toleration of views that are antipathetic to human and non-human respect: racism, sexism, homophobia, colonialism. But this move anticipates a commitment to understanding and critiquing those material factors on which these views thrive. To act in this way is, for example, to take the racist off-guard. It also implies consistent inquiry into motivations and uncovering sources of power, which obstruct praxis. But there are far more everyday aspects with which teachers struggle, e.g., disruptive behaviour – truancy, lack of resources, bullying, which can test the best of intentions in classroom practise.

Marilyn Cochran-Smith (1991) describes a strategy of "collaborative resonance", an opening up of democratic spaces, where pre-service teachers work with more experienced teachers and educationalists whose own practise resists dehumanisation of the school system. Through collaboration, discussing problems and critique, a process of inquiry is made possible which allows a constant reflection on the theory-practice continuum and to problematise their own practice through their social, professional and political contexts. The difficulty, however, is that the performative dominance of schools today is greater than it was 25 years ago, and there are precious few spaces where pre-service teachers can work within the relatively hospitable and encouraging environments that Cochran-Smith describes.

Nonetheless, the point about building on possibilities of resistance is a potentially promising one. In a research project funded by the British Academy – UK and Brazilian colleagues – explored the possibilities and challenges for science practitioners we identified as "teaching against the grain". The criteria for choice were those teachers who were known to us as mentors of pre-service teachers, with at least five years teaching experience, who were well-respected by their colleagues, took an active part in school and national science projects, were admired and loved by their students, and who were known to articulate and problematise issues of social justice through their own practise. Through a narrative approach (Goodson, 2008) in which they were encouraged over a series of sessions to tell the story of their professional development, the stories of the UK teachers demonstrated how recent managerial innovations in schools had conflicted with their own practise. Of the four UK teachers we worked with, two had since left teaching because they felt alienated by developments in their own school, which conflicted with their sense of communality and pedagogy. One was still teaching in his own inspiring way but was marginalised from effective organisational control.

The fourth teacher, Don, with twenty years' experience, not only was still teaching but also was gradually finding ways of influencing national professional bodies. He worked in an inner-city school, multi-ethnic, drawing mainly from areas of high social deprivation. Don was fortunate in working in an institution that historically allowed teachers considerable autonomy but also encouraged professional collaboration and pursuance of higher academic qualifications:

> There were all kinds of freedoms. There was the freedom to develop the kind of curriculum which suits the kids, and I think that was the culture of the place, that whole structure of team meetings, what you were going to teach anybody and being involved in the staff meetings where clearly decisions were made to some degree. There seemed to be a structure not only over which had you had control and also responsibility. For the curriculum you felt responsibility as well. Yes, there was that. But also the other freedom you've got was, and what I was involved in was the freedom to get involved outside … and at the same time I'm reading quite a lot of philosophy and so through the history of science I'm coming across a lot of people like Foucault and really sort of disturbing my philosophical outlook and I'm involved in Marxist politics so that's also having an influence … Theories of knowledge. So I think probably always looking for a secure foundation for what I'm doing. Probably not finding it … (laughter). But you know clearly not comfortable with merely a pragmatic approach.

Don's pedagogy was based on telling stories from within, to encourage students to experience what it was like to encounter new and unexplained scientific phenomena:

> … upfronting the Aristotelian approach and then contrasting that with Galileo's views of motion, say. It's still a method which is used in physics teaching and so having read Aristotle and Galileo it helps. I did have a notion which perhaps grew out of the primary school teaching about narrative and telling stories and that being useful, not only in teaching young people, but teaching, actually having a story to tell helped people relate to the conceptual material that was underlying what you were trying to teach, and also gave it a form. And at the time in the history of science in the early modern period I was studying Latin… and as part of the course I decided to do a summer school in ancient Greek and we studied the Odyssey and the Iliad and I got interested in how those epic poems were designed and so they got a structure and…the beginning and the end reflect each other…and I sort of formed the notion about lesson planning and long term planning based on the structures of poetry.…And if you give this kind of structure to your lessons and your year it may help both you and the kids or the adults to put a pattern to the learning. And so, I'm borrowing from everything that's coming at me.

His problem was not the institution, which encouraged creative disturbance, but the reactions of his students who wanted to know how his pedagogic philosophy would help them pass their examinations. They were seemingly uncurious about how they could explain the world but more concerned with being able to produce the right answers to help them get their university places:

> … particularly young people who really are increasingly instrumental about these two or three years of education … getting them through that and… their examination results which are going to get them on to the next stage and change their lives…and coming from backgrounds where they come from, workless households, this is useful, and so the degree to which you're treating it other than a body of knowledge which you need to get to grips with has to be tempered with the need to get them through the examinations…increasingly being pulled into that, isn't the most efficient method of getting through. So there are a number of tensions creeping in particularly when you become aware of them…which is when they start asking you what's in the exam.

And his approach to this problem of focus only on the examination results leads to negotiation:

> So the notion about being explicit … about why I'm taking this approach and so … I think certainly in the past few years I've been really clear with kids about talking to them about why it is I'm teaching them the way I'm teaching them and engaging them with that. I try not to burden them. I find that quite helpful.

Don's solution to the problem is to explain his pedagogy to the students, so they come to trust his judgement. The refusal to submit only to teaching to the examination is combined with a willingness to be open with the students, to expose his thinking to their consideration, so that authority for learning is shared. Like Tom and John, he refuses to impose his own knowledge but to trust to the students' judgment and their realisation that they are involved together in the project of learning and making sense of the world. Rejection is always a possibility but one which is based on openness and trust.

CONCLUSION

If those concerned by STEM wait on national and international political changes, we are likely to have to wait for a long time. But change can be encouraged from bottom-up as well as top-down, particularly through collaboration, problematisation and shared goals. The point about Don's practise, like those of John and Tom, is that his pedagogy reflects a democratic and collaborative approach. In the performative world, risk and the possibility of failure, cannot be endured but the debate can be opened up and is always sensitive to the contexts of practise. The teaching that stimulates transformative approaches holds a tension between the Said – knowledgeable teaching – and the Saying, that which is open to the Other. The most appropriate term for this seems to be *knowledgeability* – openness to change is encouraged by those whose knowledge of their subject is *non-presumptive* and built upon a notion of social justice at the very core of teaching.

REFERENCES

Adams, R. (2013, December 3). UK students stuck in educational doldrums, OECD study finds. *The Guardian*. Retrieved from https://www.theguardian.com/education/2013/dec/03/uk-students-education-oecd-pisa-report

Arendt, H. (1998). *The human condition*. Chicago, IL: University of Chicago Press.

Balch, O. (2011). Coca-cola in green bottles. *The Guardian*. Retrieved from http://www.theguardian.com/sustainable-business/coca-cola-green-plant-bottles

Ball, S. J. (2003). The teacher's soul and the terrors of performativity. *Journal of Education Policy, 18*, 215–228. Retrieved from https://doi.org/10.1080/14748460902990344

Bauman, Z. (2000). *Liquid modernity*. London: Polity Press.

Bencze, L., & Carter, L. (2011). Globalizing students acting for the common good. *Journal of Research in Science Teaching, 48*, 648–669. Retrieved from https://doi.org/10.1002/tea.20419

Bernal, J. D. (1969). *Science in history*. London: Penguin.

Blades, D. W. (2006). Levinas and an ethics for science education. *Educational Philosophy and Theory, 38*, 647–664.

Caprile, M., Palmén, R., Sanz, P., & Dente, G. (2015). *Encouraging STEM studies for the labour market*. Brussels: European Union. Retrieved from http://www.europarl.europa.eu/RegData/etudes/STUD/2015/542199/IPOL_STU(2015)542199_EN.pdf

Chinnery, A. (2001). Levinas and ethical agency: Toward a reconsideration of moral education [Featured essay]. In L. Stone (Ed.), *Philosophy of education 2000* (pp. 67–74). Urbana-Champaign, IL: Philosophy of Education Society Philosophy of Education Society

Cochrane-Smith, M. (1991). Learning to teach against the grain. *Harvard Educational Review, 61*, 279–310. Retrieved from https://doi.org/10.17763/haer.61.3.q671413614502746

Dominiczak, P. (2013, July 8). Michael Gove: New curriculum will allow my children to compete with the very best. *The Telegraph*. Retrieved from http://www.telegraph.co.uk/education/educationnews/10166020/Michael-Gove-new-curriculum-will-allow-my-children-to-compete-with-the-very-best.html

Hodson, D. (2010). Science education as a call to action. *Canadian Journal of Science, Mathematics and Technology Education, 10*, 197–206. Retrieved from https://doi.org/10.1080/14926156.2010.504478

Hutchinson, J. (2012). Career-related learning and science education: The changing landscape. *School Science Review, 94*(346), 91–97.

Kermode, F., & Richardson, J. (Eds.). (1997). *Wallace Stevens: Collected poetry and prose*. New York, NY: Library of America.

Kind, P. M., & Kind, V. (2007). Creativity in science education: Perspectives and challenges for developing school science. *Studies in Science Education, 43*(1), 1–37. Retrieved from https://doi.org/10.1080/03057260708560225

Kuhn, T. (1970). *The structure of scientific revolutions*. Chicago, IL: University of Chicago Press.

Layton, D. (1973). *Science for the people*. London: Allen & Unwin.

Levinas, E. (1993). *Outside the subject* (M. B. Smith, Trans.). London: Athlone Press.

Levinson, R. (2014). Undermining neo-liberal orthodoxies in school science: Telling the story of aluminium. In L. Bencze & S. Alsop (Eds.), *Activist science and technology education* (pp. 381–388). Dordrecht: Springer.

Llewelyn, J. (1991). *The middle voice of ecological conscience*. London: Macmillan.

Mayer, R. (1989). Cognitive views of creativity: Creative teaching for creative learning. *Contemporary Educational Psychology, 14*, 203–211. Retrieved from https://doi.org/10.1016/0361-476x(89)90010-6

McCarthy, T. A. (1978). *The critical theory of Jürgen Habermas*. Cambridge, MA: Massachusetts Institute of Technology.

Osborne, J., & Dillon, J. (2008). *Science education in Europe: Critical reflections*. London: King's College.

Owen, R., MacNaughten, P., & Stilgoe, J. (2009). Responsible research and innovation: From science in society to science for society, with society. *Science and Public Policy, 39*, 751–760. Retrieved from https://doi.org/10.1093/scipol/scs093

Rorty, R. (1989). *Contingency, irony and solidarity*. Cambridge: Cambridge University Press.

Rose, S. (1998). *Lifelines: Biology, freedom, determinism*. London: Penguin.

Schreiner, C., & Sjøberg, S. (2007). Science education and youth's identity construction: Two incompatible projects? In D. Corrigan, J. Dillon, & R. Gunstone (Eds.), *The re-emergence of values in science education* (pp. 231–248). Rotterdam, The Netherlands: Sense Publishers.

Strike, K. A., & Posner, G. J. (1982). Conceptual change and science teaching. *European Journal of Science Education, 4*, 231–240. Retrieved from https://doi.org/10.1080/0140528820040302

The Royal Society. (2009). *Hidden wealth: The contribution of science to service sector innovation*. London: The Royal Society.

Tobin, K. (2014). Using collaborative inquiry to better understand teaching and learning. In L. Bencze & S. Alsop (Eds.), *Activist science and technology education* (pp. 127–148). Dordrecht: Springer.

ABOUT THE AUTHOR

Ralph Levinson is Reader in Education at University College London – Institute of Education. After teaching in science in London schools for twelve years he became a teacher educator. His research interests include chemistry education and biology education, teaching and learning science in the context of social justice, the role of the arts in science and science education, and the nature of explaining.

MITCH BLEIER AND ASHLEY N. MORTON

5. ADVENTURES IN CHEESEWORLD

Learning in the World and on the Job

ABSTRACT

Formal schooling in the United States, particularly when publicly funded, tends toward a one-size fits-all transmission model. Although this "works" for some, the decontextualized nature of classrooms and other formal learning environments often does not provide the type of support that many learners need to produce the kinds of knowledge that afford individual satisfaction and fulfillment, and the achievement of personal and professional goals. In this chapter we document a phenomenological, hermeneutic exploration of one learner's efforts to forge and follow a self-directed professional education project that employs a dynamic bricolage of approaches to achieve knowledge production and participation in a community of practice. The authors explore the birth and development of a cheese professional – cheese maker, affineur and cheesemonger – as she learns in the world and on the job.

Keywords: phenomenology, hermeneutics, interpretive research, cheesemaking, apprenticeship, situated learning, formal education, schooling, informal education

The shortest distance between two points is not necessarily a straight line, especially when one of those points is in constant motion. When we set off on a leisurely walk in the woods, our purpose is not to get "there" nor to do it quickly. Yes, we often walk marked trails, appreciating the guidance of those long past who mapped out the routes, and those more recent guides who clear the brush and maintain the trails. But the real joy often materializes when we stop to explore something unusual or unexpected, or depart from the path to follow a set of tracks in the mud. These ruptures of the usual and expected, these surprises and contradictions are the complexities that define and three-dimentionalize both our physical and emotional journeys; they are the substance of the stories we tell when we get back home. Even if it *is* the shortest distance between two points, the straight line is not necessarily the route we want to take.

SCHOOLING VS EDUCATING

I have been a teacher of science and science education for over thirty years. I have seen both students and schools struggle to define and carry out their missions. Students may

DOI:10.1163/9789004389663_005

be labeled: deficient; gifted; Level 1, 2, 3 or 4; low-performing or high performing; at-risk; on-track or off; SPED; pre-vocational; proficient or emerging; eagles or crows. Teachers are probationary or tenured; effective or ineffective; good or bad; masters, novices and more. Schools are models, magnets, basket cases, "under registration review," in or out of compliance, world-class, last resorts, zoned, charter, public, private, parochial, vocational, for-profit and more. But none of these designations really get at the issue of what brings learners and teachers together in these institutions (formal or otherwise), and what should be happening at this nexus for any and all of the individuals involved, and for the families and various layers of governmental structures concerned with the success (however that may be viewed through the epistemological and axiological lenses each employs) of these costly and ubiquitous arrangements.

We all are painfully aware of schools that "just don't work," of students that won't "make it," and of teachers who should be anywhere else but where they are. Most of us in academia also know that "the system" allows some of us (*particularly* those who have ended up in academia) to succeed and even to thrive – to make our way into the world of work and career and livable income. In this chapter (and in my research in the Urban Education Program at the CUNY Graduate Center), I explore a single learner's ongoing navigation of the educational terrane upon which she finds herself. It includes smooth sailing, and good tailwinds. It also includes storms, portages, bushwhacking, lost and faulty navigational instruments and a great deal of trailblazing. The journey isn't over. We join it in-progress, and we leave it in-progress. My hope is that, at the end of our time together, each of us is transformed, at least a little, and finds that this tale can inform our historical views, our visions of learning and teaching, and our decision making around these issues for ourselves and for those whose lives we can affect.

This is the story of a young cheesemonger and nascent cheese professional. She is not the subject, but, in fact, a full partner in the investigation – a co-researcher. Ashley's story is one that in itself is worth telling and being told. As I take on this task, I find myself asking, "Whose story am I telling – mine or hers?" I want to honor her experience and present her voice, her point of view and her understanding (interpretation) of the story. But, equally important, is my voice, my point of view, and my interpretation of her experience.

What attracted me to this story is as meaningful as the story itself. This (in the following pages), in fact, is my story too – a story whose ontological, epistemological and axiological dimensions are mediated by over fifty years of engagement with American education as a student and an educator. It is not *the* story, but *a* story of the social, emotional, and cultural landscapes upon which our lives play out. The story's richness and depth will be found in its aspects that resonate with our own experiences, but to a much greater extent, in the contradictions, conflicts and dissonances that both complicate and illuminate each of our efforts to make sense of the events, descriptions and interpretations that inhere.

My hope and intention is that this story of one person's journey in constructing and immersing herself in a self-designed professional education program that is an alternative to, but, at the same time is inextricably entwined with the prevailing educational paradigm(s), will inform and support individuals' and organizations' efforts to provide environments that support each person's efforts to educate herself both for making productive contributions in the world of work and for leading a satisfying full life.

This is a phenomenological, hermeneutic exploration – it explores a part of one person's journey and presents, to the extent that my abilities allow, not repeatable, generalizable data and inferences, but a unique and singular story that, along with what the reader brings to the table, surrounds the narrative with meaning from which each of us – participants in the described events as well as the readers – will make meaning that, if we are successful, will improve our lives and our work, and benefit all of us.

WHOSE STORY IS IT?

A story cannot exist independently of its oration, and there is no telling in the absence of a tale. A *story* and the *telling* of it, exist in a dialectical relationship, each presupposing and dependent upon the other. Consequently, no story exists independently of the teller, the receiver and the moment in time. Stories are, in fact, re-constructed and transformed with each recounting. Our *own* stories change both in substance and in meaning with each telling as they are mediated by audience, time, place and who we are at the moment.

The gravity and responsibility inherent in the telling of stories – our own, and *particularly* those of others – looms so large in my work that I hope the reader will tolerate and find value in a brief side-trip here to explore this idea.

A Journey of Discovery and Rediscovery

Recently, a friend of mine, Janet, who had lost both parents within a few months of each other decided that she would embark on a road trip through both time and space to retrace the family's histories and geographies. She planned to visit places and people who were integral to her parents' individual and joint histories during their long and full lives. Her plan was to spend about six months on the road documenting her experiences and her insights as she travelled from her home in western Massachusetts to Virginia, the Carolinas, Florida, Alabama, Oklahoma, New Mexico and just about everywhere in between. As she travelled, she would blog with some regularity so that those of us for whom her parents were an important part of our lives could experience the trip with her, respond to her thoughts, ideas, and emotions, and be part of the discovery as well. She loaded the car with maps, camping gear, writing supplies and several cases of wine to be used as gifts for long unseen friends and relatives who would serve as her hosts and partners in this undertaking of documentation and discovery.

Her first stop was to be New York City – the last place her parents lived before retiring to rural Western Massachusetts twenty-five years earlier. She stayed with me as she visited familiar places and people from her parents' history and her own childhood, often having to explain along the way who she was, and why security guards, apartment dwellers and others should let her in to their churches, office buildings and residences.

Janet was very concerned about telling her parents' stories – she wanted to "get it right" and not misrepresent them. As my research makes claims of polyphony and polysemy, I also was struggling about how to honor and represent the voices of the participants in my own work.

Because Janet and I both were reticent about misrepresenting others' stories, we talked long and hard about our own histories. It is the rarest of tales that involve only one person. All other stories are transactional – they belong to *all* of the participants – the original actors as well as those who come to be part of the story when they receive it through listening to, or reading it as told by others. Janet's parents' histories and their meanings are, and always have been, continuously negotiated and renegotiated. They are *our* stories as well as theirs. Now that they are not here to assert their own memories and interpretations of their narratives, we continue to preserve and evolve them as our own stories.

We talked about our family stories including the ones we held in common after over forty years of friendship. Quickly, it became apparent that events we shared either by having lived them together or through myriad tellings of family history were different for each of us. The details, the meanings, even the protagonists were different in each of our versions. But our common story is honored in each of our tellings as well as in the contradictions that drive us to probe and transform the tale with each telling and with the perspective of all that has happened since. Our stories – those we live and those of ourselves and others that we tell – all are interpretations mediated by memory, perspective, time and the complexity of the world that surrounds them.

None of our stories are *the* Truth, but in them live meaningful *truths* that we can access and use for guidance and communication better than any ostensibly "objective" telling would allow. Each of us surrounds our "facts" with meaning. When our facts are shared with others, we each construct different stories characterized simultaneously by resonance, difference and contradiction. This trio exists even within our own stories (which change with time and the context in which we focus on them). Furthermore, we negotiate and co-construct these stories together as we share them.

While our discussions gave *me* a way to look at the story I was telling, I could tell it did not sit well with her. The burden of representing others who could no longer speak for themselves was having an almost paralyzing effect on her mission. On a morning that she had no appointments, she took a break from her undertaking and we paid a visit to the Tenement Museum.

At the Museum

The Tenement Museum is located in an apartment building on the Lower East Side of Manhattan. Its mission is, in part, to introduce and educate visitors to the lives of the residents of this building over the past one hundred and fifty years.

We met our guide and, after a brief orientation out on the sidewalk, we entered John and Caroline Schneider's restored 19th-century Lager Bier House at 97 Orchard Street. From the moment we stepped inside, our guide made the Schneiders, their customers and their associates come alive. She engaged us in conversation, provided us with copious amounts of information (including photographs, artifacts and stories), and guided us through the methods and thought processes that historians and archaeologists use to uncover, reconstruct and interpret the past. We spent nearly an hour chatting as we sat around a table in the tavern and viewed storage areas and other rooms including the Schneiders' bedroom and living areas. By the time our guide informed us of Mrs. Schneider's sudden and untimely death as we stood inside of the kitchen that was so much a focus of her daily life, the shock and emotion of the announcement was as real and immediate as if a member of our own family had been taken from us.

Our guide's reconstruction of the past via evidence and her own personal ideas about what the lives and the people we were exploring were really like combined to make our experience deep, meaningful and very personal. She always was explicit about where the evidence ended and the speculation or interpretation began but for all of us, it was clear that the truths and meanings generated during this experience lay no more in the "facts" than in the interpretation.

Personal connections (both our guide's and our own) to the experiences of the residents of 97 Orchard Street and the surrounding neighborhood were woven into the tour both personalizing and reifying the life experiences of all of us. Our ninety-minute journey through history, social change, and complex emotional landscapes as well as the exploration of the diligence, perseverance, hopefulness, losses, pain and resilience of the residents of 97 Orchard Street engendered, empathy, emotional resonance, and self-reflection. This visit to the Tenement Museum was truly transformative. It caused each of us to consider and reconsider our own lives in the context of the Schneiders' and of the independent and common parts of our own and each other's lives.

Discussing this experience later, we recognized that what emerged from this "common" experience was different for each of us. Each one of us brought our own lives into that tavern at 97 Orchard Street and emerged transformed with new understandings of our own stories.

* * *

Janet had new resolve and was impatient to hit the road to both explore and construct the contingent and dynamic truths about her parents' lives, and, of course, her own. My commitment to presenting both my story and Ashley's story around the same set of events was strengthened but my approach needed some work. First, it was clear that I would have to focus on telling *my* story, the only one I am "qualified" to tell. In fact, the only story I am telling regardless of my intentions. Ashley's story would have to assert itself via a process of reflexive and recursive dialogs around my writing. Second, an ontological shift now had me, more than ever, convinced that a somewhat fictionalized account of events would better enable me to explore and communicate valid and useful small-t truths that both were reflective of what was happening and would resonate with Ashley's, my and the readers' grapplings with our own experiences.

My introduction to the idea of "objective" or "accurate" memory or history of lived experience in dialectical relation to one's intentional or *de facto* fictions was in Peter Waldman's (2015) *Educating Desire: Autobiographical Impressions of Addiction in Alcoholics Anonymous*, a book with a semi-fictionalized narrative of an AA meeting at its core. While reading this book and in a number of discussions with Peter, I began to be intrigued with both the role and the inevitability of fiction in our individual constructions and our social co-constructions of reality.

We are compelled to fictionalize our stories so that our narrative conveys the meanings that the events alone – even if we *could* provide an objective account – cannot communicate. Yet we also bear the responsibility (to ourselves, our stories, our collaborators and our audience) to be truthful.

Of course, fifty plus years inhabiting an epistemological landscape dominated by the hegemony of Enlightenment-driven thinking and nurtured by an educational and professional life largely characterized by scientistic, positivistic, Western-science imperatives made me resistant to these dangerous ideas to which I was increasingly attracted. However, unpacking of my recent Tenement Museum experiences, both with Janet and with my colleagues in the Learning Sciences strand of the CUNY Urban Education Program, finally allowed me to give myself permission to employ semi-fictionalized narrative in my own work – the work described in this chapter.

WELCOME TO CHEESEWORLD

The caller ID reads "Ashley." A wave of dread (*What is it this time?*) passes and I pick up the phone. "Hello?"

"I have some good news."

Most of our conversations lately begin in this way – or something very close to it. After years of setbacks, obstacles and other difficult experiences, Ashley has taken control of her life appropriating available resources to achieve an emergent and dynamic set of goals and seeing in her world opportunity in every situation. I, having been an observer over the years, have not yet become accustomed to *this* Ashley – the agentic participant in a supportive community of practice.

"What is it?"

"I'm going to the Cheesemonger Invitational in New York. I sent you a text with the link. Scroll down and you'll see me."

I do. "You're competing, not just attending"?!

"Yep. I feel like I have to challenge myself if I'm going to get anywhere."

The Cheesemonger Invitational (CMI) is a yearly international competition for cheesemongers and a limited number of other related professionals. It is very selective, usually limited to about fifty participants. It is a friendly but intense competition combined with a sort of conference including educative sessions on techniques, equipment, and industry issues as well as interactions with representatives of dairies and other cheese-related businesses. The public part of the competition occurs on day two. It is a boisterous, sometimes chaotic event that, by the end of the day is mediated by the free-flow of alcohol from participating vendors. Ashley's decision to compete as a novice, alongside so many seasoned professionals – central Cheeseworld participants at the top of their games, is a bold and confident act.

At CMI, Ashley worked alongside of, together with, and in friendly competition against cheesemongers and other cheese professionals for two days. During this time, she learned from, stood in awe of, and shared her own expertise with peers, as well as heroes and deities of Cheeseworld. She reveled in merely being in proximity to, and even more so in being accepted by these people whom she respected. The fact that she was also respected by them did not dawn on her until much later.

European cheese professionals I have spoken to have expressed awe, joy, and admiration for the welcoming, supportive, and largely democratic world of American cheese – not the little orange squares, but the growing small-scale and artisanal commercial cheese industry in the United States. They have told me that, in Europe, one is more likely to encounter cheesemakers and affineurs who are (perhaps understandably) very secretive and protective of their work. Many of them appreciate working in this environment and, in turn, provide valuable expertise to American cheese producers, and affirming, educative experiences for new cheese makers. How the cultures of the European and American regions of Cheeseworld may ultimately influence each other is not clear, but for Ashley, the way things are right now has fueled her continued enthusiasm and engagement.

Several weeks after returning home to North Carolina, Ashley received a message that she had come in fourteenth out of fifty competitors in the Cheesemonger invitational. To her the entire experience was a complete triumph.

* * *

Less than a year and a half before this phone call, Ashley was directionless, desperate and unsure that she would have a future at all, much less what that future might look like. Cheeseworld, as she would come to know it, did not yet exist for her.

Near the end of a year's break from a college experience that was fraught with indecision, multiple changes of major, emotional turmoil and more, Ashley was preparing to return to her life in a small but lively North Carolina college town. She had used her time in New York to decompress, work, pay accumulated bills (including back tuition), save money for the next year's living expenses so she wouldn't have to work and attend classes at the same time, and forge a plan to finish college and get on with her life. She accomplished many of these things, but in many ways was not much different from when she arrived in New York for her self-designed, self-imposed renewal process. She appeared to those around her to be headed back into a not-very-transformed set of circumstances from the one that sent her in search of a new life. She was returning to finish school.

As she web-surfed for an apartment and a job Ashley came across an opening for a cheesemonger at a local North Carolina gourmet grocery store, “the only requirement is a love of cheese.” She didn’t know what a cheesemonger was, but her inordinate love of cheese made this ad jump off the page at her. She researched cheesemongering and became even more determined to get this position. She called the store and spoke to the head cheesemonger who informed her that the job had been filled, but that she should come to see him when she returned to North Carolina, because he knew *everybody* in cheese in the area and would help her get situated. She continued to work toward reentering the college for the Spring semester just a few months away. However, she knew that she had only the promise of help to find a job and that the money she had saved over the past year would not be enough to guarantee a smooth school year without at least some full-time work and the concomitant conflicts between work, school, social life and sleep.

Within a day or two of this phone call, Ashley’s aunt Alexis happened upon an ad for an unpaid internship at a Greenwich Village cheese shop. It would begin in a few weeks, at the beginning of November, and would last for three months. She suggested Ashley delay her return to North Carolina for a few months, continue working and saving money so she could, at the very least, return to North Carolina both richer and more marketable.

Ashley agreed to this with some reservations, formally applied for the internship and bought herself a notebook in preparation.

It was advertised as an unpaid internship in cheesemongering at an iconic cheese shop in Greenwich Village. When the call came to inform Ashley that she had been accepted, she was told that she would be working in the caves in a warehouse in Long Island City.

“What do they mean by ‘caves’”? and “Where is Long Island City”?

Emerging from the subway among warehouses and auto-repair shops, Ashley, notebook in hand, made her way to the address she was given and found, above a door located between loading docks in one of the low commercial buildings that characterize the neighborhood, a small oval sign reading “This is Murray’s Cheese.”

She was greeted by what came to be the first two mentors who would afford a path to legitimate, peripheral participation in a community of practice (Lave & Wenger, 1991) (more accurately, a community of communities of practice) that would, before long, come to be a major focus of her life. Brian and Tess were affineurs – agers of cheese. Their workplace was built around five caves – actually cinder block-walled rooms – each with its own, very particular ambient conditions designed to nurture and support the development of different kinds of cheese. Before entering these caves, and in fact, before entering the production part of the warehouse where the caves are located, Ashley and the other interns were given hairnets, white coats and white rubber boots. They were instructed to wash their hands thoroughly and finally directed to walk through a disinfecting "moat" of bleach solution as they crossed the threshold into a world that almost instantly triggered powerful memories, emotions and associations that would cement Ashley's resolve to somehow become a part of whatever was going on here.

Among the most positive and affirming experiences Ashley had during her college career were two work-study positions: one in a medical clinic; and one in a research lab at the University Medical Center. These positions had her working in a team involved in scientific/medical research – employing sterile technique, monitoring experimental conditions, maintaining biological cultures, collecting and recording data – all in service of work that was both real and important, and very different from the simulated reality of her college lab classes.

White lab coats; procedures designed to protect samples from contamination; cheese in wheels, blocks and truncated pyramids, stacked on incubator (cave) shelves in neat rows and columns; clipboards where data were recorded with meticulous care; sampling schedules and apparatus; corners set aside for testing new formulations and techniques. All of these and more resonated with her successful college work-study experiences in medical research facilities. But in this "lab" the sampling included tasting the product! Ashley commented after one of these early days in the caves, "I feel like I am at home here."

This was the environment that carefully cradled and fanned the embers that grew into a fire inside Ashley and nurtured and grew a nascent cheese professional.

A DAY IN CHEESEWORLD

Of Ashley's many roles in Cheeseworld, her primary function at present, and the one that occupies most of her time is "cheesemonger." Simply put, Ashley sells cheese, but this tells only a small and misleadingly narrow part of the story. Following are three events that unfolded in a single day of cheesemongering:

Cheesemonger as Consultant

A mother of an eight-year-old wanting to expand her son's limited and repetitive diet approaches the cheese counter. She and Ashley discuss the pickiness of children. Ashley shares her own pickiness with the customer, and through a deft, deliberate, but also genuine display of proxemic and prosodic communication, includes the child in the discussion: turning to include both mother and child in her field of view; bending almost, but not quite, imperceptibly to better match the height of the child; increasing the volume of her voice to accommodate the distance within which both mother and child stood; and using facial expressions and arm and hand gestures to include child, mother and the surrounding cheese in the conversation; she recommends a gruyere blend as a substitute for cheddar in her macaroni and cheese, explaining how adding complexity to a very familiar and loved food of the child's – one that clearly from the conversation has deep roots in this mother's relationship with her son – would enable him to "educate his palate," she turns to the child, "make your mouth smarter" while still in familiar and comfortable culinary territory. Turning to purchases for the adults in the house, the mother asks, as do many customers, "What do you like best?" Occasionally Ashley simply tells the customer what she likes, but more often, a simple question: "Do you like it to be more sweet or more smoky?" or "What will you be having it with?" or "How many people will be there?" allows Ashley simultaneously to engage the customer in the decision and educate her. Many of these are questions that have been posed to Ashley by mentors – in the caves, at the dairy, or behind the cheese counter. She invites the customer into a partnership in which each party is respected and each party has expertise that, with the other, is used to co-construct a meaningful selection and, later, deployment of cheese knowledge. Ashley views each purchase as more than an exchange of money for a bag of product. She sees almost every transaction as a relationship built around common and complementary goals from which each participant emerges enriched emotionally, and intellectually as well as materially.

Expert to Expert

Shortly after the episode with the mother and the once "picky" but now "discerning" eight-year-old, a cheesemaker – a colleague from the dairy farm where Ashley also works, comes to the counter to shop for a party. This colleague is one of Ashley's mentors, someone she respects and reveres as a professional, as a teacher, and as a friend. Their conversation in tone, language and content differs markedly from most interactions at the counter, but it still is a customer-expert interchange. Textures, ages, under/overtones, palate feel and more are concerns here. This knowledgeable cheesemaker needs advice on products from dairies other than his own. She helps him put together a varied array of cheeses that work with each other as well as the foods and drinks the cheesemaker plans to serve. Their conversation frequently compares flavors, textures, and other qualities of the cheeses in the store with those that both participants know together from their common experiences at the farm/creamery. Ashley frequently slices small tasting sample – sharing some of each cheese's back story as she cuts: "This was made back in March so the cows were not yet eating grass, we have a younger one that I'll let you try next." or "Legally, we can't sell this one after this week, but if you get it now and let it ripen for another few weeks, it will be perfect." Here we see Ashley as the employee and representative of a merchant, bound by law and regulation, informing her customer of the government-required expiration date stamped on the cheese. At the same time, she functions as a cheese expert displaying knowledge and experience about both the product and her current customer, a fellow cheese professional. They taste the samples together communicating through nods and facial expression as much as through their words. This cheesemaker is mentoring even now. While he genuinely needs guidance and values Ashley's expertise, he is at the same time probing and allowing Ashley to practice the same kind of instructional talk that characterizes their exchanges at the dairy, but in this case, with Ashley as the more knowledgeable partner. Ashley also is deploying cultural capital from her work with one of her earliest mentors, Cielo, the master cheesemonger she worked with in New York. Reading the customer, anticipating needs/wants that the customer may not even be aware of, and educating that customer to make decisions *of her/his own* with which s/he will be pleased.

Marketing and Educating

At the end of the work day, Ashley shared, "When he came in this evening, Michael [her boss at the cheese counter] called me aside and asked if I would pick a cheese or other item that I really like and write 100 words on it for the next store newsletter." In her voice were the excitement of being able to share her ideas about cheese and educate the public, pride in representing a cheese department of which she was proud, and appreciation of the recognition and validation by an authority figure. "I chose *Alpha Tolman* from Jasper Hill Farm because it is not one of the everyday multi-rack cheeses, but it is not so high-end that it is inaccessible to most people. It is a mid-range cheese that people can use to expand their cheese knowledge. Also, it has certain qualities that make it accessible to a less-than-expert palate, but will move that palate to a deeper level of sophistication." Again, channeling Cielo, she works a complex interaction of economics, food-knowledge and experience with her own agenda for raising the lactic consciousness of her local community and customer base.

These three vignettes illustrate a kind of code-switching behavior that is simultaneously seamless and apparent to an outside observer as Ashley moves among customers and colleagues who are cheese experts, cheese novices and everything in between. Ashley easily moves between farm, caves, store and table and deploys knowledge production from each field in all of the others. She often expresses surprise when this is pointed out to her. As we saw in her relationships with her mentors, her explicit deference to both experience and authority reflects a degree of unawareness of the depth and breadth of her own expertise and her genuine (even though sometimes misplaced) respect for elders, experts and the credentialed that is so valued in formal educational settings. At the same time, it offers her access to the human and other resources made possible by those for whom her veneration – primarily a genuine honoring of their accomplishments and knowledge – both strokes egos and minimizes the perception of threat and rivalry.

Formal-Ish Classes

Another Cheeseworld venue in which Ashley practices and sharpens her craft is a regularly scheduled two-hour class that is open to all and attracts a group characterized by a diversity of experience, knowledge, interests and purpose. Ashley teaches or coteaches some of these classes and attends the ones she does not conduct.

6:45 pm; the third Friday of the month. Several tables of the Whole Foods customer service area are occupied by people…waiting. Several more are filtering in through automatic doors from the street or from various areas of the store. They are in groups of one, two and three. Around this scene, the business of the store is continuing as usual.

Conversations about work and life are briefly, almost indiscernibly interrupted by nods, acknowledgements, and occasionally warm greetings, embraces and filling-in about events since a month prior when this scene last played out.

It is the monthly cheese class at Whole Foods – a regular part of approximately three dozen lives since nine months earlier when this series of classes began back in January. These classes are structured and conducted in a way that blurs the lines between learner and teacher, and broadens and re-delineates, in a more inclusive way, membership in the community of practice that is Cheeseworld.

At seven-thirty, one of the cheesemongers signals to the waiting "students" and they follow him through the produce department, around inviting and creative displays of cheese and prepared foods and through a door behind the cheese counter into the classroom where the sessions take place. Before the session even begins, participants pass through displays of cheese about which they have become quite knowledgeable over the past eight months, and are ushered into a usually-off-limits area of the store. This activates prior positive emotions and fuels anticipation of another positive experience.

Class participants include a core of "regulars." Intermittent attendees and newcomers often come along with a regular or have some other connection to the community. This session, for example, is attended by several people from the Chapel Hill Creamery, the dairy farm where Ashley sometimes works – the two owners, the farm manager, and three cheese makers. They have come to support Ashley and to provide some insights into their own cheeses which are part of tonight's "Southern Cheeses" theme. The attendees also include a baker and author of a book on the science of baking, a group of foodies who, among other things, regularly cook with and for each other, and a visitor from the west coast – a former chemistry professor (There is a notable number of chemists, microbiologists and other scientists and academics who enter this world either in addition to their professional work, or as a life-changing escape from their work) – who has retired and has taken a great interest in food – particularly fermented foods.

During the course of the evening, several cheeses from the Chapel Hill Creamery are featured and, although she is very familiar with their history and production, she turns the floor over to the dairy owners and workers. They, of course, talk a little about the cheeses and the dairy, but all of them redirect our attention to Ashley, indicating her expertise and knowledge. While this reflects their genuine assessment of her, there also is a political component to this. They know that Ashley is sometimes underutilized and her abilities not fully acknowledged and they want to support and promote her work.

One of the regularly attending foodies, Peter, is a former professional wine buyer. He uses his wine understandings to make sense of cheese and how it is made. Every session prominently features Peter's questions which guide the conversation. He has taken a kind of paternal interest in Ashley's career. His questions are disruptive – they often interrogate areas that the presenters only touch on. They move the conversation away from the planned trajectory and into areas that interest him. His questions often clearly are grounded in wine knowledge. His questions, or at least the way he asks them, are impish and playful. He directs questions and comments directly to Ashley or steers the discussion into areas he perceives as her areas of expertise. He manages to simultaneously challenge and champion her. This is a role that Peter has spontaneously assumed. It breaches but does not derail the intended curriculum. Space is made for this kind "interruption," which encourages increased participation and, ultimately enriches the experience both for attendees and presenters.

One of the features of this (and, in fact, most of these) sessions is that the pairing of each of the cheeses with wine, beer, condiments and charcuterie is explored. This is among the most interactive aspects of the event. By the end of the first hour, enough alcohol has been consumed to loosen up the participants. The "interactions" become more frequent, more independent of the instructors and, often, somewhat less salient to the purpose of the session. The presenters consciously drink far less than the attendees and deftly keep enough order to accomplish their goals. Even (perhaps, especially) this element of chaos strengthens the community, increases the learning and positive emotions generated during the class which are reactivated during subsequent sessions.

No accountability. No formal evaluation or assessment. Merely a group of people committed to being educated about something about which they care deeply – and, in the process, educating each other, broadening the scope of the intended curriculum, surrounding the topic around which they are gathered with knowledge, emotional context, and deep, broad meaning.

Ashley – not yet a year into her existence in Cheeseworld is embraced by this community. Social, symbolic, and cultural capital surely support her in appropriating available resources to achieve her evolving ambitions and aspirations, among which are career, security, belonging, satisfaction, and happiness. This is an overwhelmingly welcoming community. Emotional resonance and entrainment, fueled in part by food – and drink – are apparent. This is at once a community of practice, an affinity group, a social event, and a more efficacious learning context than most institutional learning environments I have experienced over the past thirty years. This is true of these classes, but is pervasive and characteristic of Cheeseworld in general.

Ashley had been a participant (and contributor) in cheese classes during her internship/apprenticeship in New York. She brings this knowledge, along with her experiences in the caves, at the cheese counter and within the formal hierarchies and structures in her internship/apprenticeship in this field, recognizing resonant structures and appropriating resources both unconsciously and deliberately to make her place and move toward centrality in this professional community of practice.

Difference

In Cheeseworld especially in the United States and even more particularly in specific pockets like the one in which Ashley lives and works, one encounters a noticeable abundance of practitioners who are tatted, pierced, hard-drinking and bearded – attributes that likely set them apart in the parts of their lives outside of cheese. Ashley, who often refers to herself as "tat-free since 1989," is unpierced, unbearded, doesn't drink much and, is both female and black. This constellation of attributes *is* her quirkiness. It makes her different and allows her to fit in as a fellow non-conformist among a community of proud and assertive non-conformers. To be sure, race, gender, age and other factors can be sources of conflict and withheld opportunity, but rarely is this manifested among the more central participants – the makers, and affineurs – in Cheeseworld.

A BROADER VIEW OF EDUCATING/LEARNING EMBEDDED IN EVERYDAY LIFE

Ashley's internship in the caves is a model of educative practice built around learning embedded in daily life. No matter what formal education we obtain, ultimately, we learn through practice. Some work requires "training," but any creative endeavor demands a dynamic interplay between practice, practitioner and community.

Much of my formal school experience as a learner and, at least in the public schools, as a teacher has been characterized by a pervasive, hegemonic transmission model of learning and teaching that privileges a largely unquestioned canon and set of assumptions about gender, race, class, authority and myriad other areas. In addition, the "more positivistic forms of educational science [all too commonly encountered in classrooms, approach] the world in a way that isolates the object of study [and] abstracts it from the contexts and interrelationships that give it meaning…ignoring the notion that "to be in the world is to operate in context" (Kincheloe & Tobin, 2006, p. 5). The supremacy of Western science with its claims of objectivity, universality and generalizability rarely is challenged.

Furthermore, epistemological assessments of knowledge production that challenge the "knowing mind–useful body" dichotomy and recognize the importance of physical knowledge, embodied knowledge, the role of the senses, emotions and social relationships (Crawford, 2009; Rose, 2004) are not central (nor, sadly, even peripheral) to much discussion, practice or policy making in education – even in many "innovative," "progressive" education communities.

Physical labor, crafts-personship, sports, the arts, and Eastern "medicine" and wellness are among the areas where a broader, situated view of knowledge production is valued. Apprenticeship, mentoring, "learning at the elbow" of another, all implicitly recognize a mind | body dialectic that embraces a broad, encompassing epistemology.

Expanding this epistemological frame, we can place the mind | body in a social context. Learning occurs around, with, through and from others in the doing of the work. Involved are all the senses, attitudes, emotions, as well as exchange via language, non-verbal communication, visual signals and physical contact. As we look beyond simplistic transmission or training models of learning and teaching, we quickly see a complexity often masked by the labels teacher and learner. These are roles rather than titles. As with many crafts, professions or occupations, in communities of practice organized around cheese, each participant occupies both of these roles often simultaneously and often without an awareness (or even an interest) in which she or he is at any moment. Lave and Wenger (1991) reject the notion of learning as a precursor to participation in communities of practice and center social practice as "the primary, generative phenomenon, [with] learning [as] one of its characteristics" (p. 34). They further emphasize that "learning is an *integral* and *inseparable* act of social practice" (p. 31). The implications of this for formal schooling cannot be overstated.

Lillian Weber (1991) often described the role of a teacher in the teacher-student relationship as one of "the more educated partner" "noticing, joining with and following after" the less educated partner. Her view is very similar to what the research described here demands – a multilectal relationship (Fellner, 2015) between researcher(s), researched and reader(s) where each presupposes, depends upon and mediates the other two. In *I, Thou, and It* (1965/2002), David Hawkins presents a broader model of the relationships between learner, teacher and the world that identifies the thing learned as a partner in the teacher | learner relationship. He explores the interactions of learner (I), and teacher (thou) mediated by the content or curriculum (It) around which they have come together. I think that the idea can be reframed (see Figure 5.1), and perhaps Hawkins has this in mind when he represents

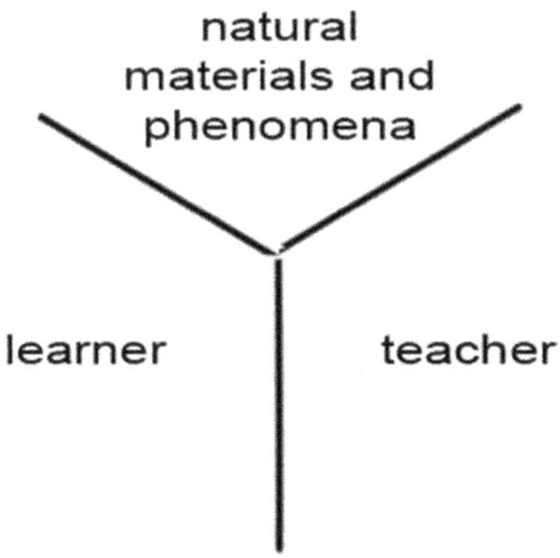

Figure 5.1. Hawkins' model viewed as a multilectical relationship between learner, teacher and the natural world

the situation as a triangle. In language that elevates the role of the curriculum/content to a partner rather than simply a mediator, Hawkins' model may be reconceived as a multilectal relationship of teacher (I), learner (Thou) and the natural world's materials and phenomena (It). Who among teacher and learner is I or thou is not fixed, and reflects the perspective of each participant at any particular moment in time. Natural materials and phenomena may not change as a result of learning and teaching, but perceptions and understandings of them certainly do.

THE SCIENCE OF CHEESE

Milk in the cow (or sheep, or goat, or any mammal) is in a constant state of flux. Conditions both in and outside of the animal leave their stamp on the milk over time. Terroir, time of year, age of the animal and countless less identifiable factors mark the color, odor and other attributes of what, from our standpoint, we think of as the product.

Once outside of the animal (and, to a lesser extent, before), the "natural" conditions that shape the milk are joined by the conditions imposed by the dairy farmer, the cheese maker and other human intercedents/interferents in the continuous and ongoing processes of change. The knowledge and understandings needed by humans to enter into these natural processes and bend them to their desires are immensely complex. Milk and to an even greater extent cheese are not unlike ecosystems with myriad and changing biotic and abiotic components in a complex system of interaction and interdependence. Observing the processes of cheesemaking, affinage and preservation/storage, it becomes apparent that the cheese professionals' work is to harness rather than control the course of cheese production. Their task is to enter into a partnership with the natural processes already in progress.

The success of humans' engagement in this partnership is attested to by the many and varied cheeses and other dairy products that can be produced with reasonable consistency and predictability. But what really reveals the true partnership nature of the interaction is when crisis occurs.

Visiting a number of farms, creameries and ageing facilities, I noticed that blue cheeses were not commonly among the variety of cheeses available. Inquiring about this I would almost always hear some variation of "When you make blue cheese, you end up making *only* blue cheese." *Penicillium*, the mold that makes blue cheese, tends to contaminate the entire facility. This can be devastating for a small operation with small profit margins. Larger operations' facilities can be designed with greater safeguards for containing blue cheese mold.

I asked Ashley about this.

> Under the microscope, blue mold looks like a dandelion [gone to seed], and like a dandelion, it spreads very easily. Bloomy rind cheeses (e.g., Camembert, brie) are our most gentle, sensitive cheeses. We watch them carefully and, whenever we see a little blue mold on them, we watch them even more

> carefully. Blue mold is less of a problem and less of a concern in the other caves, but the bloomy rind cheeses are like kids – they get dinged up a lot and they're delicate.

Despite the care and the consistency of hygienic procedures in the caves and surrounding facility, during Ashley's internship there was one major blue mold contamination incident.

> Leading up to 'the discovery,' we began noticing more and more blue mold in the bloomy rind cave. We thought that a visitor or even one of *us* might have contaminated the room accidentally. But, when we opened up the ventilation system, we found that the air filter was *full* of blue mold. Brian immediately shut everything down. We salvaged what we could and moved it to the drying room [a kind of multipurpose cave used for storage and other affinage activities]. The rest – and there was a lot of it – we threw out.

I asked, "Was the contaminated cheese inedible"?

> That's interesting. In the US it can't be sold. In Europe, there are fewer regulations…or the regulations are different. There, it is not unusual to see non-blue cheeses riddled with blue mold. People here just won't buy it. Maybe because cheese is so much a part of the culture in France, Italy…all over Europe, they understand that mold is naturally part of the process. Here cheese production is much younger. Government regulations are already in place. The problem is a combination of FDA regulations and consumer demand.

The field did not develop over centuries, even millennia, the way it has in Europe. The explosion of artisanal and small-scale commercial aged cheese production is relatively recent. Federal and state regulations govern it as they govern all commercial food production in the United States. In recent years, a number of conflicts between cheese importers, cheese makers and cheese sellers on the one hand and the FDA on the other have arisen around the use of raw milk, the ageing of cheeses on wooden shelves, and limits on the ageing duration for certain cheeses. Some of these clashes between producers and regulators may seem understandable, like the complete ban on the import of *Casu Marzu*, a pecorino-like cheese whose flavor and texture depend on the presence of live maggots at the time of consumption. In 2013, importation of Mimolette, a cheese whose distinctive qualities rely, in part, on the work of cheese mites, fell victim to proposed FDA regulations that would limit the presence of the mites to six or ten per square inch. This restriction, if implemented, would virtually eliminate the importation of Mimolette. As it was, the regulation left quantities of *Mimolette* quarantined in transit to the United States and the "Mimolette War," which lasted over a year, rose to notice in the popular press both in the United States and in Europe (NPR, May 11, 2013; Save the Mimolette, June 11, 2015).

The debate over the use of wooden shelving on which cheese is aged was a less dramatic, but ultimately more serious crisis as it threatened to block the import of almost all aged European cheeses. The bacteria that process types of cheeses and that characterize the cheeses of different producers reside in the wooden shelving and are part of the process of affinage. However, they seemed unclean and unsafe to regulators in the United States. Ashley speculated on some of these issues:

> Here in the US, blue and green are colors people associate with rotting food. Americans fear and are disgusted when blue and green molds grow on cheese or any other food. But they eat brie! And brie is covered in mold … but it's white. As far as I'm concerned, a little color never hurt anyone.
>
> I think the issue with Mimolette occurred because there already were problems with the import of raw milk cheeses and mimolette *also* has mites. It's a double whammy. I think that these two things combined made it a target for regulation. Most of us think that the problems with wooden shelving, raw milk, organisms involved in making cheese – things that aren't a problem in Europe, where they have been making cheese forever – might be a result of the big American cheese manufacturers trying to protect themselves from a new market. This has happened with craft beer also. The big breweries first tried to put the small craft producers out of business, then they began buying them up. I think that 'big cheese' is able to put a bug in the FDA's ear and cause them to crack down on the smaller importers and producers.
>
> After [the contamination crisis and subsequent clean-up], our policies and procedures changed dramatically. I'll have to go back to my notes to give you more details [Ashley's notebooks continue to be a resource to her], but we continued the handwashing, we paid more attention to cleaning our boots. We were a lot more careful. (A. Morton, personal communication, December 30, 2016)

The frequent cleaning, changing gloves between rooms, and using the bleach bath to clean boots all are visible signs that the more stringent regime was still in place. That the procedures are embedded in the culture of the caves was evident on a visit to the facility a year or so later. We were being shown around by new affineurs as both Brian and Tess had moved on to work elsewhere. I remember being the last of five of us leaving one of the caves and reaching for a doorknob to close the door behind me. One of the new interns quickly moved between me and the door and pushed it shut with an elbow. This subtle, automatic, almost balletic maneuver demonstrated a constant and consistent vigilance around the processes and procedures as well as an awareness of the lack of such in the outsiders who pass through the caves. Adjustments had been made and a new equilibrium in the symbiosis between the mold and the affineurs had been achieved.

Embodied Knowledge

"Check out these guns." Ashley took to showing off her biceps and could even be seen admiring them herself frequently during the day. It was not simply the muscles, but the shared symbolic capital that identifies her as a citizen of and central participant in Cheeseworld. "Cave arms" as they are known to "cave dwellers," as affineurs often refer to themselves, result from the regular washing and turning of large wheels of cheese that can weigh as much as forty pounds in the caves. Cave arms are not merely symbolic, however. Their development is accompanied by the ability to create and employ more efficacious cheese-care techniques and concomitant higher rates of production. Mild competitiveness in this and other areas, among both seasoned and apprentice affineurs, support community building, mutual respect and the development of cultural capital that served Ashley as she began to practice her craft in other fields within Cheeseworld.

Changes to her body accompanied changes to the cheese she cared for and guided toward readiness. And it was not just her arm muscles. Less visible than cave arms, but no less significant changes involved embodied knowledge in the form of increased sensory sophistication. Taste, smell, sight, touch and even hearing began to enable Ashley to identify and monitor subtle changes in the cheeses with which she worked. Part of the apprenticeship, because it was part of the work of affinage, was regular sampling of the developing product with the master affineur. A corer, not unlike the tool a forester uses to examine growth rings of living trees was used to extract a millimeter-thick cylinder of cheese that provides an edge-to-center view of what is going on inside of a cheese wheel. Rind, creamline and paste can be seen, smelled, tasted and felt – all of these are done with seriousness and concentration that leaves the outside observer feeling like he is in a land where he doesn't speak the language and doesn't fully understand the customs. The professionals – newbies and old-timers alike – vocalize (*hmmm*, *um-hmm*, *nice*, *wow*, etc.), which they accompany with knowing nods as they look the cheese over. Frequently, one of the master affineurs will provide direct instruction, crumbling some of the product between his fingers and offering the core for the apprentices to do the same. Sometimes an affineur will produce a pocket knife and cut into a wheel, placing the slice on a board and dividing it into as many pieces as there are people present, scooping each small sample up with the knife and silently advancing it toward each member of the group huddled in a semicircle around this activity. Talk revolves around the age of the cheese, comparisons to other wheels (both past production runs and current stock) time of year, practices of the dairies of origin and personal reactions. During these conversations, masters' and apprentices' insights, opinions and questions are received attentively, seriously and without judgement, often elucidated with facts and connections borne of experience and of the affineur's microbiological background. All of this, along with personal reflections, drawings and other information made its way into Ashley's ever-present red cheese notebook.

The Notebook

Taking notes is a "school" practice that Ashley employed from the beginning of her apprenticeship. The only preparation she could think to do before the start of her apprenticeship was to purchase a new notebook. This was symbolic of a new beginning and the readiness for a new "page" in her life. The notebook became a major part of her identity and a signifier to the community in the caves (and later, in other fields) of her enthusiasm and commitment to this work. In it Ashley kept copious notes on techniques and processes illustrated with sketches, diagrams and tables. Beyond this, there are reflections, personal thoughts, plans and ideas. It is a uniquely three-dimensional window into, and record of the interior and the exterior; the public and the private; the deep and the shallow; the emotional and the tangible. Although the book often was the subject of jokes and mild teasing, it is clearly part of what earned Ashley respect as a newbie in Cheeseworld. On those occasions that the book was misplaced or temporarily lost, all hands mobilized to find it. In time others began to carry notebooks as well. This first notebook and its successors have accompanied Ashley as she has travelled in Cheeseworld. They are frequently accessed reference volumes and generally are regarded with admiration by colleagues in the various fields she inhabits in Cheeseworld.

FOR EDUCATORS TO CONSIDER

Ashley's education history – from infancy and pre-school, all the way through college and professional life is a patchwork of success, failure, completion, false starts, validation, judgement, reward, punishment, recognition, blame, joy and sorrow.

In nearly sixty years as a learner/student and over thirty years as an educator at all levels, I have seen educational practice that is largely a language-based, knowledge-transmission affair. Even science courses that directly address physical phenomena which can be experienced, measured and engaged with, often are offered as lecture classes with separate laboratory sections where students complete directed, simulated, "hands-on" exercises designed to illustrate and illuminate the content they encounter in the lectures. Of course, educational practice is quite variable from institution to institution, but the above described situation is overwhelmingly common.

"Non-traditional" educational models, including coaching, apprenticeship and learning on the job, as well as self-directed programs that we often associate with hobbies and personal interests also sometimes reflect the epistemological and axiological stances we see in formal educational settings. However, among these less-traditional approaches we find far more direct engagement with the materials and phenomena of the natural world. Trevor Marchand (2008) in his research on craft-based learning ponders the "nature of embodied communication and knowledge." He asserts that, the learning of skilled physical activities, is affected via largely non-verbal communication and negotiation between practitioners and that "learning is achieved primarily through observation, mimesis and repeated exercise" (p. 247). Therefore, learning – perhaps better conceived as knowledge production – is inextricably bound to doing, and doing is a full-engagement activity. Emotions and senses as are engaged in working closely with, learning (from/alongside/at the elbow of), teaching, and even merely being among members of the community of practice.

In Ashley's experiences in Cheeseworld, both in the earliest stages described here and in subsequent professional settings, resonance with her lab work at college activated emotions that encouraged and supported the achievement of a set of rapidly evolving personal and professional goals. Although much about the caves was new to her, the cultural capital engendered by her lab experiences resulted in a level of acceptance among the central figures in this new world that made success (and, perhaps more important, her recognition of that success) a more prominent feature of her social and emotional landscapes. Her identity previously dominated by struggle and frustration began to be characterized by greater and greater levels of satisfaction, success, (sometimes uncontrollable) joy and the agentic pursuit of possibilities. In contrast to her experiences in the medical research labs where she had worked and studied, in the caves, she was *in* the work, data were not merely numbers read and carefully recorded from the LCD screen of a scientific instrument (although such instruments were part of the processes in the caves), but data were experienced with the body – the senses – and subjected to a much more subjective, interpretive analysis. Ashley's knowledge production and her path toward centrality were (and are) far more entwined in emotion and social relationships than she had previously encountered in the formal institutions of school and work. Without fully grasping it, she revealed an emerging recognition of this difference early on when she said, "Brian isn't a teacher, but he is the best teacher I ever had." And, "He includes you and really listens to your opinion, even though he knows so much more."

Hegemony of Formal Education

Ashley has returned to school full-time to obtain, at a minimum, her undergraduate degree. Many of those around her have encouraged her to get this degree in spite of her success in Cheeseworld: "You have to finish." "You can't get anywhere without a college degree." Her own deference to authority and to societal "norms" magnifies these admonitions. She is determined to earn her credential. Both self-doubt and the very real gatekeeping function of a college degree raise the possibility of unforeseen and insurmountable obstacles in her pursuit of a life in cheese. Considerable effort and sacrifice accompanied her return. She had to give up her secure full-time job and most of the time she spent at the farm/dairy. Although several faculty, advisors and counselors have been very supportive, the college has provided her with a number of requirements, some seemingly very arbitrary, to return to full-time study.

She is in undergraduate classes learning more and getting better grades and feedback than ever before. She approaches each assignment in almost every class, no matter what the subject, in the context of her recent years in Cheeseworld. She recently said, "I don't understand why they [the other students in the class] do so little work and don't seem to care about it. It's their major"!

For her first go-round at college, she was prepared for a continuation of high school where she was told what to do and when to do it and then reminded over and over that it had to be done; she was not prepared for the responsibility to manage time, social life and sleep without a safety net. A couple of years of work and development of purpose have changed everything. What the college offered, often seemed to her to be arbitrary and irrelevant, but has been imbued with significance in the context of her in-the-world experience. Now the completion of her degree has become merely a minor obstacle, and, at the same time, taking the classes has come to be a truly enriching experience. Of course, setbacks, major *and* minor, often invoke patterns of behavior and thinking that center the degree/obstacle in place of her success in Cheeseworld. The hegemony of the academy and concomitant pressures, both external and internal, experienced by Ashley are most apparent at these times.

During a particularly stressful finals week, Ashley was struggling over a major paper upon which she felt her grade, and her status as a full-time student on track to graduate by the end of the summer was in jeopardy. She had had more than a few nights of little or no sleep and was feeling a bit hopeless. The following exchange reveals that actual progress or success does not always outweigh perceived or, at most, possible failure:

ME: Don't worry if it takes a little longer to get your degree. You are in demand for your expertise in the work that you do. People seek you out, not because they like you, not because of your connections, but because they value and respect your work.

ASHLEY: I'm tired of going to school. I can't take it anymore. I need it to satisfy the gatekeepers. I need it to move on with my life. I need it to validate my own worth. You have no idea what it is like.

Ultimately, however, at these times, she generally has been able to re-balance these forces and re-focus herself on attainment of short and long-term goals. A few days after our exchange, Ashley, still unsure of the outcome of her paper and the rest of her finals, was able to recognize that, in the past three years, she had jettisoned many of the attitudes and behaviors that proved to be impediments to her progress through college and obstacles in other aspects of her life, and that she had since been remarkably successful in most of her endeavors both academic and professional. But it is clear that a lifetime of measurement and evaluation by other's standards – even when apparently irrelevant – looms large in Ashley's estimation of herself and her accomplishments. In addition, that these attitudes prevail among most of the people in and around her life serves to amplify and center the doubt and fear that camouflage the remarkable story she is (and I am) constructing.

As far as her professional work, Ashley's studies, as she completes her undergraduate degree, are likely largely beside-the-point. Their completion will neutralize her worries about gatekeepers, and she is bending the coursework toward surrounding her professional knowledge production with additional meaning and context. But apparent from discussions with her is that she still feels a need for the symbolic capital associated with a degree from a respected educational institution. This is especially acute as she lives in the state and the town where the institution is located and where association with the school (and its athletic teams) is an almost sacrosanct imperative.

Beyond this, however, Ashley is exploring the possibility of spending her last semester as an exchange student in France and combining this with engagement with cheesemakers and affineurs in what, even in the American part of Cheeseworld, is seen as the gold standard.

[Note: as of the date of publication, Ashley has completed her degree and has relocated to France.]

In a Nutshell

In November of 2013, Ashley took her first steps on a path to a meaningful, fulfilling future. This was the result of her discovery of a community of practice, a somewhat informal educational situation, a self-selected affinity group and an obsession – the Caves Internship at Murray's Cheese. This turned out to be a portal into Cheeseworld.

After a lifetime of formal education that afforded her access to skills, information, intellectual tools and practices with which to make sense of the world, but ultimately did not serve her well as a means to a productive and meaningful life, she entered (and was welcomed into) a community characterized by a diversity of interests, backgrounds, ages, motivations and goals that functioned to produce high-quality results that are valued by a broad range of people. It was a world that recognized, appreciated and exploited her enthusiasm so much that intrinsic motivation unlocked the best qualities of her intellectual, social, ethical, moral make up.

While the world unfailingly impinges on all of our lives, the kinds of experiences and the foundation in this world and the lasting personal and professional relationships ensure that, for Ashley, as well, I expect, for the other members of this community, there will be opportunities to build a life based on her own interests rather than simply economic considerations. Science, culture, social life and business intersect in Ashley's experience in a way that more closely emulates human social situations than it does our compartmentalized way of life that separates work, play, family life, and, most glaringly, education from each other.

The dissonance between the affordances presented in Cheeseworld and the societal pressures around success, security, and family apparent in the other fields in which Ashley moves, make the part of her story not yet told and, not yet lived obscure and unpredictable.

Making Space for Difference

For some, perhaps for most, readers, this non-prescriptive call to action will be unsatisfying. However, the story presented here demands not merely identifiable changes in the way we educate or in the way we bring people into the world of work, but rather the terraforming of a world/landscape with a variety of affordances, ways of access, and pathways that reflects the diversity of would-be participants/learners, and the diverse paths that people follow/forge as they find their way in the world. A world that provides space even for those whose differences and uniqueness cannot be imagined or anticipated. I am calling for a change of mindset among learners, teachers, parents, families, employers, and policymakers that will welcome all comers, a change that ultimately opens the possibility that each of us will be able to realize our evolving dreams and goals, one that allows all of us to derive the benefits that our fellow citizens' diverse and varied talents, skills, interests, proclivities, and quirks may provide for us.

REFERENCES

Crawford, M. (2009). *Shop class as soulcraft: An inquiry into the value of work.* New York, NY: Penguin.

Fellner, G. (2015). Multilectics and its methods. In C. Milne, K. Tobin, & D. DeGennaro (Eds.), *Sociocultural studies and implications for science education 12: The experiential and the virtual* (pp. 33–58). Dordrecht: Springer.

Hawkins, D. (2002). *The informed vision: Essays on learning and human nature*. Flemington, NJ: Agathon Press. (Original work published in 1965)

Kincheloe, J., & Tobin, K. (2006). Doing educational research in a complex world: Preface. In K. Tobin & J. Kincheloe (Eds.), *Doing educational research: A handbook* (pp. 3–13). Rotterdam, The Netherlands: Sense Publishers.

Lave, J., & Wenger, E. (1991). *Situated learning: Legitimate peripheral participation.* New York, NY: Cambridge University Press.

Marchand, T. (2008). Muscles, morals and mind: Craft apprenticeship and the formation of person. *British Journal of Educational Studies, 56*, 245–271. Retrieved from https://doi.org/10.1111/j.1467-8527.2008.00407.x

Prichep, D. (Writer), & NPR (Producer). (2013, May 11). Tiny mites spark big battle over imports of French cheese. In *NPR's Weekend Edition Saturday*. Retrieved from http://www.npr.org/sections/thesalt/2013/05/11/180570160/tiny-mites-spark-big-battle-over-imports-of-french-cheese

Rose, M. (2004). *The mind at work: Valuing the intelligence of the American worker*. New York, NY: Penguin.

Save the Mimolette. (2015, June 11). *Here's an interview I gave about Mimolette for a documentary about the "Mimolette war". They called me "the Joan of Arc" of Mimolette!* [Facebook post]. Retrieved from https://www.facebook.com/SaveTheMimolette/

Waldman, P. (2015). *Educating desire: Autobiographical impressions of addiction in alcoholics anonymous.* Rotterdam, The Netherlands: Sense Publishers.

Weber, L. (1991). *Inquiry, noticing, joining with, and following after.* New York, NY: City College Workshop Center (CUNY).

ABOUT THE AUTHORS

Mitch Bleier is a science teacher and teacher educator. He has attended New York City public schools and universities for his entire student career. As an educator, he taught K-16 science and graduate and undergraduate science education classes at CUNY for over thirty years. He has served as a science education consultant for schools and school districts in New York State, Nevada and California. He recently received a PhD in Urban Education from the Graduate Center of the City University of New York.

Ashley N. Morton LOVES cheese. By happenstance, luck and persistence, she has made cheese her life – she is a cheese professional, practicing cheese making, affinage and cheesemongering. She was educated in the public schools of New York City and Mecklenburg County, North Carolina as well as at the University of North Carolina – Chapel Hill. But the most important parts of her education occurred in the caves, pastures, milking parlors, cheese production facilities, and retail counters at Murray's Cheese, Chapel Hill Creamery, Whole Foods, and other venues in the United States and in Europe.

SONYA N. MARTIN AND DA YEON KANG

6. EXPLORING THE TRANSFORMATIVE POTENTIAL OF EXPERIENTIAL LEARNING

Fostering Positive Attitudes towards Inclusive Science Education for Special Education Needs (SEN) Learners in Korea

ABSTRACT

We offer socio-historical context for understanding current challenges for implementing inclusive education for students with special education needs (SEN) in the Korean education system. We present current barriers to SEN students' learning, including limited access to physical spaces, material resources that can accommodate impairments, and teachers who are qualified in special education and content areas. These factors, as well as negative attitudes towards people with disabilities in Korean society, prevent SEN learners from being physically and academically integrated into regular education settings. To address these challenges, we describe a course designed to engage pre-service secondary science teachers in experiential learning with SEN students during authentic science teaching activities. This course supports teachers to confront unexamined biases about people with disabilities and inclusive education and to consider how teachers' beliefs shape their perceptions about SEN learners and their attitudes about the value of science education for students with disabilities. Using autobiographical writing and dialogues in small-group and whole-class discussion to promote reflection and to generate intention for future action, we demonstrate how experiential learning courses can support teachers to develop more positive attitudes towards inclusive education and SEN learners. We discuss limitations and the need for research to improve education for diverse learners in science.

Keywords: Korean education, special needs, science education, inclusive education, equity

INTRODUCTION

For researchers concerned about equity issues in science education, ensuring that all students have access to a high-quality science teaching and learning experiences is critical. Educating students in inclusive science education environments is an important way to reduce barriers for learning science for students with special education needs. Educational policies that legislate the practice of educating

 DOI:10.1163/9789004389663_006

SEN students in inclusive learning environments generally seek to promote free, equal access to quality education in age-appropriate classrooms in schools geographically near to students' homes. While such laws have been passed in many countries, many factors continue to prevent SEN students from receiving the benefits of learning in truly inclusive environments. Advocates for equitable education contend that until the practice of inclusive education is extended beyond laws and mandates to become an essential philosophy integrated into all aspects of teacher education, the potential for improving education for all students will remain limited in scope.

In every country, there are barriers preventing students and teachers from realizing this goal. In this chapter, we describe challenges to inclusive education in the Republic of Korea (hereafter, Korea) and introduce an approach to pre-service science teacher preparation that aims to transform teachers' classroom practices. Korea's education system has experienced considerable change in the last seven decades related to policies and practices promoting inclusive education in K-12 classrooms. To provide context for our discussion about transforming teachers' practices, we first offer a brief overview of the history of special education and inclusion education in Korea and highlight the current challenges facing students, teachers, and teacher educators – especially in the context of science education. Following this introduction, we describe an experiential pre-service science teacher education course designed to address some challenges researchers have identified as presenting the greatest obstacles to inclusive education in Korea. Through this overview of the course, we highlight some principles of experiential education that underpin our approach to working with pre-service teachers that we think are useful for promoting changes in how teachers view inclusive education in general, as well as their personal roles and responsibilities for creating inclusive learning environments for all students.

Drawing from qualitative data collected in the evaluation of the course, we discuss the potential for transforming inclusive education in Korea through teacher education initiatives that engage pre-service teachers in authentic inclusive teaching experiences and share findings that suggest this course supported pre-service teachers to adopt more expanded views of what science education "for all" could look like in their future classrooms. We discuss implications for teacher education and professional development informed by a philosophy of inclusion that aims to expand science learning opportunities for marginalized students in Korea and other countries. We conclude with suggestions for future research.

A MOVEMENT TOWARDS INCLUSION EDUCATION IN KOREA

Historical Advances in Educational Opportunity for All Learners

Challenges facing SEN students and teachers in Korea are both historically and socially constituted, so it is important to provide some context for understanding what factors limit the implementation of fully inclusive education today. While there are some historical records documenting educational efforts for people with disabilities in Korea dating back nearly 600 years, modern educational programs for children with disabilities were not established until the 1890s (Kim, 1995; Ryu, 2009). The majority of these schools were established by newly arriving missionaries from Western countries and they provided services mainly for students with visual, auditory, and physical impairments (Park, 2002). During the Japanese colonial occupation of Korea from 1910–1945, many of the missionary established schools were maintained, but they functioned more as orphanages with limited focus on providing academic learning for students (Kim, 2014). However, there was some advancement in this time period, including the development of the Korean Braille system in 1926 and the establishment of both a finger-spelling and uniform signing system based on Hangeul, the written system for the Korean language (Byun, 2005).

It was not until the post-war establishment of Korea's National Constitution in 1948 that the government extended educational opportunities for all citizens. The constitution mandated free and compulsory elementary education "according to their capabilities" (Kim, 2014). For the first time in Korea's history, all people, regardless of gender, religion, social class, or financial status, were allowed to pursue an education. By 1959, student demand was outpacing the school systems' resources with as many as 90 students in one classroom and more than 40% of schools using a "two-shift" model, holding both morning and evening classes to accommodate students (Korean Educational Development Institute [KEDI], 2007). By 1965, only 35% of students were continuing to secondary education due to the selective examination process and the required costs associated with it. This meant that only high-achieving students from families with the ability to pay for secondary education were able to continue beyond the elementary level (Morris, 1996). However, government-supported system-wide changes that reduced families' financial burdens and abolished the high school entrance examination policy helped to expand access so much that by 1986, student enrollment in secondary school had risen to about 95% of the entire population of secondary school-aged children (Martin et al., 2014).

Today, 95% of Korea's students complete secondary level education with the vast majority (78%) graduating from general education programs (Organisation for Economic Co-operation and Development [OECD], 2016a). Of these graduates, about 51% enroll in tertiary education programs, the largest proportion of people to pursue post-secondary education among all countries in the OECD (OECD, 2016b). Recent estimates suggest that by 2025, about 80% of all 25- to 44-year-old Koreans will have participated in a post-secondary education program, making it the most educated population in the world (McNeill, 2011).

These advances in education are amazing, especially considering that in less than seven decades, Korea transformed from being one of the poorest, pre-industrialized, agrarian nations in the world to the 15th largest economy in the world (The World Bank, 2017). The post-war devastation of Korea's physical infrastructure, agricultural lands, and economy left few resources from which to rebuild society. Investing in education allowed the government to maximize the potential of Korea's human resources, which many attribute to Korea's economic success today. In addition to expanding general education, the government has provided considerable support for developing gifted education programs aimed at maximizing human capital. Special legislation, including the Gifted and Talented Education Promotion Act (영재교육진흥법, 2000) and the Gifted and Talented Education Promotion Enforcement Decree (2002), have provided significant funding for the development of gifted schools, in- and out-of-school classes and programs, research centers, teacher development institutes, and departments of gifted education, all of which together serve more than 120,000 students (Lee, Kang, & Lee, 2016). Unlike in some countries, such as the United States, in Korea gifted education is not categorized as special education – so these funds do not come from monies earmarked for special education. This information about the many educational opportunities available to general education and gifted students in Korea will provide a contrast to the limitations of the services and possibilities available for students with special education needs.

Advances in Special Education in Korea

Initially, the same 1949 Education Law that provided wide-ranging benefits for general-population students also sought to support SEN student populations. The law provided provisions for the establishment of special schools dedicated to providing instruction for basic living skills for students categorized as being blind, deaf, or physically or intellectually impaired. This law ensured that special attention was given to the educational needs of these children (Kim, 2014). However, the same law also allowed for the establishment of "special classrooms" within schools, which effectively separated students into general education and special education learning environments. Additional provisions allowed for exemptions and postponement of compulsory education for some students due to disabilities, disease, under-development, or other unavoidable reasons (Kim, 2014) – which allowed for the exclusion of many children with special needs from receiving any formal school experience.

During the 1950s, many families struggled, for a variety of reasons, to provide educational opportunities for children with disabilities. By 1962, only 10 special education schools were established, and by 1971, there were only 30 special education classrooms within general education settings (Hwang, 2015). It was not until the late 1960s and early 1970s that policy makers and advocates made some advances in special education. In 1961, the first professional programs for training personnel to support SEN students were established at what is now Daegu University. In 1971, Daegu University offered the first special education certification program. In 1967, the first national curriculum for visually and hearing impaired students was released, followed by a national curriculum for students with intellectual disabilities in 1974. Special education advocates built momentum for change and saw the first significant advancement in educational rights with the passing of the 1977 Special Education Improvement Act (특수교육진흥법), which included stronger provisions for ensuring nine years of free, compulsory education for all students and to legislate protections of educational rights for all students with disabilities.

Additional amendments passed throughout the 1980s and 1990s saw regulations that expanded access to integrated educational settings and ensured parents had improved capacity to enroll children with mild disabilities to be educated in special classrooms situated within general education schools. Revisions in the laws continuously contributed to improvement in the quality of special education services available. The 1994 re-authorization of Special Education Improvement Act first clearly defined "inclusion" to mean educating SEN students in either regular classrooms or in special schools, but in each case, students were to be instructed using the national curriculum to develop both their academic and social skills (Kwon, 2005). Additional provisions included federal mandates to develop teacher education programs to certify teachers in special education. However, it would more than a decade before more expansive legislation targeting inclusive education was passed.

The Development of Inclusion Education

The 2007 Special Education Law for Disabled Persons (장애인 등에 대한 특수교육법) expanded the categories used to designate students for services. Currently there are eight protected categories of students, including students with impairments related to vision, hearing, cognition, language, physical ability, health, students with emotional/behavioral disturbance, and learning disabilities. The 2007 law increased the number of students identified as needing special education services so that between 2004 and 2016, the total number of students receiving services increased by 37% (from 55,734 to 87,950). In the same time period, this law also increased the number of students being educated in regular education school environments by nearly 50% (from 31,612 to 61,989).

However, a close look at these numbers show that while fewer students were being educated in separate special schools, the majority of SEN students were still being placed in special classrooms within general education schools (see Table 6.1).

Table 6.1. Changes in student population receiving special education services

Year	*Special school*	*Special class in general school*	*Inclusive class in general school*	*Total*
2004	23,762	28,002	3,610	55,374
2008	23,400	37,857	10,227	71,484
2012	24,720	44,433	15,647	85,012
2016	25,467	46,645	15,344	87,950

Currently, less than 18% of students experience inclusion in the same classrooms with general education peers, and even then, inclusion tends to be limited to several class periods a day and is generally available only to students with mild disabilities. Therefore, even while the 2007 Special Education Law has allowed for the physical inclusion of SEN students in the same buildings as their general education peers, it has failed to ensure either social or academic inclusion for these students.

CHALLENGES TO INCLUSION EDUCATION IN KOREA

In this section, we discuss the philosophy of inclusive education and we briefly describe some of the challenges limiting inclusive education as a system-wide practice in Korean schools. Building from this discussion, we introduce our experiential education approach to preparing pre-service science teachers to be better able to meet these challenges in their future classroom teaching. We share some findings from the evaluation of our program that suggest that this innovation in Korea's teacher education programs could positively transform inclusion education for SEN students.

Inclusive Education: In Principle

As introduced earlier, an evolving goal in special education has been to integrate SEN students within general education settings using the "mainstreaming" approach. However, a main challenge to achieving this goal has been the lack of flexibility in the organization of general education schools, especially with regard to curriculum and a reliance on standard teaching and learning strategies. A response to this challenge has been the drive to reform schools through the implementation of inclusive pedagogy, which advocates the celebration of individual differences as a means for enriching the learning of all students. The core principle of inclusion and inclusive education is a positive view of difference, whereby all students are considered to be valued members of the school community and where educators work to actively combat discriminatory practices towards learners who are exceptions from the norm. Inclusion requires that educators respond to diversity by making necessary modifications in content and utilizing teaching approaches that expand participation and achievement by improving the overall quality of education for all learners – not only students with special education needs (United Nations Educational Scientific and Cultural Organization [UNESCO], 2005). For inclusion to be successful, it requires both a society and an educational system that espouses the belief that education is a human right, that all students are capable of learning, and that pursuing an education is valuable and meaningful for all learners (Evans & Lunt, 2002). These ideals are not easily realized in all educational contexts for a variety of reasons.

Inclusive Education: In Practice

The provisions of the 2007 Special Education Law attempted to address some of necessary structural changes that would strengthen the responsibilities of the local and federal governments to improve special education service delivery. Specifically, the law sought to (1) increase the numbers of teachers certified in special education to ensure a lower teacher-to-student ratio (with 1:4 being the aim) in the classroom, (2) expand training for care professionals and classroom aides, (3) stress early childhood intervention services, (4) expand compulsory education to secondary levels, and (5) develop a system-wide program to support the development of appropriate educational plans for students with SEN.

Although the provisions of this law have been instrumental in moving forward the fight to more fully integrate people with disabilities into the Korean education system and society, there are still many limitations to its implementation. While researchers, special education advocates, and teachers around the world continue to debate the educational benefits of inclusion learning for all SEN students, there are some clear inequities in the Korean education system that must be addressed before inclusive education practices become more widely and equitably implemented.

Lack of Physical and Human Resources to Support Inclusion

Examples of system-wide limitations include an inadequate number of special education certified teachers and a continued lack of appropriate and accessible facilities for students with disabilities beyond the elementary level. Currently there are only 18,000 special education teachers hired to support 88,000 students, which means schools cannot meet the federally mandated 1:4 teacher-to-student ratio (Ministry of Education [MOE], 2016). In addition, rather than integrating SEN students into general education classrooms, since the law's passage in 2007, there has been a 42% increase in the establishment of special classrooms within schools. The majority of these classrooms are housed in elementary school settings, meaning there are not enough secondary school placements for SEN students to advance beyond elementary school levels (Kim, 2014). More than 70% of SEN students are separated into these special classrooms, which limits their capacity to develop relationships with general education students.

This policy of educating SEN students in special classrooms within schools and a lack of these special classrooms in secondary schools has also resulted in reduced access to secondary education. As recently as 2016, only 19% of special classrooms were located in high schools (MOE, 2016), which could account for the very small percentage (25%) of SEN students who continue their education at the secondary level (Kim, 2013). This means that students who want to pursue secondary education courses are forced to transfer to special schools. Data show that SEN students in middle and secondary grades are three times more likely to transfer to a special school setting than SEN students at the elementary level (Hwang, 2015). Another reason students may transfer while in the upper grades is that the separate classroom policy means that SEN students with a range of physical, cognitive, and behavioral disabilities all receive instruction in the same room at the same time – regardless of each student's ability level, age, or individual special needs. As a result, many families and children make the decision to transfer to special schools to access more resources to support academic learning. SEN students may also transfer to special schools to escape bullying from general education peers. Finally, being with peers who face similar challenges may allow students to develop friendships, but limits potential for relationships with students without disabilities.

However, even the transfer process can be a challenge as there are only 170 special schools in the entire country and many are not easily accessible to families living outside of cities. Even in Seoul, a city of 10 million people, there are only 29 special schools (Hwang, 2015). A recent study found that more than half of all SEN students travel more than 1 hour a day to access educational facilities and one quarter travel as much as 2 hours a day (MOE, 2017). As a result, they are not easily able to develop and maintain relationships with general education students at the neighborhood school or the SEN peers in their classrooms who may also be traveling from many distant neighborhoods. Access to physical resources is critical, but SEN students' limited access to highly qualified teachers also has a negative impact on their learning – especially in content areas such as math and science.

Need for Teacher Preparation and Professional Development

Currently special education-certified teachers are not required to be certified in additional content areas, such as math or science, so SEN students educated in either special classrooms or special schools generally lack opportunities to learn specialized content from highly qualified teachers. For the 17% of SEN students who are actually included in general education classrooms, they can expect to interact with teachers who have little or no training to support their learning. While the government has recognized the need to increase general education teachers' professional competency in special education, in 2016, only 63% of general education teachers in inclusive classrooms had completed the minimum 60 hours of training required (MOE, 2016). However, this is significantly improved from 2013 when less than 25% of inclusive classroom teachers reported having any special education training (Kim, 2013).

Plans to improve future teachers' knowledge about special education have been limited in scope and impact. For example, pre-service teacher programs currently require only one two-credit course in special education, which is commonly completed in the second or third year of their programs. The course is not content specific, but does provide information about special education practices in the Korean education context. Additionally, secondary pre-service teachers typically complete only 5 weeks of teaching practicum during their entire program. During practicum, there is no guarantee they will be placed with an in-service teacher who has knowledge and skills about how to support inclusive classroom teaching, and they also have no guarantee of being placed in a classroom setting that actually serves SEN students. Thus, the majority of pre- and in-service general education teachers are inadequately prepared to support SEN learners and special education teachers may lack content knowledge expertise.

These issues tend to be compounded at the secondary levels where fewer teachers are exposed to SEN learners. Kwon (2016) focused on special teachers and secondary general education teachers' beliefs about SEN students and found that teachers believed that inclusion would be successful if SEN students' learning was aligned with the regular curriculum and monitored through the individualized education plan (IEP) process, but they acknowledged this rarely happened in practice and that this process was the exclusive responsibility of special education teachers. IEPs are formal written statements created by teachers, care givers, and SEN students to design an educational program to specifically meet the individual needs of the student. The IEP offers a road map for targeting gains in academic, social, physical, and vocational growth for SEN students by following an agreed upon set of strategies to support the learner to meet their goals during the school year.

Teachers in the same study noted that SEN students should only be instructed by special education teachers because they felt they lacked the knowledge and strategies needed to support these learners. A case study (Kim, 2012) examining general education teachers' perceptions of inclusion revealed that these teachers did not consider SEN students would be an appropriate addition to their classrooms. Rather, they believed that simply being in a special classroom being taught by a special education teacher within the general education school setting was sufficient for meeting the learning needs of SEN students.

Lack of Societal Acceptance for Inclusion Education

Although the Ministry of Education announced plans to build 21 new special schools by 2019, as of fall 2018, no construction for new schools had been started. This is due in large part to citizens' opposition to building special schools in residential neighborhoods. Citizens regularly protest the development of special schools for fear that property values will decrease and that the school will present a negative image of the neighborhood (Lee, 2016). It is well documented that individuals with a disability in Korea are likely to experience discrimination and encounter negative interactions with others (Kwon, 2005). This is true for social interactions occurring both inside and outside of the school. Some researchers (You & McGraw, 2011) have suggested that Confucian ideals requiring children to reciprocate the support given by their parents in later life positions people with disabilities as burdens because they cannot fulfill these filial obligations. Not conforming to societal ideals of success, people with disabilities threaten societal norms, which can bring shame to families. For some families, the stigma associated with having a special needs child prevents the parents from having their child tested in order to identify what kind of supports are needed. Official diagnoses require that parents register their children with the local administrative district office as being disabled. This designation will follow children into adulthood and can compromise not only their own future chances for university entrance and employment, but also marriage and even the marriage of their siblings and family members.

Several researchers (Kim, Kim, Beach, & Han, 2016; Kim, 2013, 2014) have suggested that because Korean people's cultural and historical orientations towards education are rooted in Confucian traditions, society members value an education system that promotes a highly competitive and disciplined approach to achieving prescribed standards of excellence. This view means that the Korean education system does not readily recognize and value individual learning differences. As such, teachers are encouraged to maintain an equal distance with all students. In addition, providing differentiated materials to accommodate individual students' needs would be seen as unfair and antithetical to the notion that all students are competing to meet the same standard. When SEN students are placed in general education classrooms, parents worry that SEN children will receive a disproportionate amount of instructional time, which could disadvantage general education students who are all preparing for competitive, high-stakes exams. These negative attitudes are not only found among general education parents, as studies also have found opposition to full inclusion from parents of SEN students (Kim, 2014), especially at the secondary level, because they fear the negative reactions to their child by other students, teachers, and parents.

TRANSFORMING TEACHER PREPARATION PROGRAMS

As science teacher educators, we are limited in our ability to transform societal structures that prevent SEN students from gaining access to learning opportunities that can improve their life trajectories beyond elementary school. However, by developing an experiential pre-service science teacher education course aimed at raising future teachers' awareness about current challenges to inclusive education in Korea, we sought to effect positive change in the science teaching and learning of our teachers and any future SEN students whom they might serve. In the sections that follow, we offer an overview of our course and highlight principles of experiential education and inclusive education that we used to help our teachers adopt a more inclusive mindset, which enabled them to create a positive science learning environment for SEN students who participated in our Science Fair for All Students.

Experiential Learning

Experiential learning has its origins in interdisciplinary and constructivist traditions of "learning by doing", where students are supported to apply knowledge to experience and then encouraged to reflect on their experiences in order to develop new skills or ways of thinking (Lewis & Williams, 1994). In educational settings, students may engage in experiential learning through experiments, field trips, or apprenticeships. Chapman, McPhee, and Proudman (1995) assert that simple participation in a prescribed set of learning experiences does not make an activity experiential; rather, it should be built around several key elements.

Chapman et al. (1995) advocate for activities that integrated theory and practice by engaging learners in purposeful tasks that encourage learners to consider a big picture perspective by incorporating reflection that supports the learner to gain insights into how their participation in various activities influences their own beliefs and practices. They reason that experiential activities should also create emotional investment such that learners see their engagement as meaningful for themselves and the community with whom they are engaged. They also advise that experiential learning is most effective when learners are accountable for their actions, such that learners' poor performance could result in negative consequences. Finally, they suggest that all experiential learning activities must provide explicit opportunities for learners to reflect on their growth and development over time. As such, this kind of learning experience is meant to be transformative in nature.

Learning that seeks to be transformative must be conceptualized as more than a product or single outcome. It should be seen as something that occurs as individuals engage in experiences that cause them to change what they think and that support them to develop dispositions that may lead to changes in practice or thought (Baglieri, 2008). Characterized as a process, Kolb (1984) described experiential learning as that which takes place through interpretation and reflection on activity that gives meaning to the experience. Meaning-making requires a process of critical reflection in which people form intentions to learn from experience by thinking about the nature of their thoughts in order to consider possibilities for change (Baglieri, 2008).

To support pre-service teachers to engage in critical reflection, we incorporated autobiographical writing in this course (Martin, 2005) by having teachers respond to their experiences with personal reflections. In teacher education, proponents of experiential learning have advocated for expanded opportunities for participating in extended teaching practicums (Hascher, Cocard, & Moster, 2004) and the use of field-based methods courses that integrate reflection on practice to promote deeper understanding of how theory from the classroom can inform decision making in the classroom (Siry, 2011). We also drew from research advocating for collaborative dialogue about authentic classroom teaching experiences for pre- and in-service teachers (Siry, 2011) with a goal of promoting reflection on individuals' experiences within a professionalized forum where pre-service teachers were encouraged to reflexively consider the ways in which their changing attitudes and beliefs may shape their teaching with SEN students.

TEACHING SCIENCE TO DIVERSE LEARNERS COURSE

Origins for the Course

To provide context for the origins of this course, we briefly describe our experiences with inclusive science education and how we came to co-create and co-teach the course.

Sonya is a former public school science teacher from the United States where she had experience teaching science in inclusive classrooms with English as second language (ESL) and special education teachers. Living as a researcher and science teacher educator for seven years in Korea, she has had numerous opportunities to work with teachers who have struggled to support SEN students' science learning. To improve science learning opportunities for SEN students and to better support future science teachers, she created an elective course for undergraduate secondary pre-service science teachers at her university. Da Yeon was a student in the first course, which was conducted as a typical lecture course with an emphasis on exploring effective science teaching methods for SEN learners. Specifically, students worked to develop science lessons and resource materials with specialized modifications for learners with different impairments. Da Yeon, who is an earth science education major, created a model for visually impaired students to demonstrate the theory of continental drift (Pangea).

The following semester, Da Yeon engaged in an international exchange program to complete her student-teaching practicum in the United States. She student taught in an inclusive science classroom and focused her undergraduate research thesis on the examination of the effects of teacher and school support on the science learning of SEN students. Sonya served as her advisor and after entering graduate school the following year, we decided to revise the original course to be more responsive to the needs of future science teachers in Korea. Our re-designed course is offered as a 3-credit (45-hour) elective course that meets for fifteen weeks and utilizes an electronic course platform to support asynchronous information sharing, communications, and project development. This course has been offered three times and has had a total of 61 participants. This semester, a former teacher-participant in the course (Kieun Eom) has become a third co-teacher in the course.

Revised Course with Experiential Learning Focus

In alignment with earlier described features of experiential learning, the entire course is framed around a central goal of having teachers design and implement an "incursion" experience for secondary students whom we host for a full-day of science learning activities at our university. We partner with administrators, teachers, and students at a local private school for students who are deaf and hearing-impaired. These students present a wide variety of hearing, visual, and cognitive impairments, for which our teachers are tasked with creating highly interactive, inquiry-based activities that meet the learning needs of these individual students. The activities are offered as separate learning stations, which participating SEN students elect to visit. We have named this set of activities the Science Fair for All Students,

To prepare our teachers to meet this goal, our course focuses on three general goals: increasing teachers' knowledge about effective inclusive education strategies for teaching science, expanding teachers' awareness and understanding about how personal and societal beliefs and attitudes towards difference results in the disabling of people, and providing authentic teaching and learning experiences coupled with explicit reflection through autobiographical writing and collaborative dialogue aimed at developing each teacher's personal and professional growth. The course employs a combination of course readings, lecture, and workshops to introduce teachers to inclusive education principles and practices and to promote discussion and reflection on prior and new experiences.

After each activity, students are assigned to write an autobiographical reflection. The reflection should include a brief summary of what they experienced during the activity, a discussion about what they felt they have learned about themselves and about the topic, and a reflection on how this experience may impact on their beliefs and future teaching practices. In addition to the five reflective writings, students also participate in small group dialogues with one of the course instructors to debrief about what they have learned and they are asked to critically reflect on how their experiences may impact on their perceptions about SEN students and inclusion education. Below we offer a brief description of the experiential activities in our class. Following this description, we share data from our evaluation to discuss what impact these activities have had on teachers in our class.

Challenging Personal Bias towards Disability

In addition, we designed a series of interactive workshop activities that encourage students to explore their own attitudes towards and beliefs about people with disabilities and inclusive education. Topics explored in workshops include reflecting on how people with disabilities are depicted in Korean film, an interrogation of terminology used in Korean society to describe and label people who have different forms of impairment, and a discussion about how these terms are used in everyday life. During these activities, we encourage students to challenge their personally held biases and unexamined assumptions about people with disabilities. These discussions become the foundation for our exploration of education for SEN students as a human rights issue that affects individual children and their families, as well as, the greater society.

Understanding Disability in Society

Another feature of the class is an intensive question-and-answer workshop between our teachers and the parent of a child with a disability, which offers a unique and intimate space for learning about the discrimination people with disabilities and their families face in Korean society. We partner with a parent advocacy organization for people with disabilities to identify a parent speaker. The dialogue is structured in part by questions our pre-service teachers generate while role-playing what kinds of questions they would like to ask of a parent of a newly arrived student with special needs who has transferred to their classroom. The questions touch on various academic and social inclusion issues, some of which are very personal. Our partnership with a parent advocacy group provides a professional space for engaging in discussion about many taboo topics. Teachers are eager to understand how having a child with special needs impacts on familial relationships, personal finances, parental expectations for their child's future, and experiences with discrimination. In this workshop, we want our students to consider what role and responsibility teachers need to play to effectively support these children and families.

Field Visit to Special School and Meeting with SEN Students

Our teachers also make a field visit to our partner school, where they have a chance to tour a school designed to meet the learning needs of children who are deaf and hearing impaired. Our teachers learn about the benefits of open-classroom designs, which remove walls, doors, and other visual obstacles that can prevent teachers and students from effectively communicating via sign and voice. We observe classrooms of students engaged in math and science lessons, and we tour specialized library facilities, music rooms, physical education spaces, outdoor gardens, and art spaces.

While at the school, our teachers meet the secondary students who will attend their science fair. During the meeting, the SEN students share their science interests and learning goals with our teachers. They also give some suggestions to our teachers about teaching strategies that work well for them. This provides the SEN students an opportunity to contribute to the co-development of the curriculum and design of the science fair activities. The SEN students communicate with our pre-service teachers through the school's teachers who act as interpreters. They also have direct dialogues with students who read lips or have a cochlear implant. During this meeting SEN students are encouraged to give their opinions about what they enjoyed at the previous fair and what changes would be beneficial for their learning.

Interview with Special School Teachers and Administrator

As part of the school site visit, our teachers also conduct an interview with a special education teacher and with the principal of the school. Again, many of the questions are generated in a workshop held prior to the field visit. The workshop focuses teachers' attention on issues relevant to professional and administrative knowledge necessary for effectively co-teaching with a special education teacher and on the specific learning needs of the students who will attend the fair. Our teachers are also encouraged to ask personal questions about job satisfaction, the challenges and benefits of working with SEN students, and how to support parents and families.

Science Fair for All Students – Preparation and Implementation

The above activities all take place in the first 6–7 weeks of the class. The remainder of the course is spent working in small groups, based on science content interests, to design 20–30-minute inquiry learning activities that SEN students can choose to participate in during the science fair. Teachers are provided with a budget for purchasing materials to prepare their inquiry stations. They work in groups and are supported by us to access resources necessary for addressing appropriate curriculum standards. Teachers need to provide evidence of differentiated instructional strategies that will be used to meet the needs of students who will visit us and include necessary assistive technologies, adaptations, and modifications that will allow each SEN student to successfully engage in their inquiry station. In addition, our teachers work with volunteers from the university's Korean Sign Language club to learn signs and fingerspelling for simple aspects of their lesson. The KSL club members also volunteer as interpreters, so our teachers must also collaborate with the KSL club members to be able to provide support to sign specific terminology and instructions to SEN students during the science fair.

The Science Fair is a full day event that includes an opening ceremony with college administrators, a tour of the campus, visits to science laboratories, picnic lunch, participation in science activities, and a closing ceremony with a presentation of student certificates from the university. These certificates are evidence of extra-curricular activities that can be used in college admissions submissions. All of these events are planned and implemented by teachers in our course. The final week of the class is spent reflecting on all that was learned, de-briefing from the events of the science fair, and discussing what teachers will be able to realistically implement in their future teaching careers.

TRANSFORMATIVE POTENTIAL OF EXPERIENTIAL COURSES

In this section, we share findings from excerpts from interviews, autobiographical reflections, online discussion posts, and field notes collected over two semesters to highlight the experiential education and inclusive education design principles that we believe were salient in helping teachers to develop and grow.

Expanded Awareness about Inclusive Instructional Strategies

A central goal for our course was to develop experiential activities that could integrate theory and practice while engaging our teachers in purposeful tasks. Our teachers indicated that while other pedagogy courses had required them to create lesson plans, none were actually implemented with real students. Having a chance to actually teach SEN learners allowed them to put into practice some of the strategies they learned about from the course – including learning to modify instructions and materials to accommodate students' needs.

> I was able to think a lot about SEN students while designing the lessons [and learned by] finding ways to communicate information more effectively and organizing diverse experiments. I also realized the difficulties that can happen when teaching a class. (PST 3, Autobiography 5, June 2016)

> I found that I had to go very slowly when teaching the class. Because most of the students tended to rely on their eyes, I found I needed to point to the objects

> while speaking. Lastly, I learned to minimize [simplify] the number of letters, lines, and words on one slide so that students could easily read the text. (PST 9, Group interview 3, June 2016)

Being able to implement the lessons was important, but reflecting on and evaluating the effectiveness of their planning and teaching allowed teachers to have an improved understanding for how their knowledge about learners and instructional strategies could be used in practice. We wanted these future teachers to develop an appreciation for the wide range of diverse abilities that all SEN students have, regardless of what disability label is used to describe them. Engaging in meaningful activities supports teachers to become emotionally invested in their work. We found not only that the course supported teachers to see the value of providing science learning experiences for these students, but it also helped them find the experience rewarding as educators.

> I thought it was touching to see that the students were so excited that they were talking a lot. Even though they had unclear pronunciation, when we gave students time to share what they learned, I felt thankful to them [for their comments]. (PST 5, Autobiography 5, June 2016)

> Participating in the science fair, I felt and learned a lot of things. In particular, I felt rewarded when students expressed joy in learning. (PST 6, Autobiography 5, June 2016)

Experiential learning requires participants have a sense of ownership and accountability for what they are learning. To help foster this notion in our teachers, during the science fair preparation and implementation period, we worked mainly as supporters and facilitators for teachers' discussions and helped provide guidance about where to locate needed resources.

> [In the future, if I have a SEN student in my class] I will think about various ways to overcome challenges. Since I learned a lot from this experience, I even want to try some of these methods. If I become a teacher [of a SEN student], it will be a rewarding and proud year in my life as an educator. Of course, I must try my hardest, because if I fail, the student will be disadvantaged. (PST 2, Post-interview, June 2016)

> I wish the science fair could be even more open and include everyone if possible. After seeing how excited the students were for the fair, it felt really rewarding. I want to have more students enjoy this kind of experience. So, if I have a chance in the future to do something like this, I want to do it again. (PST 7, Group Interview 3, June 2016)

We believe that this experience helped our teachers to recognize the challenges that will face them in the future should they have an opportunity to work with SEN learners. We also believe that supporting our teachers to find success through their own efforts and ingenuity allowed them to build the confidence they needed to be open to trying to teach these students in the future. Their success in identifying and implementing appropriate strategies to accommodate different learners demonstrated that it was at least possible to support these students to do science. This realization led many students to assert that in the future, they would be open to trying to find ways to support SEN learners in their classrooms.

Expanded Awareness about SEN Students as Learners and Teachers' Expectations

Based on teachers' reflections about prior experiences with people with disabilities and with SEN student peers, we found evidence that our teachers had become more critical about their beliefs and attitudes towards SEN students as learners. Several teachers recognized they held negative expectations for what the students in our fair would be capable of doing. While many pre-service teachers indicated that their observation of and interactions with the SEN students and teachers during the school site visit helped to challenge some of those assumptions, their interactions with SEN students during the science fair were most compelling.

> Honestly, I had a deep perception that SEN students might have lower scientific thinking skills than general students. Even though I knew there were differences in [disability] levels among them, I thought it would not make much difference and these students would achieve at lower levels than general [education] students. However, through the science fair activities, my negative perceptions were strikingly changed. This was not just due to their enthusiasm or concentration while learning. When I looked back at students' questions and practices in the activities, I observed that some students had really high level scientific and logical reasoning. (PST 2, Autobiography 4, June 2016)

In an education system that places considerable value on preparing students to compete on a once-per-year college entrance exam, there is limited discussion about the inherent value of learning science for everyday life or fostering scientific literacy. Because there is no ability tracking in the Korean education system, there are no general education science courses designed to meet the needs of students who will not pursue a post-secondary education.

> [Before] I really questioned the importance of science education for students with disabilities. However, I [now] realize that education, especially science education is important and can offer great possibilities for students with disabilities. I found that science content [may be] more appropriate for students with disabilities [than other subjects]. Students easily become interested [in science] and approach it more than other subjects. (PST 4, Autobiography 2, March 2016)

Convincing teachers that there is some value to learning science for SEN students who have limited likelihood of entering university was a significant challenge we identified during the first time the course was offered. Of the nearly 10,000 SEN students who completed secondary education in 2016, only 54% transitioned into tertiary education or employment (MOE, 2016). Teachers can expect roughly 98% of all general education students will continue their educational training beyond the secondary level. Our desire to have teachers really work with SEN students was born from this need to encourage teachers to be able to see the potential benefits of inclusive science education for SEN students. Several teachers indicated they had been able to expand their view about not only what was possible for SEN students to do but also what was important for them to learn.

> The most important thing [for SEN students' learning] is that the teacher must believe science is possible for SEN learners. If teachers do not think it is possible, they will not even try. But by thinking that *it can be possible*, the teacher can try something and discuss with other teachers [to get support]. (PST 6, Post-interview, June 2016)

Not only does this remark show that the teacher is aware of her role in making science accessible to SEN students, she also indicates that she could work with other teachers to try to provide instruction. Many participants echoed these sentiments across both courses.

> Now I know a person with a disability can dream about doing various things and education can help them to achieve it. The biggest change [in my thinking] from this course is now I know that SEN students are not incompetent, they just have difficulty to access new information. If they can successfully access the information, they can have the same learning outcomes. (PST 3, Post-interview, June, 2016)

Importantly, our teachers demonstrated a growing awareness of ways in which limited expectations for what SEN students can do also impacts on their opportunities to learn. We found this to be one of the most important outcomes from teachers' participation in our course.

Expanded Awareness about Disability Issues in Society and Role of Teachers

For inclusive education to be tenable in Korean public schools, teachers must be convinced they have the skills to be effective for SEN students and that educating SEN students in inclusive settings is appropriate and desirable for all students. We believe that for teachers to adopt this mindset, they must view the education of SEN students as more than a privilege to be bestowed by a benevolent government; rather, it should be viewed as a human right to be demanded for all people with disabilities. To help our teachers reflect on their roles and responsibilities as advocates for an equitable and just educational system and society, we tried to provide teachers with experiences that could help them see connections between their lives and the lives of disabled people they may not know or understand.

> From this lecture, I came to realize that a teacher should be a protector of SEN students during school time. The teacher needs to teach [SEN and general education] students not only content knowledge, but also personal character and how to behave in life. Teachers must intervene in peers' conflicts and prevent discrimination against SEN students. (PST 1, Autobiography 2, March 2016)

> I met an advocate fighting for the educational rights of SEN students. I could understand the difficulties of family members (parents and siblings of SEN students) and also the greatness of [women as] mothers. This lecture changed my perspective on those struggles. (PST 4, Autobiography 2, March 2016)

Having teachers engage with a parent advocate who could speak about her personal struggles to raise a child with disabilities and also educate students about the legal discrimination that faces families and adults with disabilities, was an effective way to have teachers develop greater empathy families and SEN students.

> I could not accept what I had learned in the special education textbook because of my prejudiced beliefs that SEN students cannot be educated and there is nothing they would want to learn, unlike general education students. Now I think that I should improve my awareness about rights of [people with] disability. (PST 11, On-line discussion 2, March 2016)

We hoped by reflecting on their experiences in this course, teachers could gain a big picture perspective of how people with disabilities are positioned in Korean society and to reflect on the ways in which they could individually work to disrupt inequities through their own actions. Several teachers began to imagine what kinds of future roles and responsibilities they could take on to help create a more equitable and just society for these children and their families.

CONCLUSIONS AND DISCUSSIONS

Numerous studies have shown that pre-service teachers' attitudes toward inclusion cannot be positively affected by coursework alone. Fieldwork has increasingly been seen as a necessary component for challenging teachers' notions about disability (Varcoe & Boyle, 2014). Other than previous volunteer experiences with organizations supporting people with disabilities or being a student in a classroom where SEN students were physically included, our teachers began the course with limited knowledge and understanding about SEN students as individuals or as science learners. Current practices regarding inclusive education in Korean schools make it nearly impossible for us to identify inclusive science classrooms where our pre-service teachers could gain similar experience. While this means our teachers have limited possibilities for seeing positive examples of inclusive classroom practices in real school settings, our teachers' experiences in the science fair offer evidence of their own potential for providing positive science learning experiences for SEN students.

In this course, teachers were responsible for identifying students' needs, developing content appropriate lessons, and providing accommodations to support individual students to meet the learning goals for each activity. This placed teachers at the center of their own learning, as they each needed to access different kinds of knowledge and resources to support their individual science fair activities. Designing and implementing lessons and then reflecting on shared teaching experiences positioned participants as teaching professionals rather than pre-service teachers. Teaching and interacting with these students help teachers to challenge preconceived notions about people with disabilities. Our teachers were surprised to find that SEN students were as diverse in their personalities, interests, and abilities as students who they considered to be "normal or regular". To prepare teachers to have positive self-efficacy for teaching SEN students, it is important to focus attention not only on the resources and knowledge teachers need to plan and implement effective lessons, but also to foster positive attitudes about SEN learners and the value of science education for all students.

If we want teachers to effectively engage SEN learners, they need to feel comfortable interacting with people with disabilities and be able to embrace a philosophy of inclusion in their classrooms. Our work builds on previous studies in the field of special education (Forlin, Loreman, Sharma, & Earle, 2009; Sharma, Forlin, & Loreman, 2008) that has shown that having direct teaching experiences with students with special needs can help future teachers to build more positive attitudes towards these learners specifically and inclusive education in general. We believe that structured experiential learning activities with SEN students, framed by individual and whole group reflection, is effective for providing pre-service teachers important opportunities to connect theory and practice in an authentic science learning environment. Without autobiographical writing tasks that challenged teachers to consider how their past lived experiences shaped their current attitudes, beliefs, and dispositions to act, many teachers in this study reported that they would have been less inclined to struggle with their unexamined prejudices and assumptions. For example, it was not uncommon for teachers to recount painful and embarrassing childhood memories of having participated in the teasing (or failing to stop the teasing) of a classmate with a disability. Having a community of teaching professionals with whom to honestly share their thoughts helped our teachers to gain some perspective on their experiences and, more importantly, consider how they might choose to act in the future – both as a teacher and a fellow citizen. We believe the transformative potential of this course may come from preparing teachers who feel empowered to advocate for the educational rights of all students, including students with disabilities or students with different learning styles or life goals. Teachers have the potential to positively disrupt societal norms regarding what it means to be *dis-abled* by modeling inclusive education practices in their own classrooms.

IMPLICATIONS FOR INCLUSIVE EDUCATION IN KOREA AND BEYOND

Making science learning accessible and meaningful for diverse groups of learners has increasingly been a goal of science educators and researchers over the last 30 years. A notion of *science for all* students has generally been directed towards addressing disparities in achievement and engagement in science for girls, economically disadvantaged students, and students who are culturally and linguistically diverse from dominant, mainstream student populations. Publications focused specifically on science teaching and learning for SEN students have been fairly limited. A recent review (Villanueva, Taylor, Therrien, & Hand, 2012) of articles published in top-tier international science education journals from 1990–2012 found that only 13 articles examined the science learning needs of students with disabilities. However, at a time when more research attention should be focused on this issue, there are increasingly loud voices advocating for reinstituting separate educational policies for students with disabilities (Evans & Lunt, 2002; Zigmond, Kloo, & Volonino, 2009). Proponents of policies seeking to diminish inclusive education efforts argue it is too difficult to effectively differentiate instruction for all students and that better educational oversight can be achieved through separate educational systems and teacher preparation programs for special education teachers.

However, we assert that inclusive education is critical not only for the academic benefits it can provide SEN students, but also because teachers and general education students can gain a greater appreciation for the underlying principles of inclusive education, which seek to promote tolerance, diversity, and equity in society. Korea's educational system is currently working under a separate/inclusive education model in which SEN students are physically included in general education schools, but fail to be integrated academically and socially. We can see the effect of this segregated education policy in our pre-service teachers' lack of understanding about SEN students and the challenges facing people with disabilities in society. This model significantly contributes to limiting opportunities for SEN students to be successful in school and severely limits these students' chances for being independent in adulthood. One recent study examining poverty and disability in Asian countries (Economic and Social Commission for Asia and the Pacific [ESCAP], 2012) found that Koreans with disabilities had significantly lower employment and educational attainment rates than both their Korean peers and disabled people living in other Asian countries. More than 70% of those surveyed indicated their school provided no vocational training, and less than 30% of the participants indicated that they were presently holding jobs (ESCAP, 2012). Of those who were working, only 6% indicated they were able to meet their livelihood needs without assistance from their families or the government. These findings suggest a grim future awaits most SEN students, especially if their families cannot support them.

Currently, only about 1% of Korea's student population has been identified as needing special education services. However, because Korean women tend to delay childbirth into their mid-to-late 30s, the proportion of SEN learners is expected to increase over time. Later-in-life births tend to increase the number of premature and low birth-weight babies (Tough, Newburn-Cook, Johnston, Svenson, Rose, & Belik, 2002), which can lead to an increased incidence in the number of children born with developmental delays and cognitive and physical challenges (Boyle et al., 2011). In addition, as the field of special education continues to develop in Korea, we can expect more children will be identified as needing support due to improvements in diagnostic tools. At the same time, Korean schools, are experiencing increased enrollments of culturally and linguistically diverse students (Park et al., 2016), which means general education teachers will need to be able to differentiate instruction for a variety of learning needs.

Such changes suggest the need for expanding inclusive education, not only to meet the needs of diverse student groups, but also to promote equity and tolerance in a society that is becoming increasingly diverse. To help expand opportunity, we need innovative teacher education programs and more research in Korea and in the larger education community that focuses attention on improving inclusive education. While our approach is limited in potential to create systemic change or impact, we believe it is an important first step towards establishing research-driven teacher education initiatives to promote authentic inclusive education. We believe similar initiatives are necessary in many countries where there is a trend towards increased diversity of students. We hope to see more researchers and educators addressing this issue in the future.

ACKNOWLEDGEMENTS

This work was supported by the Ministry of Education of the Republic of Korea and the National Research Foundation of Korea (NRF-2016S1A3A2925401).

REFERENCES

Baglieri, S. (2008). 'I connected': Reflection and biography in teacher learning toward inclusion. *International Journal of Inclusive Education, 12*, 585–604. doi:10.1080/13603110802377631

Boyle, C., Boulet, S., Schieve, L. A., Cohen, R. A., Blumberg, S., Yeargin-Allsopp, M., Visser, S., & Kogan, M. D. (2011). Trends in the prevalence of developmental disabilities in US children, 1997–2008. *Pediatrics, 127*, 1034–1042. Retrieved from https://doi.org/10.1542/peds.2010-2989

Byun, H. K. (2005). Historical understanding of special education. In H. G. Shin (Ed.), *Understanding of special education* (pp. 19–42). Seoul: Kyoyookbook.

Chapman, S., McPhee, P., & Proudman, B. (1995). What is experiential education? In K. Warren (Ed.), *The theory of experiential education: A collection of articles addressing the historical, philosophical, social, and psychological foundations of experiential education* (3rd ed., pp. 235–248). Dubuque, IA: Kendall/Hunt Publishing.

Economic and Social Commission for Asia and the Pacific [ESCAP]. (2012). *Disability, livelihood, and poverty in Asia and the Pacific.* Bangkok: United Nations Publications.

Evans, J., & Lunt, I. (2002). Inclusive education: Are there limits? *European Journal of Special Needs Education, 17*(1), 1–14. Retrieved from https://doi.org/10.1080/08856250110098980

Forlin, C., Loreman, T., Sharma, U., & Earle, C. (2009). Demographic differences in changing pre-service teachers' attitudes, sentiments and concerns about inclusive education. *International Journal of Inclusive Education, 13*(2), 195–209. Retrieved from https://doi.org/10.1080/13603110701365356

Hascher, T., Cocard, Y., & Moser, P. (2004). Forget about theory – practice is all? Student teachers' learning in practicum, *Teachers and Teaching, 10*, 623–637. Retrieved from https://doi.org/10.1080/1354060042000304800

Hwang, J. W. (2015, December). Status of special education in Korea. *Education Development, 194*, 1–5. (Published by Korean Educational Development Institute (KEDI)).

Kim, J., Kim, M., Beach, E. M., & Han, A. (2016). A study of health perception, disability acceptance, and life satisfaction based on types of leisure activity Koreans with a physical disability. *Applied Research in Quality of Life, 11*, 791–804. Retrieved from https://doi.org/10.1007/s11482-015-9397-8

Kim, M. (2012). Qualitative case study on school principal and general education teachers' experiences of inclusive education and classroom practices at college-bound high school. *Asian Journal of Education, 13*(4), 59–79.

Kim, S. G. (1995). *Special education.* Seoul: Yangjisa.

Kim, Y. W. (2013). Inclusive education in Korea: Policy, practice, and challenges. *Journal of Policy and Practice in Intellectual Disabilities, 10*(2), 79–81. Retrieved from https://doi.org/10.1111/jppi.12034

Kim, Y. W. (2014). Inclusive education in South Korea. *International Journal of Inclusive Education, 18*, 979–990. Retrieved from https://doi.org/10.1080/13603116.2012.693402

Kolb, D. (1984). The process of experiential learning. In D. Kolb (Ed.), *Experiential learning: Experience as the source of learning and developmen*t (pp. 20–38). Englewood Cliffs, NJ: Prentice-Hall.

Korean Educational Development Institute (KEDI). (2007). *Understanding Korean education: School education in Korea* (Vol. 3). Seoul: KYUNGHEE Information Printing.

Kwon, H. (2005). Inclusion in South Korea: The current situation and future directions. *International Journal of Disability, Development and Education, 52*(1), 59–68. Retrieved from https://doi.org/10.1080/10349120500071910

Kwon, H. (2016). Secondary inclusive school teachers' perceptions about inclusion, inclusive classroom, and teachers' role for inclusion. *Journal of Special Education: Theory and Practice, 17*, 263–282.

Lee, H. J. (2016, April 19). Disabled students suffer under reluctant education system. *The Korea Herald.* Retrieved from http://www.koreaherald.com/view.php?ud=20160419000855

Lee, J., Kang, B., & Lee, D. (2016). Law for gifted and talented education in South Korea: Its development, issues, and prospects. *Turkish Journal of Giftedness and Education, 6*(1), 14–23.

Lewis, L. H., & Williams, C. J. (1994). Experiential learning: Past and present. *New Directions for Adult and Continuing Education, 62*, 5–16. Retrieved from https://doi.org/10.1002/ace.36719946203

Martin, S. (2005). Not so strange in a strange land: An autobiographical approach to becoming a science teacher in an urban high school. In K. Tobin, R. Elmesky, & G. Seiler (Eds.), *Improving urban science education: New roles for teachers, students, and researchers* (pp. 225–243). New York, NY: Rowman & Littlefield Publishers.

Martin, S., Choe, S.-U., Kim, C.-J., & Kwak, Y. (2014). Employing a socio-historical perspective for understanding the impact of ideology and policy on educational achievement in the republic of Korea. In J. V. Clark (Ed.), *Closing the achievement gap from an international perspective: Transforming STEM for effective education* (pp. 229–250). Dordrecht: Springer.

McNeill, D. (2011, November 27). After decades of building colleges, South Korea faces a lack of students. *The Chronicle of Higher Education* (Global News Section). Retrieved from http://chronicle.com/article/article-content/129896/

Ministry of Education. (2016). *2016 annual report on special education.* Seoul: National Institute of Special Education.

Ministry of Education. (2017). *2017 special education statistics.* Seoul: Ministry of Education.

Morris, P. (1996). Asia's four little tigers: A comparison of the role of education in their development. *Comparative Education, 32*(1), 95–109. Retrieved from https://doi.org/10.1080/03050069628948

OECD. (2016a). *Indicator A2: How many students are expected to complete upper secondary education? Education at a glance. 2016: OECD indicators.* Paris: OECD Publishing. doi:10.1787/eag-2016-8-en

OECD. (2016b). *Indicator C1: Who participates in education? Education at a glance. 2016: OECD indicators.* Paris: OECD Publishing. Retrieved from https://doi.org/10.1787/eag-2016-23-en

Park, J. (2002). Special education in South Korea. *Teaching Exceptional Children, 34*(5), 28–33. Retrieved from https://doi.org/10.1177/004005990203400505

Park, J., Chu, H. E., & Martin, S. (2016). Understanding challenges for inquiry and language based teaching to culturally and linguistically diverse students in Korea. *EURASIA Journal of Mathematics, Science and Technology Education, 12*, 1799–1841. Retrieved from https://doi.org/10.12973/eurasia.2016.1536a

Ryu, S. Y. (2009). Special education and social services in Korea: Past, present, and future. *International Review of Research in Mental Retardation, 38*(1), 125–146.

Siry, C. A. (2011). Emphasizing collaborative practices in learning to teach: Coteaching and cogenerative dialogue in a field-based methods course. *Teaching Education, 22*(1), 91–101. Retrieved from https://doi.org/10.1080/10476210.2010.520699

Sharma, U., Forlin, C., & Loreman, T. (2008). Impact of training on pre-service teachers' attitudes and concerns about inclusive education and sentiments about persons with disabilities. *Disability & Society, 23*(7), 773–785. Retrieved from https://doi.org/10.1080/09687590802469271

The World Bank. (2017). *Overview: The World Bank in republic of Korea.* Retrieved from http://www.worldbank.org/en/country/korea/overview

Tough, S. C., Newburn-Cook, C., Johnston, D. W., Svenson, L. W., Rose, R., & Belik, J. (2002). Delayed childbearing and its impact on population rate changes in lower birth weight, multiple birth, and preterm delivery. *Pediatrics, 109*, 399–403.

United Nations Educational, Scientific and Cultural Organisation (UNESCO). (2005). *Guidelines for inclusion: Ensuring access to education for all.* Paris: United Nations Educational, Scientific and Cultural Organization.

Varcoe, L., & Boyle, C. (2014). Pre-service primary teachers' attitudes towards inclusive education, *Educational Psychology, 34*, 323–337. Retrieved from https://doi.org/10.1080/01443410.2013.785061

Villanueva, M. G., Taylor, J., Therrien, W., & Hand, B. (2012). Science education for students with special needs. *Studies in Science Education, 48*(2), 187–215. Retrieved from https://doi.org/10.1080/14703297.2012.737117

You, H. K., & McGraw, L. A. (2011). The intersection of motherhood and disability: Being a "good" Korean mother to an "imperfect" child. *Journal of Comparative Family Studies, 42*, 579–598. Retrieved from http://www.jstor.org/stable/41604469

Zigmond, N., Kloo, A., & Volonino, V. (2009). What, where, and how? Special education in the climate of full inclusion. *Exceptionality, 17*(4), 189–204. Retrieved from https://doi.org/10.1080/09362830903231986

ABOUT THE AUTHORS

Sonya N. Martin is an associate professor in the Department of Science Education at Seoul National University in Seoul, Republic of Korea. Sonya holds a bachelor's degree in Biology from Bryn Mawr College, and master's degrees in Elementary Education and in Chemistry Education from the University of Pennsylvania in the United States. She also holds a doctoral degree in Science Education from Curtin University in Australia. Her research focuses on identifying science teacher practices that promote learning for students with special education needs (SEN) and for culturally and linguistically diverse students.

Da Yeon Kang is a doctoral student in Science Education at the Seoul National University in Seoul, Republic of Korea. Da Yeon earned a bachelor's degree in Earth Science Education and a master's degree in Science Education from Seoul National University. Her research focuses on equity issues in science education. Da Yeon is particularly interested in improving teacher education programs for pre- and in-service teachers by providing experiential learning opportunities to expand teachers' exposure to and awareness about diverse learners in Korean K-12 science classrooms.

LEAH D. PRIDE

7. TAKING A RISK

Having Mindful Discussions about Race, Privilege, and Social Justice

ABSTRACT

In my experience, science teachers traditionally avoid whole-group discussions as they relate to topics considered to be "social" issues like whether Black Lives or All Lives Matter; or whether it is appropriate to allow the use of the N-word in the learning environment. Compartmentalizing issues of race, privilege, and social justice are not appropriate. Whenever and wherever thorny issues arise, they are to be problematized with great care and not dismissed in haste or relegated to classrooms belonging to the social sciences and humanities teachers. The purpose of this chapter is to take a risk and engage in an unorthodox, mindful discussion about race, privilege and social justice. Unique in academia, I use the authoring of auto-ethnographic, impressionist tales and writing prompts derived from reflections of the "Thorny Issues Heuristic" as mindfulness tools for discussing thorny topics with science teachers enrolled in a teacher preparation program at Brooklyn College of The City University of New York. The informal writing style of this chapter invites the reader to participate in an active dialogue with the purpose of sharing emotional narratives in their rare and authentic nature. The following is what I have learned from my students and how some narratives were painful yet encouraging; welcoming yet alienating.

Keywords: race, privilege, social justice, thorny issues, impressionist tales, heuristics, mindfulness

INTRODUCTION

> Leah: When encouraged to write a chapter on race, I became overwhelmed by a surge of emotions and metacognitive thoughts and questions.

Who am I to discuss race, equity and science education? Humbly, I am a Black American woman and a doctoral candidate of the learning sciences in an urban education program. I am also a biochemist. I am an anomaly, given the history of Black Americans (politically termed as African-Americans) in the United States and I write knowing and having experienced the histories of others who struggle(d) to make their voices heard.

 DOI:10.1163/9789004389663_007

Why do I identify as a Black American and not as an African-American? I capitalize the term, Black, and use it as a noun rather than an adjective to make transparent a worldview belonging to a group of people who originate from the continent of Africa within their ancestry and who, as a cultural group, were cosigned to family histories of involuntary and forced immigration, so as to fulfill the expectations of servitude throughout the African diaspora (Ogbu & Simons, 1998). "Black," as a political moniker, is used to marginalize a group of people who share a common experience of racism, subjectivity, and (mis)representation as it relates to assumptions (often void of discourse) about culture, values, beliefs, and politics within the Black community (Hall, 1996). Blacks or African Americans, as a demographic are numerically underrepresented in the United States, making up thirteen percent of the total U.S. population (United States Census Bureau, 2016). From a power structure analysis, overwhelmingly Black Americans and other non-White minority groups possess a dearth of political, social, cultural and economic capital compared to their White counterparts (Bourdieu, The Forms of Capital, 1986). This chapter entails *my* experience as a Black American and is not to be generalized, even though some narratives may resonate with other oppressed people(s). The Black American experience contrasts others, particularly those Blacks who have immigrated to America voluntarily. Blacks who immigrated to America from the diaspora often distance themselves from Black Americans, even though they are racially similar (Shaw-Taylor & Tuch, 2007). Intersecting, contextually embedded, and often times contradictory, "Blackness" transcends skin color as a biological descriptor and represents a cultural shift for Blacks of the diaspora to seek and connect to their African heritage through racial consciousness. Blackness does not belong to any one sub-group within the diaspora, but is appropriated by individuals to usurp societal, prescribed cultural boundaries (Johnson, 2003). As a socio-political construct on an individual level, Blackness is dependent on location, cultural identity negotiations, and Black identity politics (Godfried, Sekimoto, & Brown, 2016). A post-colonial, cosmopolitan Blackness is not universally defined; however, it is generally accepted that Blacks should resist the Eurocentric burden of assimilation and celebrate diversity of what it means to "be Black" (in whatever context the individual appropriates) (Thiam, 2014). Blackness as a construct is often criticized for probing what it means to "act Black" juxtaposed against what it means to "act White," ignoring the perspectives that *any* other non-Black or non-White brings to the dialogue.

How will my narrative contribute to existing bodies of literature written by Black Americans? Historically, slave narratives throughout the African diaspora represent a genre of writing that offered Blacks in bondage an opportunity to name their oppressors and describe the harsh realities of plantation life. The slave narrative was usually crafted for the benefit of white abolitionists who had not experienced the throes of slavery, but wanted to read first-hand accounts as a means to authenticate the realism of bondage. Written with physically and emotionally excruciating details, the slave narrative subjugates the reader into an existential crisis where the hardships of exhaustive labor, beatings, sexual exploitations, and denial of inalienable human rights: life, liberty and the pursuit of happiness, and the illegalities of existence beyond servitude become a temporary reality in much the same way that a work of science fiction transplants the reader to an imaginary, alternate reality (Sherryl, 2007). Neo-slave narratives are contemporary literary works that capture the oral traditions of slavery as recounted by descendants of slaves or sympathizers of antebellum American history. Neo-slave narratives may also originate from a third person's pontification about what slavery life might be like or the narrative may represent a fictional or faux biographical life-story of a slave (Rushdy, 2004). Contemporary neo-slave narratives have been written as fictional novels with the author's objective being to "come to terms with [the] American past, and salvage the vestiges of a distant African past." (Ouimet, 2002, p. 133). Examples of contemporary, neo-slave narratives are William Styron's *The Confessions of Nat Turner* (1967), Ernest J. Gaines' *The Autobiography of Miss Jane Pittman* (1971), Octavia E. Butler's *Kindred* (1979) and Toni Morrison's *Beloved* (1987) to name a few prominent works. Unique to the genre of the neo-slave narrative is that it allows the author to explore the emotions of an unanswered past and questions about an unsettled identity: Black Americans not belonging to colonial America as authentic citizens and not being recognized as *human*, but rather, as property.

Who may author neo-slave narratives? At issue with this genre and notable works, particularly *The Confessions of Nat Turner* and Harriet Beecher Stowe's *Uncle Tom Cabin*, is whether non-Black counterparts have the authority to appropriate Black American slavery history and author their own slavery narratives or neo-slave accounts (Neo-slave narrative, 2005). A critique of historical slave narratives is that the narratives of Black males overwhelming populate the genre. There are examples of seminal works by Harriet Ann Jacobs (*Incidents in the Life of a Slave Girl*, 1861), Sojourner Truth (*Ain't I a Woman,* 1851), and a small constituency of notable female writers. Neo-slave narratives allow a space for Black feminists to have voice.

Structurally, my neo-slave narrative emerges throughout the chapter, beginning with intimate details about my family history and the emotions that surround having questions about one's ontological citizenship. While teaching, I often bring in the histories of my lived experience and ask my students to share their ways of being, knowing, and experiencing the world. As we incorporate diverse voices (polyphonia), alongside varied ways of knowing (polysemia) as a resource, discourse is enriched and potentially transformative (Tobin, Learning from a good mate, 2011). The neo-slave narrative usually continues with grotesque descriptions of oppression: both of mind and spirit, as the body is no longer legally bound to shackles and servitude. In this chapter, there will be many places where the emotional entanglement of my life as a person of color in America will be centralized. The emotions shared are not to be generalized, in much the same way the slave experience was not the same for all persons of color. Authored slave narratives share an impassioned quest for freedom: of the spirit, of the body, and most of all, the mind. My neo-slave narrative is an evolving awakening that comes about through authentic inquiry and the sharing of others' perspectives and differences in safe spaces. I write from a place of responsibility, an expectation that my narrative(s) and those of others serve as counter-narratives (Delgado, 1989): a tenet of critical race theory which offers writers of color the agency and responsibility to share their experiences with race and racism, coupled with the hope that we may be able to communicate to our non-Black counterparts (e.g. White), matters that they are unlikely to know and/or understand because their skin color gives them certain "privileges" in America.

During the Great Depression, in the cotton-fields of Oklahoma, my poorly educated, Black-American father learned how to persist through the days of segregation, discrimination, and the emotions that emerged in response to the struggle for existence. This unorthodox schooling was rooted in an oral tradition that has traversed the annals of history beginning with slavery into the contemporary, post-Barack Obama presidency. In "post-racial" America, I am a cotton-picker's daughter who became a scientist and science educator, despite a presumed deficit model that was forecast by society based on my family's lineage. I too, have learned to persist through the challenges of "-isms" (racism, sexism, classism, body size-ism). Today this oral tradition, which I have received, is referred to in the contemporary classrooms of urban, New York City where many of my students and I are faced with debilitating and emotionally jarring macro-structures: segregation, fear of being detained by authorities or killed given one's place of birth or the color of one's skin, and discrimination. Persistence in the context of this chapter, is the "inexplicable" determination to push through racial battle fatigue: psychological and physiological responses of stress relating to being a person of color and having feelings of frustration, shock, anger, disappointment, resentment, anxiety, helplessness, hopelessness, and fear in response to discrimination and racism (Smith, Allen, & Danley, 2016). Persistence is consciously embedded in my unofficial curriculum (as it relates to sensitive topics like, race, gender, privilege).

In writing narratives, how do I theorize beyond an egocentric form of activism and social justice? At a research forum, I shared my perception of activism and social justice in America, as being on a spectrum. At the individualized level, my theoretical position suggests individualized participation in social justice movements relies on constituents valuing their lives *and* the lives of others, concurrently, while also appreciating differences, not fearing them. Persons of color (including myself) learn to love and respect themselves by focusing introspectively. On the social justice spectrum, individualized self-love as an activist standpoint is gradated towards more collective activist efforts, which may include but are not limited to protestations against injustice of any kind. After sharing thoughts with the audience, rooted in my interpretation of the contemporary works of bell hooks, a counterpoint was offered by a member of the meeting. The scholar, who happened to be White and male, suggested that my thoughts were too "egocentric" and activism must grow beyond that of the self. Keeping in mind that my voice is only one view offered as a counter-narrative (alongside that of students in my class), I challenge the readers to see this work as a notch on the broad spectrum of activism.

GETTING TO AN EMOTIONAL "PLACE" WHERE I CAN THEORIZE RACE

I listened to a still voice inside and wrote during a meditation. The following is what I heard:

In a vast sea of emotion
there is me;
The place I seek,
Is Race.
A buried plea,
to unearth a systemic disgrace.
Set sail,
An anticipated odyssey deep within;
Prevail!
America's original sin-
The mariner,
Fatigued
Yet persistent-
Where Race lies,
I approach with haste.
Silence,
I find.
In the wake of violence
Prevail!
A barrage of micro-aggressions
at this place,
Race.
Resignation was the intended lesson.
The act of oppressors-
Piracy of valuable goods,
"Inalienable" rights.
Life, liberty, the pursuit of happiness,
Set Sail.
Give chase I should,
A promised land is in my sight.
Prevail.
In a vast sea of emotion
there is me.
The place I seek,
Is Race.
Set sail.
To speak,
my voice,
is a tempest to the commotion.
Prevail.

The informal writing style of this chapter invites the reader to an active discussion with the purpose of sharing emotional narratives in their rare and authentic nature. Below is an example of questions you may ask and answer, following the reading of a narrative:

> *The reader is invited to an active and mindful discussion*: What are your own emotions as they relate to the reading of the above narrative? Considering the emotions shared, what do you think the author wants you to experience?

SILENCE AS A PLACE FOR THEORIZING

bell hooks writes:

> I have been working to change the way I speak and write, to incorporate in the manner of telling a sense of place, of not just who I am in the present but where I am coming from, the multiple voices within me. I have confronted silence, inarticulateness. When I say, then, that these words emerge from suffering, I refer to that personal struggle to name that location from which I come to voice – that space of my theorizing. (1996, p. 49)

In order for the authoethnographic words that follow in this chapter to remain authentic and vulnerable, they emerge from within the captive walls of my heart and mind without judgment. Silence was difficult to accept in the writing of this chapter dealing with race: a part of my life (not just a topic to discuss) that is painfully isolating at the intersections of who I am, what I believe, and how I want to be in the world. Like hooks, I found the personal struggle to name the place of theorizing an arduous and transformative feat. I've appreciated how this writing process has made me more mindful, confident, and critical. As I've allowed my inner, more reflective self to have voice, I've grown into a more complete *person*.

During a writing period, when I encountered silence (in the context of this chapter, defined as the lack of a word or utterance in speech or writing), I was insatiate. I wanted an output but had to come to realize that silence is a valued output at times. It may be the best answer to tapping into a theoretical discourse:

Silence?
No,
a pregnant pause.
I smile.

Mikhail Bakhtin proposed a "word or utterance" as a means to express one's relationship to the other (a social entity). My interpretation of Bakhtin's work is that the utterance (or lack thereof) is contingent on *who* is engaged in the discourse and the exchanges (written, spoken, action – enacted or symbolic) between the participants. The pronoun, *who*, is an amalgamation of the self in relation to the other and is inclusive of historical contexts and situations that may or may not be nuanced at the time of the discourse (Bakhtin, 1981). Silence, or the act of silencing, is the enactment of emotional dissonance as a result of disrupting or omitting one's turn to "talk."

UNOFFICIAL CURRICULUM AS THE PLACE FOR THEORIZING

In science education, there are few teacher preparation programs with the courage and support to explicitly confront issues like race, privilege, and activism for social justice. From experience in the "hard sciences," these types of conversations are: non-existent, out of place, too emotional, and too subjective in STEM classrooms. Layering difficult social issues, like race, with technical terms strips away the humanistic context. The types of conversations that need to happen when discussing these issues are raw and subdural, honest, and transparent. Having the connotation of being "non-academic" and not belonging in the same space that experiments are being run, issues of discrimination, activism, and the like are often avoided and ignored; leaving those who live with them feeling violated, oppressed, and powerless.

The "Historical, Philosophical, and Sociocultural Foundations of Education and Science" course at Brooklyn College (of the City University of New York) is unique in that the official curriculum deals with the broader implications of history, contemporary philosophies, and the sociopolitical-cultural intersections all have had on American schooling, in particular science education (Alexakos, 2015). The course is meant to rouse critical reflection and authentic inquiry into one's way of "being" in the world and to help teachers become more aware of their own ways of teaching. The official curriculum of the course arms teachers with theoretical frameworks often excluded within scientific and teacher preparation programs.

As an instructor of the course, alongside pre-and in-service science teachers, I encouraged students to speak and write honestly and passionately, irrespective of how they may perceive it will affect me, emotionally, as a person of color. Implicit in the construct of this class was an expectation that we listen radically to each other. Radical listening, in the context of this work, is a way of listening with the purpose of learning from the other by addressing and accepting differences (Tobin, 2011). There is an inherent expectation that coherency expands when differences and contradictions are nuanced and culture is transformed through the act of dialogue (Sewell, 2005). Void of radical listening, discussions between participants can lead to confusion and contention; misunderstandings and feelings of anger or guilt will result when great care isn't taken to value what others are saying and lack of conformity between what we want to hear and what is being said.

> Our discussion took a different turn when we started to think of some of the ways we phrase our words and how they could become personal when they hurt another person. Someone from the [cogenerative dialogue] mentioned, "we should stop using the phrase, I agree with you" or "I disagree with you"

> and instead build on each other's conversations. By building on each other's conversations, we will be acknowledging what someone else contributes and let our opinions be heard, as well. Professor Leah contributed by mentioning that she didn't like when others used phrases, such as "I think you are trying to say this" when we are having a discussion because many times during our discussions many students rephrased what another person was saying. This discussion took a little turn, when I noticed some anger from some of the participants and they weren't even trying to include everyone else in the discussion. Few voices became dominant and this was my first experience in which I saw no safe space. I realized many of us became quiet and the ones who were speaking didn't even acknowledge that many of us weren't even talking or weren't even given a chance to talk (a reflection from Zara who is a Muslim woman, in-service science teacher).
>
> *The reader is invited to an active and mindful discussion*: What are your own emotions as they relate to reading the above narrative? How could the statement, "What I think you are trying to say," strip agency from marginalized voices? How do you define safe spaces? What lines in the narrative cause you to question the role of silence?

Implicit in the structure of the course and the physical classroom environment in which we met, was a need for a "safe space" for dialogue. A safe space not only refers to the need for physiological safety and a place where basic needs are met, but also a place that promotes dialogue, respect, and inclusion of *all* members, including those who may be underrepresented (Gayle, Cortez, & Preiss, 2013). In our course, not all members of the group were comfortable with participating in discussions on all topics that relate to race. As the instructor, I had to respect that some voices would be silenced from our conversations because it is the enactment of one's agency and convenience to exclude him or her-self when there is fear of potential danger or violence:

> Unfortunately, the conversation in our class took an uncomfortable path. I just couldn't bring myself to participate in the discussion about the use and origin of the "N-word." I just felt like I didn't have a place in that discussion and looking back, I feel upset and sad that it was that way (a reflection from Pamela who is a White female, in-service science teacher).
>
> *The reader is invited to an active and mindful discussion*: In terms of your own response to silence, what emotions are you most aware of while experiencing it? Do non-Blacks have a place in discussions that probe issues of race and Blackness (e.g., the appropriateness of the N-word, writing neo-slave narratives)? There are Black individuals, who are uncomfortable discussing race in mixed-racial company and homogenous-racial company, where does their voice fit in? Is a safe space attainable and is so, how is it protected?

Unique to our course, science teachers open each class meeting with a five-minute breathing meditation before coteaching units on the foundations of the history of science education using the lens of social life and the varied socio-political factors that shaped it. The breathing meditations are important, so as to discharge emotional distress, incorporating the principles of mindfulness, whenever it arises as a result of engaging in this authentic work (Powietrzynska, Tobin, & Alexakos, 2015).

I practice deep breathing while discussing and writing about race, which is emotionally laden with pain and causes me to feel vulnerable and unsafe. In writing this chapter, there will be times where I will insert the breathing prompt, so the reader will not only read, but also share in my emotional turmoil.

> Breathe in (deep, abdominal inhalation through the nose).
>
> Exhale (count down from 10 as the breath is released through pursed lips – in a controlled, slow manner).

The breathing prompt has served as my pain reliever and is now available to the reader to utilize throughout the reading of this emotionally heavy chapter for me to read and write.

THEORETICAL FRAMEWORK: NARRATIVES AND IMPRESSIONIST TALES AS A FORM OF AUTHENTIC INQUIRY

My opening narrative is written in a genre unlike the style of writing expected in a doctoral dissertation, peer-reviewed journal, or academic book chapter, yet it is an artifact of authentic inquiry. My work incorporates authentic inquiry, which centers on four-authenticity criteria adapted by Kenneth Tobin (2014) from an interpretation of Guba and Lincoln's work on authenticity in the research process (Guba & Lincoln, 1989):

> *Authentic inquiry is ontological*; the researchers | participants are open to change based on what they learn in the research process. The Sheffer stroke (|) in the context of this chapter denotes a dialectical relationship. For example, in the following dialectical relationship, self | other, the two elements are recursive in nature. If one element is changed (e.g., self), the other is transformed in the process as well. In the scope of this work, the opening narrative is autoethnographic | ontological and nuances the emotional multi-voicedness of the self | other. (Bakhtin, 1981)

> *In a vast sea of emotion*
>
> *there is me;*
>
> (Self: I am speaking as myself)
>
> *The place I seek,*

Is Race.

(Self | Other: I want to be critical of how race and racism, a social construct of the other, affects me)

A buried plea,

to unearth a systemic disgrace.

(Other: Racial categorizations and the racism that ensues are systemic, but function to suppress and oppress others)

Authentic inquiry is educative; all participants are encouraged to understand and learn from the other. By analyzing the written discourse within the dialectic of self | other, the researcher in this work (me) can learn from the voiced "other" within, which manifests throughout the narrative. The process of considering *who* is speaking, line by line, allows space for teaching and learning for both the reader and me. In autoethnographic narratives, my voice has agency and is open to inform and transform those willing to read and hear it.

Authentic inquiry is catalytic; stakeholders are expected to benefit from the research and the research process is meant to catalyze change, not stand in the way of it.

Authentic inquiry is tactical; the *teacher | researcher* addresses any disadvantages that may have emerged throughout the work, particularity for those who may not have the power to address them or are disadvantaged. Sharing my narratives alongside those students enrolled in my course, reveals inquiries into our ways of being which may be painful to some, yet encouraging to others; welcoming, yet alienating.

Autoethnographic narratives are grounded in mindfulness: a contemporary framework that embraces reflexivity and authenticity, becoming aware of the unaware and embracing difference as a resource (Tobin, 2011). During discussions involving issues like race, privilege, and social justice, there is an inherent responsibility to be kind to oneself and others, while also an expectation that the discussants are compassionate, incorporate radical listening, are non-judgmental, and do not get stuck in negative emotional spaces. In a mindful place, there is an awareness that some social issues, like discussions about race in mixed racial company, are emotionally unsafe and may be too sensitive to discuss without concern for the wellbeing of those engaged (Powietrzynska, Tobin, & Alexakos, 2015).

In response to unraveling the emotional entanglement with race and what it means to be a Black American, the opening narrative takes on the methodological framework of an impressionist tale: a "story" that is ethnographic and ontological in nature. The emerging tale privies the reader to creative and often underappreciated details about the vulnerabilities of a character(s) or an event, that must be told, with the purpose of being educative. As the impressionist tale is exempt from the constraints of peer-reviewed, academic language in its most elemental form, the author has unabridged license to tell a story how he/she wants and/or needs (Van Maanen, 2011).

The autoethnographic impressionist tale takes on whatever structure that emerges, with the intentions of pushing past the boundaries and prisons of one's mind and opening the possibilities of theory and research beyond raw emotion (Van Maanen, 2011). The writing style, as it relates to my autoethnography mimics that of poetry: a discursive methodological framework used in research to marry existentialism (or the way in which one thinks about and exists in the world) with the author's emotional truths (Furman, 2007). Poetry, as a qualitative methodological tool, is unique in that it allows the use of metaphors and figurative language to explore the external and internal interplays of a person's emotions, irrespective of how they conflict, conform, or corrupt norms that may be generalized or marginalized in the social realm (Furman, 2007). Qualifying a poem for peer-review is quite subjective and frowned upon by positivists and other researchers that value objectivity and generalizability (Furman, Langer, Davis, Gallardo, & Kulkarni, 2007). However, when discussing issues that are quite emotional and often times difficult to express, allowing the flexibility of a poem (or any genre of writing) can embrace the realization that language itself is a structure that oppresses some who struggle to access it, particularly in academia. Looking at the narrative through the lens of polyphonia (speaking with many voices), the author and reader make meaning of the shared experience together (Bakhtin, The Dialogic Imagination: Four essays, 1981)

While my particular style of writing emerged in the form of a poem, so as to capture the "otherness" and emotional attachment to what was being discussed in the "Historical, Philosophical, and Sociocultural Foundations of Education and Science" course, others shared their narratives more directly.

> We are all one group of people, and we should all be held to the same standard – no more "select your race," no more "black lives matter," no more "white privilege" nonsense! My race is not a factor, all lives matter, and I have not been given this white privilege but if it's out there, send it my way (a reflection from Emma is a White female, in-service science teacher).
>
> *Breathe in (deep, abdominal inhalation through the nose).*
>
> *Exhale (count down from 10 as the breath is released through pursed lips – in a controlled, slow manner).*
>
> *The reader is invited to an active and mindful discussion*: What are your own emotions as they relate to the reading of the above narrative? How would you respond to the author?

When discussing race, the course provides readings that address the historical standpoints of Black-White relations in America. While this is not the only perspective on racism, it is one that is explored in greater context. Not all students appreciate the lens in which the course is structured:

> While going through the readings for this week, I realized that for the past two weeks we have been speaking about racism in a very bi-modal fashion. The focus has been on how African Americans have been oppressed or silenced by white Americans. It bothers me that we are discussing racism, but the sole focus is on one group of people who have felt the effects of racism as though there are no other oppressed groups. Considering the fact that many other groups are also stereotyped and stigmatized, I was hoping for a wider variety in the readings. I think too many people too often associate racism solely with the conflicts between black and whites that have spanned decades. Clearly those issues need to be addressed, but they are not the only issues that need to be addressed and I dislike how easy it is to become single minded in our views of what racism is and who it affects. I am often wary of becoming so caught up in my own suffering that I neglect the suffering of others" (a reflection from Ruby who is a Black female, in-service science teacher).
>
> *The reader is invited to an active and mindful discussion*: What is your own standpoint as it relates to the context in which race could/should be discussed (i.e. Black-White and Black-Black relations, Islamophobia, Critical race theory, Asian and Latino critical theory, Black feminist theory, Intersectionality, etc.)

THORNY ISSUES ARE EMOTIONAL AND HAVE A PLACE IN SCIENCE EDUCATION

As schools are a microcosm of social life, it is reasonable to expect difficult topics to be discussed in classrooms of *all* disciplines. During our time together at Brooklyn College, science teachers asked for opportunities to narrate their personal and emotional tales as it relates to the history of race, privilege and social justice in this country and express why it is or is not a "thorny issue." Thorny issues are sociocultural constructs (like race and colorism, gender and sexuality, implications of socioeconomic status and privilege) that are layered with deep emotions (could be positive or negative) and tend to be more closely aligned with the vulnerabilities of questioning assumptions about identity, "place" in society, and the emotional valences inextricably associated with engaging in these types of discourses (Alexakos et al., 2015).

Below is an excerpt from a narrative that was authored following a reading of "Skin Deep" a scientific article that explores the predisposition of variations in human skin color through a bio-geographical lens (Jablonski & Chaplin, 2002). The sociocultural perception that lighter skin is more desirable and lends itself to intended or unintended privileges not afforded those with skin tones of a darker hue is defined as colorism. In America and other countries, where some Whites and other lighter-skinned people of color *are* the ones with more sociocultural-political-economic power (Bourdieu, 1977), colorism often times but not always intersects with de facto racism and discrimination (Hunter, 2007). The discussion was quite thorny as we discussed the intersections of biology (i.e., skin color) and implications in social justice and perceived privilege.

> The thorniness [of a discussion that surrounds race and education] was apparent as we discussed the "elephant in the room" more deeply…I felt particularly quiet because I never felt that I could say anything since my race is considered the "privileged" one. Although I personally don't believe I fall into that category (I've worked so hard to be where I am), my skin color makes it tough for me to speak without being viewed a certain way… Somewhere along the line, skin color and societal standards overlapped. Skin color is no longer an indication of ancestral origin (how close to the equator you are), but it symbolizes and represents things like level of education, socioeconomic class, values, morals, and respect for others/authority. These societal standards extend into the classroom and unfortunately impede and hinder the progress and desire of achievement for those with an "undesirable" skin color (Pamela).
>
> *The reader is invited to an active and mindful discussion*: Can we substantiate the perception of perceived privileges associated with lighter skin color in America and/or around the world (e.g., skin bleaching)?

Throughout the semester, and as the course progressed, we science teachers asked each other weekly: How do we create safe emotional spaces in order to have the kinds of thorny discussions needed to tackle difficult issues like race and other "elephants in the room" (Alexakos et al., 2015).

> As I'm attempting to write this I'm having trouble articulating the words in my head, so I guess I'll start at the beginning of the class. We started off the lesson with a free write on how skin complexion affects your position in society. I cannot recall exactly what I wrote but to sum up society; the lighter your skin complexion the easier you have it. After writing what we had thought we did a travel talk where I had the privilege of speaking with you [Leah]. From what I remember, we both felt very uneasy sharing our thoughts and I believe it was because of two main issues. First being I don't think either of us thought we were going to share our thoughts. Second, but most prevalent to the actual topic, was the difference in skin color. After I had finished what I had written, I recall that you had verbalized your uneasiness, which actually eased my own discomfort. After both of us discussed our thoughts it was clear that we were very much on the same page, which I find troubling as a human. Why should the color of our skin, which is not something we choose, define us as a person?" (a reflection from Kenneth who is a White male, pre-service science teacher).

There were times during the semester when some voices were more privileged than others. The voices, typically of White students (with the exception of one Black female), compromised the safe space with swift interjections, fierce rebuttals, and deliberate invalidations of others' experiences during thorny discussions. This required me to address the students privately for not respecting the culture of radical listening. Sanctions of silence were imposed for the sake of keeping dialogue open for the rest of the class:

> Your [Peter] passion was not offensive, it was an indicator that you were engaged and in-tune with the emotional contagion that emerges when talking about subjects that challenge or question people's world views, and in some cases, ways of being. I appreciate hearing passion coming through one's voice; it is engaging and liberating. However, at times, it seemed as though there wasn't a reciprocal space for others' passion to emerge and that was suffocating. By the end of the session, it was clear that all the air was sucked out of the room. I have never been a debater, so it felt un-natural having to cut you off at some point. If it came off abrasive I'm sorry, was never my intention" (an excerpt from an email I sent to Peter who is a White male, in-service science teacher).
>
> Breathe in (deep, abdominal inhalation through the nose).
>
> Exhale (count down from 10 as the breath is released through pursed lips – in a controlled, slow manner).
>
> *The reader is invited to an active and mindful discussion*: How would you respond to this email? Peter chose not to respond.

Thorny discussions often include sharing personal narratives that may not only be emotionally painful, but also physiologically painful or awkward to bear. I wrote the following during a breathing and writing meditation:

I cried today.
It is an affective response to a reservoir of angst within.
Fear.
The question,
Who am I?
Fear captures and impregnates the first teardrop.
I ask,
What do you, the polyphonic you, want as an answer?
The reply,
I can't articulate clearly because the cacophony of voices are nauseating.
I hear an umbilical whimper amidst the contrast.
I connect to the lifeline its chord offers.
Physiologically, this translates into tears outside.
Fear claimed the first tears, but
The surge of finally wanting to answer the question,
Broke the dam of judgment.
Each drop has a justified place and it is reconciliation day.
I reply,
You want to know who I am?
I am human.

> *The reader is invited to an active and mindful discussion*: Take a moment to meditate. What voices are speaking? How does *your* body feel as you listen to what is being said? What are your own emotions as they relate to the reading of the above narrative?

THE THORNY ISSUES HEURISTIC AS A PLACE FOR THEORIZING

Heuristics are ontological shape-shifters; characteristics that cause the individual engaged in a reflexive process to pause and become aware of his/her ways of being (Powietrzynska, Tobin, & Alexakos, 2015).

> I believe heuristics are very helpful, because when we are presenting or talking during a discussion we need to be aware of other's feelings. It lets us keep in check what we are saying and how it could have an effect on others around us, which is very important. We usually make comments and think they are harmless. These comments could hurt those around us and it [heuristics] allows us to have a discussion. It also allows us to question if we are dominant during our discussions and presentations, which will allow others to speak as well and share our thoughts. Depending on where you stand in the discussion and presentation, you will make a change regarding your behavior in the classroom. If you don't speak as much, you will try to contribute more and if you are dominant, you will allow others to speak up. (Zara)

During the week in which we discussed race, the "Thorny Issues Heuristic" was given for students and me to author a free write about any of the characteristics. Prior to writing, we practiced a five-minute breathing meditation. Immediately following the meditation, I engaged in a free write and used the "Thorny Issues Heuristic" characteristic number 24, to write whatever came to the forefront of my mind, without judgment (Alexakos et al., 2015).

> I understand that others and I may not be able to use language to describe the emotions related to thorny issues.

Heuristics contain compassionate statements, such as characteristic 24, which engage the reader in a hermeneutic process. As a person becomes more aware of his/her emotions, change is inevitable because forgiveness and reflexivity are crucial underpinnings (Tobin, 2006). The aforementioned characteristic gave me an opportunity to engage in a research process that moved past the barrier of language, so as to probe my standpoints as they relate to race.

> Breathe in (deep, abdominal inhalation through the nose).
>
> Exhale (count down from 10 as the breath is released through pursed lips – in a controlled, slow manner).

The following impressionist tale is how my free-write emerged:

Stanza I
Pride.
The name always left my lips in sincerity, but reached my ears, mockingly.
Why did I have to have that name?
Ironic.
Leah has Pride.
Leah, full of Pride.
Pride of lions, powerful.
The metaphors were endless,
But shameful,
to me.
Inauthentic.

Stanza II
Growing up, I was not full of pride nor did I feel worthy of the name,
Pride.
Ironic, I declare!
Pride.
American.
Black.
Pride in Black America?
Black and Proud in America?
America has Pride in Blacks?
It was all confusing!
I hated myself, not White.
Sad but honest,
Too young to understand the complexities of colorism,
Why was white the idealized form?
Built like "Mammie" from the movies,
Couldn't escape my curves in this Black body.
I felt different.
Always reminded of being an(other).
My color defines how I am perceived in this country?
Judge by the content of one's character and not by the color of their skin,
But only if you are white, it seems.
In kindergarten,
I used to look at the palm of my hands and try to convince myself that the lightest color was my true color.
At age five, I was made aware that my skin was darker than the other kids in my class.
Innocence.

Pride in what skin color?
My father's hazel eyes and light skin,
The origins of my skin color were always questioned?
Too light to be truly "Black," and too Black to be "White."
Awkward.
"What are you?" people would ask.
"I am Black." was my reply.
"Then why are you so light?"
I was ashamed of the historical answer.
Nigger Pride.
Pride.

Stanza III
Pride in my lineage.
Sharecroppers who were hard working survivors;
Fruit of their labor,
Was me.
The roots of cotton and tobacco plants,
Burrow deep in the ground.
They absorb every drop of my blood.
During slavery and Jim Crow,
It was called "Nigger blood."
Nigger Pride, we are no longer called.
My blood.

Stanza IV
I've been told, "Black Americans! They are animals."
Even animals were treated better than us at times.
America.
Land of Opportunity,
America is the land of the free and home to the brave.
But, what if you are Black?
How do you reconcile history against the hopes of a future?
Breathe in (deep, abdominal inhalation through the nose).
Exhale (count down from 10 as the breath is released through pursed lips – in a controlled, slow manner).

By writing the impressionist tale, I begin theorizing out of pain; the purpose being to nuance the emotions of disgust as a person of color in America, coupled with the desire to self-identify.

The reader is invited to an active and mindful discussion: How would you respond to a person [i.e., the author] who expressed shame with being a person of color?

RACIAL DISGUST AS A PLACE OF THEORIZING

Following the struggle to contextualize this chapter and having exposed my emotional vulnerabilities in the course I taught at Brooklyn College, I authored an email to Konstantinos Alexakos, a mentor and friend. Below is my excerpt from the exchange:

> I'm unsure of what emotion you want me to capture in this paper because, generally, I feel as though I've expressed my disgust with being a Black person in America in my 2nd [doctoral] exam, in the paper I authored (Pride, 2014) and in emails. I readily identify as a Black person but I know that comes with discrimination, great challenge and pain: some pain that others will never experience or even understand. It will always be so just because I have more melanocytes than others. The history, in America, of having that "flaw" cannot be erased. I guess that is it: nothing fancy and nothing novel, it is the nature of the struggle. I am Black and that is that: the rest of the emotional attachment comes from what is projected by society towards me. So, I guess my point is: WHAT should I focus on in this piece that will be considered a good narrative – really not sure?

Digging a little deeper into the narrative and in the spirit of authentic inquiry, engaging the theoretical underpinnings of race studies and the personal and varied degrees of emotions that enmesh the social construct, the email expresses a "struggle." Considering my interpretation of Bakhtin's addressivity framework for discourse analysis (the utterance of a word or phrase with the expectation of a response), and the ontological authenticity criterion we take a closer look at the emotions within the context of the email as it relates to speaking to the self and the other. The methodological approach is to isolate each "I" statement and the verb (whatever it is) that follows (Bakhtin, 1981). In the context of this chapter, I term the voice that emerges as a result of this approach, the "I" voice which, as a methodological approach contributes to the growing body of multilevel research (Tobin & Ritchie, 2012).

> *I'm unsure of what emotion you want me to capture* (Self addresses the Other: What is it that YOU want to hear?)
>
> *I feel as though I've expressed my disgust with being a Black person in America* (Self addresses the Self: Am I being honest with myself?)
>
> *In my 2nd [doctoral] exam* (Self addresses the Other: These are publications that I authored and began to discuss my uncertainties and insecurities as a person of color in academia)
>
> *In the paper I authored* (Self addresses the Other)
>
> *I readily identify as a Black person* (Self addresses the Self: This is an epistemological, phenomenological, ontological, existential standpoint! It is the intersection from where all my theorizing and narratives emerge)

I know that comes with discrimination, great challenge and pain (Self address the Other: The ills of discrimination are historically documented in America)

I have more melanocytes than others (Self addresses the Self: This is a biological observation: innate and genetic)

I guess that is it (Other: The question is one that is emergent and contingent on the other)

I am Black (Self addresses the Self and the Other, concurrently: This statement is both a declaration and a plea for validation)

The rest of the emotional attachment comes from what is projected by society towards me. (Self addresses the Other: The statement points to an emotional conflict and a reductionism which seems saturated with more questions rather than answers. This line would be an event or worthwhile narrative to explore further in a hermeneutic research approach)

I guess my point is: (Other: There is a hesitancy in this statement, an insecurity as to whether or not it is allowed for the author to have the privilege of a point)

WHAT should I focus on in this piece (Self addresses the Other: This should read, "What do you want me to focus on?" or "What do YOU want to hear?")

The reader is invited to an active and mindful discussion: Who do you theorize the "I statements" are addressing (i.e., Self | Other)?

In this narrative, the tensions of living in America as a person of color are contrasted by the desire to "readily identify as a Black person" knowing that it will lend itself to "discrimination, challenge, and pain." Depending on the field of study, the term "identity" can be defined any number of ways. In the context of this chapter, identity is characterized as follows: a) it is fluid, contextual, temporal, and contingent b) emerges out of a dialectical relationship that involves self | other and c) it is reflexive in nature, where awareness brings about transformation and ontological shifts (Sewell, 2005).

"WHITENESS" AND THE HISTORY OF RACE IN AMERICA AS A THORNY ISSUE AND A PLACE FOR THEORIZING

Throughout the course at Brooklyn College, the history of race in America was a thorny issue for many members, including myself. If we were to take a snapshot of present-day America (Fall, 2017), there is a debate about whether it is appropriate to argue that "#Black Lives Matter"; a moniker that began on social media in the summer of 2013 after the deaths of unarmed Black men at the hands of police officials (Chokshi, 2016). #BlackLivesMatter, as a political movement, denotes the perceived historical devaluing of Black lives in America while it also stands as a mantra for awareness, social change, and justice. While many argue that it is time to focus on the injustices skin color has had on Black and Brown American citizens (while the term, "Brown," has been adopted by those who originally descended from the South Asian sub-continent, in the context of this paper the term is used to represent those of Latin@ heritage); others argue that "#AllLivesMatter" in post-racial America and the setting apart of any one racial group in protest is equally as discriminatory and damaging as the perceived persecution of said group (Carney, 2016). For those that argue that "#AllLivesMatter" there is an underlying conduit that stems from the privileges of Whiteness, defined as an identity that comes with hegemonic normalcy that remains unchallenged (Hayes, Juarez, Witt, & Hartlep, 2013). Whiteness, like Blackness (defined earlier in the chapter), does not refer exclusively to a racial group (i.e., White people or Black people, respectively), but rather denotes the *mindset* of the dominant social and political "lives that matter." In 2017, "#AllLivesMatter" is synonymous with the not so popular slogan, "#WhiteLivesMatter" except that it is palatable and less overtly racist in nature. "#BlueLivesMatter" is another moniker that has gained momentum in the wake of the #BlackLivesMatter movement and the retaliatory killings of police officers following the publicized killings of unarmed, Black men at the hands of armed officials. The color, blue, tends to symbolize the fraternal order of the police. While law enforcement bureaus around the country are inclusive of women and other racial demographics, overwhelmingly most police departments throughout the country remain predominantly White and male. Supporters of the "#BlueLivesMatter" movement, in the context of social justice, have typically maligned the #BlackLivesMatter movement as rogue. Following the political arguments in opposition of police reform segregates with the whitewashed nature of the "#AllLivesMatter and #WhiteLivesMatter" movements in contemporary America (Coates, 2014).

Additionally, reflecting on the contentious and controversial 2016 Presidential election and the current political atmosphere that celebrates the input of fundamentalist conservatives and Alt-Right leaders known to align with White supremacists and White nationalists, contributes to my struggle with reconciling the emotions that ensnare living as a Black person in America. During a political rally in Virginia, President Donald J. Trump (who was the Republican Presidential candidate at the time of this event) likened the inner cities where Black and Brown families overwhelmingly reside, to "hell" amidst claims of extreme poverty, unemployment, increased gun violence in Black-on-Black crimes, and a school-to-prison pipeline that relegates some children to correctional institutions based on their zip codes (Fausset, Blinder, & Eligon, 2016). In hopes of garnering the minority vote, Trump forced Americans of all ages and all racial groupings to reconcile an unspoken history of social injustice. How did the communities where minorities and the poor populate become incestuously contaminated with nefarious social derelictions? A painful answer is woven into the fabrics of America's, Declaration of Independence:

> We hold these truths to be self-evident: that all men are created equal; that they are endowed by their Creator with certain unalienable rights; that among these are life, liberty, and the pursuit of happiness. (The Declaration of Independence, A Transcription, 1776)

The egalitarian standard: all men (with no mention of females, Blacks, or Native Americans) were "created equal" and "endowed with unalienable rights" to freedom, as long as they bear a resemblance to the founding fathers; were of the same social class (economic and political freedom), had some higher educational status, (intellectual and civic freedom) and practiced the same religion (Tozer, Senese, & Violas, 2009). Blacks, at the time the Declaration of Independence was written, were overwhelmingly enslaved and remained so until the legal end of slavery in 1865 and were not privy to the "pursuit of happiness." According to Trump, Blacks and Hispanics living in the inner cities of today are in "hell" where there is no hope for "life and liberty." One hundred and fifty-one years after the legal end of slavery, some Whites do not readily acknowledge their unearned racial privilege, which is invisible and ubiquitous in their lives, but hyper-visible to people of color (Frideres, 2007).

The following is a narrative that was written by Emma, a student enrolled in the "Historical, Philosophical, and Sociocultural Foundations of Education and Science" course at Brooklyn College.

> I don't disclose my background typically, for several reasons: 1. I don't see why it's relevant – I should be respected regardless of where my ancestors are from; 2. I've been a victim of "reverse discrimination," which I find particularly amusing, again because if you don't know my race, how are you putting me into a category? and 3. I feel that the reason race is still an issue is because we're making it an issue. I can empathize, and I understand how certain groups feel the burdens of their oppressed ancestors, but at some point, we need to move on and take matters into our own hands. My grandparents and great-grandparents didn't sit around saying, "woe is me, I don't have opportunities because I'm [black, Native American, European immigrant, etc.]," they said, "what are my resources, and how can I make a change?" My grandparents worked terrible jobs in sweat shops – multiple jobs per day. They were at the bottom of the class-system, and somehow engrained in their children that the American Dream is what you make out of your own life. Life isn't fair, but we can't sit back and complain. Racism does still exist, I'm not denying that, but I see it on an individual level, not a national level. Another reason I don't fully disclose my ancestry is because I do not want handouts. I hate more than anything that I know a "minority" will get chosen over me any day, even if my credentials are higher. But I am even more insulted that the requirements are lowered for "minorities." That right there, is discrimination, and that is not fair, it is not what America was founded on. Equal opportunity for all, not different standards for each ethnic group. (Emma)

Emma's narrative has much emotion to explore, beginning with the quote, "How are you putting me into a category, if you don't know my race."

> Breathe in (deep, abdominal inhalation through the nose).
>
> Exhale (count down from 10 as the breath is released through pursed lips – in a controlled, slow manner).

If the racial identity of the author were not revealed earlier in the chapter, it would be a worthwhile exercise to explore whether or not the reader could predict what the racial epithet of the narrative's author might be. Emma identifies as a White, female science teacher. After reading her reflection, which this narrative was a part of, I felt silenced to respond. How does one respond to the following comment?

> I feel that the reason race is still an issue is because we're making it an issue. I can empathize, and I understand how certain groups feel the burdens of their oppressed ancestors, but at some point, we need to move on and take matters into our own hands?

> Breathe in (deep, abdominal inhalation through the nose).

> Exhale (count down from 10 as the breath is released through pursed lips – in a controlled, slow manner).

While I always wanted my students to feel open and vulnerable to share their emotional reflections about the topics we discussed, I did not anticipate how much of a challenge that would be when the topic was about race and privilege. I found that I was unable to respond to Emma because of my fear that an oppositional response to what she shared would be viewed as a failure on my part "to move on," particularly as my narratives and viewpoints are written as contemporary neo-slave narratives, as discussed earlier in the chapter. Critics of safe spaces argue that there is no safe space in race dialogues and violence (not to be inferred as, an intentional act to cause physical and/or fatal harm to an individual) is inevitable and educative, particularly for those Whites and people of color entering the dialogue with emotions of anger, hostility, frustration, and pain (Leonardo & Porter, 2010). Emma's narrative was an emotional hurdle for me to overcome and I felt unsafe to challenge how insecure many of her comments made me feel, as it directly contradicted my entire lived experience.

As Emma and I had *no* dialogue about her narrative, I am careful not to make assumptions about what she meant by the things she said. Using authentic inquiry as a methodological framework for doing research given the sharing of the narrative and my resulting silence, the next steps would be for Emma and I to have a dialogue so as to learn from our differences and this chapter may serve to initiate. As authentic inquiry is ontological, catalytic, tactical, and educative, there is an inherent space for us to learn from each other and to radically listen to our points of view without judgment, if we are *both* open to this possibility.

Being respectful of and careful with Emma's narrative, particularly as it completely opposes how I see the world, I'd like to posit the portions of the narrative, which were most jarring and suggestive of the controversial framework of Whiteness and White privilege. The narrative allows us to be critical of the privilege of invisibility, meaning some Whites inherently have the power to delude themselves into thinking that "all men and women are created equal" and have full access to the "American Dream." As this narrative reflects, there is an element of "rightness" that reinforces the privileged standpoint. In America, some Whites are privileged to control what others know of their history and can choose to discount historical events and characters that may be deemed inconsequential to how they want the narrative to be recounted (Frideres, 2007). The egalitarian standard is what is considered the "right" way to succeed in America and sitting around saying, "woe is me, I don't have opportunities because I'm [black, Native American, European immigrant, etc.]," is the immoral and perceived wrong way to live out one's citizenship.

In the general sense, Whiteness has a way of silencing visible minorities, as there is a tacit expectation of the non-Whites to agree with whatever Whites with access to power propose as right, good, and "fair" (Frideres, 2007).

> Breathe in (deep, abdominal inhalation through the nose).
>
> Exhale (count down from 10 as the breath is released through pursed lips – in a controlled, slow manner).

I had an adverse emotional reaction to the following sentence in the narrative:

> I hate more than anything that I know a "minority" will get chosen over me any day, even if my credentials are higher. But I am even more insulted that the requirements are lowered for "minorities." That right there, is discrimination, and that is not fair, it is not what America was founded on.

The impassioned testimony is an ideation surrounding the perceived "special rights" that are granted to marginalized populations (Dudas, 2005). The use of the word, "hate" in the narrative is indicative of a feeling of resentment that stems from perceptions that "minorities will get chosen" over a White person. Some Whites describe the perceived "special rights" of the marginalized minority groups, as authored in the narrative, as being "unfair' and un-American: "not what America was founded on." Resentment by some Whites, alongside the standpoint of patriotism and nationalism, reflects a skewed interpretation of egalitarianism and a perverted standard in which "fairness," actually refers to Whiteness (Caouette & Taylor, 2007). Some Whites saturate the current political climate with the fears of a declining America amidst widespread resentment over the perceived loss of privilege within the status quo (Dudas, 2005).

> *The reader is invited to an active and mindful discussion*: What is your standpoint as it relates to perceived privilege and "Whiteness"?

Structurally, without delving into assumptions about what Emma meant, the narrative, as a standalone narrative, can be theorized through interpretations of Mikhail Bakhtin's, carnival theory. A characteristic of carnival is the reveler, or speaker: idolizes oneself and structures every behavior, emotion, and discourse in a manner that boosts the self-image and self-satisfaction (Bakhtin, 1984). The author uses laughter and amusement as a form of mockery against political structures. In the theoretical context of Whiteness and how it relates to inherent or perceived privileges, the "I" voice in this narrative is preoccupied with fantasies of unlimited power, self-importance and uniqueness; demands for admiration, indifference for others, a lack of empathy, and a sense of entitlement (Matias, 2016). Without making assumptions as to what Emma hoped to convey to me, the purpose of this line by line description of the "I" voice is to name the parts of the narrative which I found difficult to respond to and rather met with silence. In the lens of Whiteness, privilege, and carnival, the following is how the narrative could be theorized:

Line 1: I don't disclose my background (the author has unlimited power to decide what parts of his/her background can be disclosed, there is a self-importance and sense of entitlement to being un-named, there is an indifference and lack of empathy for others who may not have that "right)

Line 2: I don't see why it's relevant (self-importance: the author has the privilege to make him/herself invisible in this discourse)

Line 3: I should be respected regardless of where my ancestors are from (self-importance: Historically, this respect is something that marginalized populations cannot command)

Line 4: I've been a victim of "reverse discrimination" (lack of empathy for those who are victims of de facto discrimination)

Line 5: I find particularly amusing (indifference for others: an element of carnival and a mockery of those who claim they have been victims of discrimination and racism)

Line 6: If you don't know my race (uniqueness: Is the race not suggestive in the lens of Whiteness?)

Line 7: How are you putting me into a category? (self-importance and demand of admiration)

Line 8: I feel that the reason race is still an issue is because we're making it an issue (there is an indifference for others who have no choice but to be recognized as a person of color)

Line 9: I can empathize (the narrative lacks empathy and this "I" statement is quite ironic)

Line 10: I understand how certain groups feel (another example of a statement with a lack of empathy because there is no space for other groups to share how they feel)

Line 11: My grandparents and great-grandparents didn't sit around saying (sense of entitlement because there were some ancestors who were slaves, indentured servants, migrant laborers, etc. and had no/few choices and/or freedom)

Line 12: My grandparents worked terrible jobs in sweat shops (demands for admiration: Historically, the opportunity to work select jobs were not available to every one in a manner that was equitable. The division of labor was based on race)

Line 13: Life isn't fair, but we can't sit back and complain (indifference and sense of entitlement because there is no consideration about the sociocultural and historical impacts of racism and discrimination on "others")

Line 14: I'm not denying that [Racism exists], but I see it on an individual level (This line reads as a unique standpoint and may serve as a prompt to discuss when Emma and I continue our research together, should she be open to it)

Line 15: I don't fully disclose my ancestry because (power and sense of importance – discussed in lines 1 and 2 – note that this is a recurring theme in this narrative)

Line 16: I do not want handouts (sense of importance and entitlement)

Line 17: I hate more than anything that (sense of importance: the word "hate" shows strong emotion. This line may serve as a prompt for further dialogue, but it *will* cause violence as it stirs strong emotions for me to read. Herein lies the challenge with discussing thorny issues and engaging in authentic inquiry research)

Line 18: I know a "minority" will get chosen over me any day (lack of empathy, sense of entitlement and a failure to recognize the possibilities that being White is a privilege for some Americans. I wonder why "minority" is in quotation marks. This line may also serve as a prompt for further dialogue between Emma and myself.)

Line 19: I am even more insulted that the requirements are lowered for "minorities" (lack of empathy because the perceived "lowering of requirements" seems to be used synonymously with policies of affirmative action and the notion of "special rights" as discussed earlier in the chapter)

Breathe in (deep, abdominal inhalation through the nose).

Exhale (count down from 10 as the breath is released through pursed lips – in a controlled, slow manner).

The reader is invited to an active and mindful discussion: How would you respond to "Emma"? I was unable to share my thoughts with the author, Emma, out of shame and fear. Which emotions are you most aware of as you read this section of the article?

CLOSING THOUGHTS

Other *thorny* discussions like the (in)appropriate use of the N-word, the alleged onslaught of police brutality against unarmed Black men, the politics of Whiteness, defining equity and social justice beyond a buzzword and political correctness; were all controversial subject matters discussed during our time together in the "Historical, Philosophical, and Sociocultural Foundations of Education and Science" course at Brooklyn College.

The purpose of this chapter was to open a dialogue about issues of race, privilege and social justice. Earlier in the chapter, I shared with Konstantinos, an impassioned email that expressed my struggle with writing about race and being vulnerable enough to share deep emotions (i.e., anger, fear, disgust) with readers in a potentially unsafe space (i.e., academia).

> Thank you for your note and thank you for *taking the risk* to share such personal and emotionally painful reflections. This paper is on some very sensitive topics, which adds many new layers of responsibility, caution, and risk. (Konstantinos)

Konstantinos' response was non-judgmental and welcoming of difference, providing the space for theorizing beyond the pain of my thoughts and emotional attachment to what was written and how it was articulated. As a mentor and friend, the emotional space between us is open for teaching and learning from each other, the essence of which is embedded in the structure of the course we teach. As instructors, the goal is to model risk-taking in our learning environment and push the boundaries of safety and critical thought. Emotional safeguards are in place to allow for conflict. In fact, the absence of conflict is not the gold standard for what it means to be safe in a learning environment, while at times it may be a welcomed respite for healing to take place (Gayle, Cortez, & Preiss, 2013). The challenge, as shared throughout the chapter, is to confront biases and oppressive thoughts. Reflexive inquiry tools (i.e., free-writes, breathing meditations, heuristics) allow members of a learning environment to reflect on their worldviews and (re)build knowledge about the self | other.

> Leah: This chapter closes with space for both the reader and myself to continue a dialogue. The work may feel incomplete or open-ended, but my intention was to posit the struggle that I bring to academia and any learning environments, as a person of color.
>
> A personal goal was to name my place for theorizing and to introduce myself as an emergent scholar who struggles with emotions that relate to fear, anger, and at times disgust. The active and mindful discussion prompts were included to further this work and a response is welcomed, if the reader is so inclined (lprideotubanjo@gmail.com).
>
> While engaged in mindful discussions, there is space for healing and safety, but it takes time to manifest. There is risk-taking and vulnerability. Violence will occur, with the victor being humanization for all participants. In my humanity, I was careful not to make generalizations about experiences in this chapter,

but some may be offended by this work. My hope is that the voices of others (students at Brooklyn College), as well as my own narratives, may be used to transform talks about race, privilege, and social justice. On the activism spectrum, my desire is that this work continues to push boundaries by offering additional theoretical and methodological frameworks for engaging in and analyzing race dialogues.

REFERENCES

Alexakos, K. (2015). *Becoming a teacher | researcher: A primer of doing authentic inquiry research on teaching and learning*. Rotterdam, The Netherlands: Sense Publishers.

Alexakos, K., Pride, L. D., Amat, A., Tsetsakos, P., Lee, K. J., Paylor-Smith, C., Zapata, C., Wright, S., & Smith, T. (2015). Mindfulness and discussing "thorny" issues in the classroom. *Cultural Studies of Science Education, 11*(3), 741–769. Retrieved from https://doi.org/10.1007/s11422-015-9718-0

Bakhtin, M. (1981). *The dialogic imagination: Four essays*. Austin, TX: University of Texas Press.

Bakhtin, M. (1984). *Rabelais and his world*. Bloomington, IN: Indiana University Press.

Bourdieu, P. (1977). *Outline of a theory of practice*. Cambridge: Cambridge University Press.

Bourdieu, P. (1986). The forms of capital. In J. Richardson (Ed.), *Handbook of theory of research for the sociology of education* (pp. 241–258). New York, NY: Greenwood Press.

Caouette, J., & Taylor, D. M. (2007). Don't blame me for what my ancestors did: Understanding the impact of collective White guilt. In P. R. Carr & D. E. Lund (Eds.), *The great White North? Exploring whiteness, privilege, and identity in education* (p. 84). Rotterdam, The Netherlands: Sense Publishers.

Carney, N. (2016). All lives matter, but so does race. *Humanity & Society, 40*, 180–199. Retrieved from https://doi.org/10.1177/0160597616643868

Chokshi, N. (2016, August 22). How #BlackLivesMatter came to define a movement. *The New York Times*.

Coates, T.-N. (2014, December). Blue lives matter. *The Atlantic*.

Delgado, R. (1989, August). Storytelling for oppositionists and others: A plea for narrative. *Michigan Law Review, 87*(8), 2411–2441.

Dudas, J. (2005, November 15). In the name of equal rights: "Special" rights and the politics of resentment in post-civil rights America. *Law & Society Review, 39*(4), 723–757. Retrieved from https://doi.org/10.1111/j.1540-5893.2005.00243.x

Fausset, R., Blinder, A., & Eligon, J. (2016, August 24). Donald Trump's description of Black America is offending. *The New York Times*.

Frideres, J. (2007). Being White and being right: Critiquing individual and collective privilege. In P. R. Carr & D. E. Lund (Eds.), *The great White north? Exploring whiteness, privilege, and identity in education* (p. 46). Rotterdam, The Netherlands: Sense Publishers.

Furman, R. (2007). Poetry and narrative as qualitative data: Explorations into existential theory. *Indo-Pacific Journal of Phenomenology, 7*(1), 1–9. Retrieved from https://doi.org/10.1080/20797222.2007.11433939

Furman, R., Langer, C., Davis, C. S., Gallardo, H. P., & Kulkarni, S. (2007). Expressive research and reflective poetry as qualitative inquiry: A study of adolescent identity. *Qualitative Research, 7*(3), 301–315. Retrieved from https://doi.org/10.1177/1468794107078511

Gayle, B. M., Cortez, D., & Preiss, R. W. (2013). Safe spaces, difficult dialogues, and critical thinking. *International Journal for the Scholarship of Teaching and Learning, 7*(2), 1–8. Retrieved from https://doi.org/10.20429/ijsotl.2013.070205

Godfried, A., Sekimoto, S., & Brown, C. (2016). Becoming "Black": Exploring the racialized experiences of African immigrants in the United States. *Howard Journal of Communications, 27*(4), 367–384. Retrieved from https://doi.org/10.1080/10646175.2016.1206047

Guba, E., & Lincoln, Y. (1989). *Fourth generation evaluation*. Newbury Park, CA: Sage Publications.

Hall, S. (1996). *Critical dialogues in cultural studies*. New York, NY: Routledge.

Hayes, C., Juarez, B. G., Witt, M. T., & Hartlep, N. D. (2013). Toward a lesser shade of White. In C. Hayes & N. D. Hartlep (Eds.), *Unhooking from whiteness: The key to dismantling racism in the United States* (p. 2). Rotterdam, The Netherlands: Sense Publishers.

hooks, b. (1996). Choosing the margin as a space of radical openness. In A. Garry & M. Pearsall (Eds.), *Women, knowledge, and reality: Explorations in feminist philosophy* (p. 49). New York, NY: Routledge.

Hunter, M. (2007). The persistent problem of colorism: Skin tone, status and inequality. *Sociology Compass, 1*(1), 237–254. Retrieved from https://doi.org/10.1111/j.1751-9020.2007.00006.x

Jablonski, N. G., & Chaplin, G. (2002). Skin deep. *Scientific American, 287*(4), 74–81.

Johnson, E. P. (2003). *Appropriating blackness: Performance and the politics of authenticity*. Durham, NC: Duke University Press.

Leonardo, Z., & Porter, R. K. (2010, July). Pedagogy of fear: Toward a Fanonian theory of "safety" in race dialogue. *Race, Ethnicity and Education, 13*(2), 139–157. Retrieved from https://doi.org/10.1080/13613324.2010.482898

Matias, C. (2016). *Feeling White: Whiteness, emotionality, and education* (Chapter 5). Rotterdam, The Netherlands: Sense Publishers.

Neo-slave narrative. (2005). *Columbia guide to contemporary African American fiction* (pp. 177–179). New York, NY: Columbia UP.

Ogbu, J. U., & Simons, H. D. (1998). Voluntary and involuntary minorities: A cultural-ecological theory of school performance with some implications for education. *Anthropology & Education Quarterly, 29*(2), 155–188. Retrieved from https://doi.org/10.1525/aeq.1998.29.2.155

Ouimet, L. (2002). American studies in review. *Canadian Review of American Studies, 32*(1), 133–144.

Powietrzynska, M., Tobin, K., & Alexakos, K. (2015). Facing the grand challenges through heuristics and mindfulness. *Cultural Studies of Science Education, 10*, 65–81. Retrieved from https://doi.org/10.1007/s11422-014-9588-x

Pride, L. D. (2014). Using learning stories to capture "gifted" and "hard worker" mindsets within a NYC specialized high school for the sciences. *Theory into Practice, 53*, 41–47. Retrieved from https://doi.org/10.1080/00405841.2014.862121

Rushdy, A. A. (2004). The Neo-slave narrative. In M. Graham (Ed.), *The Cambridge companion to the African American novel* (pp. 87–105). Cambridge: Cambridge UP.

Sewell, J. W. (2005). *Logics of history: Social theory and social transformation*. Chicago, IL: University of Chicago Press.

Shaw-Taylor, Y., & Tuch, S. (2007). *The other African Americans: Contemporary African and Caribbean immigrants in the United States*. Lanham, MD: Rowman & Littlefield Publishers.

Smith, W. A., Allen, W. R., & Danley, L. L. (2016). Assume the position…you fit the description. *American Behavioral Scientist, 51*, 551–578. Retrieved from https://doi.org/10.1177/0002764207307742

The Declaration of Independence: A Transcription. (1776, July 4). Retrieved December 6, 2009, from http://www.archives.gov/exhibits/charters/declaration_transcript.html

Thiam, C. (2014). Negritude, Eurocentrism, and African agency: For an Afrocentric renaissance of Leopold Sedar Senghor's philosophy. *French Review: Journal of American Association of Teachers of French, 88*(1), 149–163.

Tobin, K. (2006). Qualitative research in classrooms: Pushing the boundaries of theory and methodology. In K. Tobin & S. R. Steinberg (Eds.), *Doing educational research: A handbook* (pp. 15–58). Rotterdam, The Netherlands: Sense Publishers.

Tobin, K. (2011). Learning from a good mate. In K. Hayes, S. Steinberg, & K. Tobin (Eds.), *Key works in critical pedagogy: Joe L. Kincheloe* (pp. xv–xxiv). Rotterdam, The Netherlands: Sense Publishers.

Tobin, K. (2014). Using collaborative inquiry to better understand teaching and learning. In J. Bencze & S. Alsop (Eds.), *Activist science and technology education*. Dordrecht: Springer.

Tobin, K., & Ritchie, S. M. (2012). Multi-method, multi-theoretical, multi-level research in the learning sciences. *The Asia-Pacific Education Researcher, 21*, 117–129.

Tobin, K., & Roth, W.-M. (2006). *Teaching to learn: A view from the field*. Rotterdam, The Netherlands: Sense Publishers.

Tozer, S. E., Senese, G., & Violas, P. C. (2009). *School and society: Historical and contemporary perspectives* (6th ed.). New York, NY: The McGraw-Hill

United States Census Bureau. (2016, July 1). *United States Census Bureau*. Retrieved September 1, 2017, from https://www.census.gov/quickfacts/table/PST045216/00

Van Maanen, J. (2011). *Tales of the field*. Chicago, IL: The University of Chicago Press.

Vint, S. (2007). "Only by experience": Embodiment and the limitations of realism in neo-slave narratives. *Science Fiction Studies, 34*(2), 241.

Walsh, D., & Andersen, S. (2008, July 29). *House apologizes for slavery, 'Jim Crow' injustices*. Retrieved April 13, 2009, from http://www.cnn.com/2008/POLITICS/07/29/house.slavery/index.html

ABOUT THE AUTHOR

Leah D. Pride is a graduate of the Urban Education program at the Graduate Center of the City University of New York. She has taught secondary and post-secondary science and special education courses within the New York City Public school system and the City University of New York for over a decade. Her research interest focuses on narratives and the emotional underpinnings of race, privilege, and social justice dialogues.

JENNIFER D. ADAMS

8. CREATIVE CRITICAL INQUIRY

Transforming Our Understandings of and Engagements in the World

ABSTRACT

Twenty-first century scientific problems are complex and have multiple, intersecting etiologies that require us to work across disciplinary boundaries in order to solve them. This warrants creative scientific thinking that will lead to the new knowledge, innovations and social movements that will address existing and emergent issues and move us towards more equitable and sustainable futures. This chapter offers and approach for thinking about creativity in relation to science teaching and learning. Using the example of extreme climate events, I situate this issue in the economic, historical and social contexts that complicate the both the causes and challenges of addressing this issue. This is followed by a discussion of the Crit-Trans heuristic created to expand opportunities for all students, especially racialized and marginalized students, to engage in science in meaningful ways. I then expand on this heuristic by centering creativity in order to broaden how people approach and address scientific challenges from classroom-based tasks to the larger societal issues. I provide an example of a creative engagement with science to offer a starting point from which to build the creative practices necessary in science teaching and learning in order to imagine and advance a better society for all.

Keywords: creativity, heuristic, science education, socioscientific issues, equity

THE CRITICAL ISSUE

Despite a growing trend by critical researchers to decolonize science, including centering indigenous knowledges in science classrooms (e.g., Mpofu, Otulaja, & Mushayikwa, 2014), myths about science persist, rendering it an exclusive space dominated by a European male patriarchal view. These myths include science as objective, solely rational, and exclusive of those who are not "smart" or do not "get it." Furthermore, with education in general trending more towards neoliberalism, including market-based reforms and meritocratic practices, science teaching and learning spaces are further "enclosed" (Strong, Adams, Bellino, Pieroni, Stoops, & Das, 2016); becoming a private, profitable enterprise leaving more students at the margins of STEM fields. Moreover, with all of the work and rhetoric around

DOI:10.1163/9789004389663_008

increasing diversity in STEM, disparities still exist (Landivar, 2013) with people who are often held as spokespersons for advancing equity in STEM being, in many ways, complicit in perpetuating the disparities. For these trends to change, science must change, however the critical question is how must science change and what can we, as science educators and researchers, do to transform it into a space of authentic equity?

Space not only implies the physical spaces where science is taught, learned and practiced, but also in the conceptual spaces that define science and describe who gets to engage in science and in what ways. In a space of authentic equity, multiple knowledges are valued and multiple perspectives are centered in problem-solving; a true transdisciplinary approach. This not only means opportunities to engage in science but also opportunities to be co-producers of relevant and meaningful scientific knowledge. In this space, there are no assumptions about who is there as it is the expectation that all who are there are legitimate participants. However, if there are absences, these are critically examined in terms of why and what we can do to engender a broader sense of inclusivity. These kinds of science spaces are vital if we are to not only survive, but also thrive as a collective species – living harmoniously with non-human species and non-living components of the ecosystem e.g., rocks, air and water (Bai, 2015) that all too frequently are caught in the wake of human self-destruction.

Complexity of Scientific Challenges

Global scientific challenges are becoming increasingly complex, ill-defined and open-ended. For example, climate change, food security and scientific intellectual law and policy all have transdisciplinary dimensions (Kawasaki & Toyofuku, 2013). The recent climate events demonstrate this very well. The 2017 hurricane season was unprecedented. Hurricanes Harvey and Irma were both category 5 storms and caused vast and unprecedented damage on many Caribbean Islands, including two that I know and love very well. Irma completely defoliated the islands of Barbuda and St. John, USVI and rendered concrete structures to piles of rubble. Hurricane Maria followed and decimated Dominica and Puerto Rico. With the warmer oceans feeding the intensity of these storms, the etiologies are undoubtedly connected to climate change, with climate change itself being a complex, global problem. Storm recovery will take months, even years as many homes and livelihoods have been destroyed. If we are to actively work towards equitable solutions to address climate resilience, disaster recovery and curtail climate change, we need to move beyond the disciplinary siloes that keep the political, historical and social separate from the scientific. In most contemporary science classrooms students are rarely exposed to a "systematic analysis of the social origins, traditions, meanings, practices, institutions, technologies, uses and consequences of natural sciences" (Harding in Gough, 2015, p. 447). This contributes to dangerous ignorance around climate change, especially in a nation like the United States whose policy makers make climate-related decisions that have global consequences. The people who are living the reality of climate change, such as those in the Eastern Caribbean, including the non-presidential-voting United States Virgin Islands, have little say over climate-related policy. A few days after the devastation of Hurricane Maria, Dominica's Prime Minister Roosevelt Skerrit tearfully addressed the United Nations General Assembly, "to deny climate change is to procrastinate while the Earth sinks; it is to deny a truth we have just lived" (Brown, 2017). He went on to note:

> While the big countries talk, the small island nations suffer. We need action and we need it now. We in the Caribbean do not produce greenhouse gases or sulphate aerosols. We do not pollute or overfish our oceans. We have made no contribution to global warming that can move the needle. But yet, we are among the main victims on the frontline. (Brown, 2017, p. 1)

It is often the marginalized who are the most vulnerable to natural disasters (Chakraborty, Tobin, & Montz, 2005) and as we see in the case of the Eastern Caribbean, Puerto Rico, and the USVI, where we now have climate displaced and refugees. This is, in essence, a reproduction of colonialism, it is a neocolonialism where the economic systems and corresponding political policies of seemingly sovereign nation-states are directed by global powers, usually the so-called West, "The result of neo-colonialism is that foreign capital is used for the exploitation rather than for the development of the less developed parts of the world. Investment under neo-colonialism increases rather than decreases the gap between the rich and the poor countries of the world" (Nkrumah, 1966, p. 1). If we think of financial capital also as a metaphor for scientific capital, then neocolonialism is very present in climate science policy where the lower GDP countries and poorer areas within affluent nations are most vulnerable to climate related disasters, both in probability and intensity of effects. Planning for the climate realities of the near future that will afford safety and resilience for all people will require interdisciplinary thinking and new knowledge that goes beyond disciplinary siloes. It will require people in academic, industry, policy and lay, to think collaboratively, collectively and in novel ways about addressing and surviving climate realities.

Need for an Epistemic Revolution

In response to the storms and the subsequent musings about a disaster economy that might somehow connect to the anti-climate science rhetoric that currently exists caused us to ask who profits from the specter of increasing "natural" disasters and how does this influence how scientific policy is considered? Through an engagement with colleagues on Facebook, both sociologists, and women of color with cultural connections to the Global South, we thought about a revolution that needs to happen for the benefit of the future of the planet and how this revolution has to look very different from those of the past. We discussed the history of political education classes that were organized and taught by unions and how these were beneficial in helping people to make political and economic decisions that forefronted the needs of the community and working class. In light of this discussion, I thought about how might a redefinition of scientific literacy be folded into such an education; how would a socioscientific centered science education make citizens aware of the ways that political decisions have grave scientific consequences? Colleagues like Larry Bencze, Erin Sterling, and Lyn Carter (2012) call for enabling learners to become citizen activists, "ready and willing to take social actions to reduce risks associated with issues," to become more aware of personal, economic choices and how they are intertwined with the political and serve to reproduce the historical.

This demands an "epistemic disobedience" that seeks to "undo the systems through which knowledge and knowing are constituted" (McKittrick, 2014, p. 106); one that intellectually delinks from the knowledge systems that are taken for granted, "from the magic of the Western idea of modernity, ideals of humanity and promises of economic growth and financial prosperity" (Mignolo, 2009, p. 161). This requires us to not only do the hard intellectual work of rethinking science teaching and learning, but to actively work towards new and different ways of engaging in science towards the co-creation and co-production of knowledge. Furthermore, science education needs to center lived experiences while assuming a critical, decolonizing stance in ways that afford citizen agency in making the individual/collective decisions towards reclaiming our planet for the betterment of all – humans and non-humans alike.

CREATIVITY AND THE CRIT-TRANS APPROACH TO STEM

A Bold Vision of STEM teaching, learning and research is one that is expansive, inclusive and collective. It is one where disciplinary silos are reconsidered and the false objectivity of Western Modern Science exposed. It needs to be a space were divergent ideas are generated and condensed towards the most feasible, Earth-benefitting and equitable solutions. Levinson (this volume) advocates for creative teaching, that challenges the hegemony of STEM and "promotes ontological and with it epistemological transformations" (p. X). I extend this to incorporate a stance on creativity in STEM that will not only challenge the existing neoliberal system within which this hegemonic science is situated, but one that centers a critical stance to creativity that will not only allow ideas to flourish but also compel those ideas to confront the social, political and economic and be responsive to the historical dimensions of given issues.

In order to center creativity in science, it is important to critically examine both notions of creativity and science in ways that broaden conceptualizations of both terms. A salient starting point for this project is Wynter's science of being human (*scientia*); one that "unsettles our familiar (Darwinian, objective, racist, sexist) governing cores of scientific thought…and seeks to locate [scientific] knowledge-making as connected to the human lived experience" and "the recoding of science through representational and biological feelings; the interdisciplinary and collaborative task that allows us to think about how the creative narrative can and does contribute to what is otherwise understood as 'the laws of nature…'" (McKittrick, 2014, p. 154). Wynter's project of decolonial scientia centers the lived experience of being human or a reclaiming of humanness (for people who have been historically excluded from this term) and reclaims knowledge generation for the benefit of healthy communities where "life had priority over economic gains, growth and development…it subjects economic growth to human needs rather than the other way around" (p. 118). Scientia allows us to merge the sociohistorical, scientific and creative toward the advancement of an expansive science teaching and learning framework.

Elsewhere, the Urban Education Research Coven (Strong et al., 2015) proposed a Crit-Trans heuristic to science teaching and learning. This heuristic is aimed at countering the "neoliberal enclosure" of science education and expanding opportunities for science engagement for all students, but especially students of color who are often racialized and marginalized from school science and afterschool science designed within a school science paradigm. In creating this heuristic, we drew on both our collective experiences with STEM and environmental education and from bodies of literature that addressed critical race theory, youth participatory action research, critical pedagogy, eco-pedagogy and eco-justice. Our use of transdisciplinarity was adapted from Basarabe Nicolescu (2005) where, "transdisciplinarity concerns that which is at once between the disciplines, across different disciplines, and beyond all disciplines. Its goal is the understanding of the present world, of which one of the imperatives is the unity of knowledge" (p. 2). The key tenets of the Crit-Trans heuristic include:

- Contextualize and historicize knowledge
- Challenge assumptions of neutrality and objectivity through critical inquiry
- Decenter hegemonic notions of knowledge production
- Situate place and space
- Privilege process over product
- Promote participatory teaching, learning and research

The heuristic recognizes and values the diversity of knowledge and lived experiences that learners bring to a given context. It centers collaborative work and the co-production of knowledge along with the critique, reconstruction and transformation of educational practice. Moreover,

> An enclosed and narrow framing of science fails to capture the historically, socially, and culturally situated nature of science. Using a more polysemic view of science, we argue that all students and educators bring valuable knowledge and expertise into a learning space, whether it is a formal classroom or informal context. Recognizing everyone as experts of their lived experiences and thus creators and holders of knowledge translates into an environment in which teaching and learning is a collective endeavor. (Strong et al., 2016, p. 227)

The Crit-Trans heuristic is designed to be a living document to be both used in a range of learning and teaching contexts and transformed as agents create and engage in liberatory transdisiplinary spaces. Thus, I will use this heuristic to shape a notion of scientific creativity that moves beyond canonical science and investigate what it means to be human and what this means for our entangled relationship with the Earth, other living beings, and the technologies that we create and use that shape our planet.

Centering Creativity

Why creativity? What does this afford in (transdisciplinary) STEM learning spaces? First we need to inquire about the ways that creativity is defined and how it is conceptualized in relation to science. There are numerous definitions of creativity. Robert Sternberg (2003) defines creativity as "the ability to produce work that is novel, high in quality and appropriate" (p. 89). A common conceptualization of creativity refers to convergent and divergent thinking, with the latter being associated with being able to explore many solutions to a given problem rather than following a linear process (like the "scientific method as often taught in schools) (Madden et al., 2013). Psychometrics, cognition, social-personality and other domains have all approaches and definitions of creativity, however many of these domain-specific approaches have ignored the multiple influences underlying that which affords creativity (Sternberg, 2003); creativity is a [product] of a confluence of multiple factors. I begin with a definition of creativity as a collective endeavor; "a flow of affect between assembled bodies, things, and ideas" (Fox, 2013, p. 6) and that it is in the relations that creativity is produced "in the most unpredictable and unexpected ways" (p. 8), with the unexpectedness leading to innovation, "the implementation of ideas into practice" (Shavinina, 2013, p. xxvii) in the form of new artifacts. This perspective of creativity allows one to focus on the contexts that allow creativity to flourish rather than on designed interventions that foster individual creativity. According to Sternberg (2003), "creativity can lead to new scientific findings, new movements in art, new inventions, and new societal programs" (p. 89), this is especially important in STEM where, as mentioned above, many issues are ill-defined and transdisciplinary (de Sousa Santos, 2007).

In the science teaching and learning field, there has been a movement to reframe STEM as STEAM through the integration of the arts. This comes from the recognition that both scientists and artists use similar approaches to their work including drawing on curiosity, making careful observations, imagining and remaking the world (Sousa & Pilecki, 2013) and a call for obliterating disciplinary silos (Boy, 2013). In addition, research shows a positive correlation between participation in the arts and patents produced, citing that "artistic skills-such as analogies, playing, intuition and imagination" contribute to being able to solve complex problems (Parker et al., 2013). However, with an emphasis on design and innovation in the service of entrepreneurialism and global competitiveness (Land, 2013), in many spaces STEAM has now been co-opted into the neoliberal agenda. As one proponent notes, "design creates the innovative *products* [emphasis mine] that will propel our economy forward, and artists ask the deep questions about humanity that reveal which way forward actually is" (Maeda, 2013, p. 1). With profit being given more importance than the larger questions about the impacts of production and the resulting products on the socioecological well-being (Lloro-Bidart, in review) of our planet. This is not to say that innovation towards a healthy economy is not important, but first we need to reconsider what the "healthy" means in terms of economics; one that does not emphasize production for consumption but rather one that centers co-production and engagement. A Crit-Trans approach to science, art and creativity centers the process, allowing scientists and artists to co-collaborate and co-create knowledge, and contextualizes the knowledge production in a way that considers the historical, social, political and economic consequences of design and innovation.

Creativity and STEM

Similarly, in reviewing literature on creativity in relation to STEM, the emphasis is often on production and measured in terms of commodities, numbers of patents, awards, grants and publications. Although these products are important to advancing scientific knowledge, they render a narrow view of scientific creativity. Furthermore, a common definition of creativity that is also employed in science refers to the production of knowledge or artifact that is considered novel and useful, ignoring that notions of innovation and utility are highly subjective and culturally embedded. As Glăveanu (2014) notes, "it is rarely creativity alone that decides who is genius but institutional structures reflecting power relations between and within social groups" (p. 4). If we think about dominant institutional structures of science – universities, governmental agencies, corporate industries – we can see why this narrow view of scientific creativity is valued and articulated. This leads us to ask, what aspects of scientific creativity are necessary but undervalued and minimally acknowledged? In what ways could measures and descriptions of scientific creativity be broadened? How might science change, including science teaching and learning, if notions of scientific creativity were broadened? These are important starting questions if we are to reconsider the relationship between creativity and science, including bringing in approaches from the critical social sciences to unpack existing notions of creativity and science and expand it with new meaning and understandings. For example a recent article posits the need for diversity in science describing how gender and culture influence research perspective, including the kinds of choices scientists make about their problems and approaches (Medin, Lee, & Bang, 2014). Extending this to thinking about creativity we can ask how diversity also influences perspectives on creativity, including what is considered novel and useful, in science. This would allow us to expand our definition of creativity in science to include aspects, relevance for a just world, innovations towards Earth-friendly collaborative practices and innovative science teaching, that are not included in more canonical definitions of creativity and innovation in science.

The creative narrative allows our cultural practices to determine how we learn and practice science (Medin et al., 2014). For, it is neither the concept of science nor creativity that decides what is "genius" or even "smart" but rather the institutional structures within which it is situated (Glavineau, 2014). STEM and creative texts are simultaneously lived and assembled and allow considerations about how humanness might be reimagined thus challenging the very foundation of colonial STEM teaching, learning and research contexts. For example, African-diasporic dance scholar Yvonne Daniels (2005) notes, "Dancing bodies accumulate spirit, display power and enact as well as disseminate knowledge" (p. 59), thus allowing us to think about learning and knowledge production in different ways thus opening up the ways that science might be considered and practiced. Where STEM and creativity intersect we might find the "scientific poetics of our future" rewriting science as "necessarily a human project" (McKittrick, 2014, p. 160).

Engagement in epistemic disobedience requires imagining alternative choices and futures. Since the arts are often more closely associated with creativity than the sciences, many projects that go beyond the STEAM framework to examine learning in art/science collaborations. For example, in their EMMAS project, Ann Rosebery, Beth Warren, and Megan Bang (2015) noted that the art/science nexus fosters "boundary-crossing thinking" and "heterogeneous ways of seeing." In Bartbara Clark and Charles Button's (2011) Sustainability Transdisciplinary Education Model (STEM), creativity was central in promoting critical inquiry of environmental awareness and sustainability; "the arts promote cultural change, trigger the imaginative conscience and community action and act as a bridge towards scientific understanding and application of sustainable efforts" (p. 43). Larry Livingston (2010) urges educators in higher education to leverage the creativity that students already possess and access in their daily-lived experiences through creating collaborative, problem-solving and social justice-oriented learning opportunities. Thus having the notion of creativity as central to STEM teaching and learning requires transformative practice, which environmental artist Tim Collins (2006) describes as "related to ideas about freedom and inter-sociality; the development of critical awareness, distributed creativity and social agency" all necessary in educating the creative thinkers and STEM innovators of tomorrow.

Creativity and Crit-Trans Enacted

I belong to a research collaborative on the island of Barbuda. The Barbuda Research Complex (BRC) embodies Crit-Trans because we recognize that an integration of the hegemonic science, as practiced in the academy, and Traditional Ecological Knowledge is necessary to both understanding human ecodynamism and devising solutions to enhance sustainability and resilience, especially in relation to food security (Boger, Perdikaris, Potter, & Adams, 2016). Central to this research is an interdisciplinary collaborative approach with scientists from multiple disciplines, Barbudan scholars and the Barbuda youth as well as examining human adaptations to living on the island dating back to circa 5000 BCE. Also integral to this approach is youth learning directly about agricultural practices from the elders and then applying that knowledge in the high school agricultural program. This serves not only document traditional and historical practices but also to integrate these practices with contemporary understandings about climate and agriculture. Through these collaborations, we are able to learn about past patterns of climate adaptations and current agricultural patterns from "cultural, anthropological, geographical, geological, botanical and pedagogical perspectives" (p. 3). As the pedagogue, the Barbuda work is a chance for me to reconnect with my own embodiment of Caribbeanness, my mother's agricultural practices and the contemporary education of the youth (Adams, 2014).

As a part of engaging the young people in learning about climate resilience and cultural practices, we initiated the Barbuda Engaged Artist Residency an art/science collaboration that focused on cultural identity and climate resilience. In this ongoing project we examined the ways that students attain both visual and scientific literacy while engaged in an arts-based exploration of the natural world. Using a participatory approach, the artist and students engaged in place-dialogues and message to the future about climate change. High school aged students first engaged in different activities that focused on observation and artistic documentation and then they collaborated on producing murals and music that demonstrated their understandings of their natural and built world. This project opened a space for students to be creative and learn to "interpret things better" and to "imagine things when they are not there," both important skills in scientific inquiry. The students working in a collaborative environment, alongside artists and scientists, allowed them to begin to imagine resilient futures, which is a important for them in thinking about, adapting, and integrating the science at the research center (aquaponics, archeology and agriculture) to meet their future needs. Two of the artifacts that were co-produced in this engagement were murals at the research station. One is a scene with both natural and cultural artifacts that were important to their identities as Barbudans. This includes the magnificent frigate bird (*Frigata magnificens*), the ferry that shuttles residents between Barbuda and its sister island Antigua, mangroves and historical structures. The other is of a vintage-style postcard with "Barbuda" framed by the quote "the seas will rise, but Barbuda will survive." Although Hurricane Irma damaged ninety percent of the structures on the island, including the research station, these two murals survived and serve as a symbol for resilience. The Barbudan youth created the mural that demonstrated their understanding of their island and its relationship to climate change and presented their timely and enduring message of survival in their reality of vulnerability to climate events. In the wake of Hurricane Irma, we see evidence of this message coming to fruition with the younger generation of Barbudans first playing a key role in the survival of the animals – pets and livestock – left behind with the mass evacuation by fundraising to return and feed them and now eager to cleanup and rebuild their island with the experiential lessons learned from Hurricane Irma and knowledge of their history of resilience, from the archaic peoples to the Saladoids to African enslavement and to the current epoch of the aftermath of climate change.

Creative Critical Inquiry

Anna Stetsenko (2018) advocates for a relational stance of creativity and creative development that views creativity as a co-authored activity characterized by change, transformation and activism. It is a "bringing forth" of a new idea, innovation or artifact towards changing the world. With the sciences being fields focused on generating new knowledge and understandings of our natural and built world(s), creative critical inquiry is germane to insure that the applications of this newly generated knowledge is for the greater good. Therefore in science education the focus should be on, what Robert Kelly (2016) calls "creativity-intense" practice, "educational design must shift toward a learner and educator disposition that emphasizes creative design practice characterized by learning through engagement in original research, production and action" (p. x). He continues,

> The educational imperative is to enable the longitudinal growth and development of a learner's creative capacity to grow from innate intuitive/ adaptive creativity to the ability to engage in sustained creative practice through original research and production that takes on increasing levels of sociocultural importance. (p. 2)

Fidget Spinners

I end this chapter with a discussion about toys and what they could teach us about creativity in STEM. In a short workshop planned to get rising high school freshmen excited about science, I showed them YouTube videos of "old school" break dancers and subway "It's Showtime" acrobatic dancers. We talked about the spinning and what causes them to spin faster or slower. The students were able to make connections to friction and some even hinted at angular momentum, noting that when arms or legs were extended then contracted caused the dancer to spin faster. I then introduced a random assortment of fidget spinners I purchased in the *chotski* district in New York City. Of course the students were excited as these are toys that were, unbeknownst to me, banned from their classrooms. They were now sanctioned to play with these forbidden toys in the name of science. I had a carefully planned out activities that asked them to observe, design and experiment and then test the spinners. They would test for speed, spinning time, whatever parameters they chose. However with the excitement and the forty-five minutes rapidly ticking away, I just left it with "learn all that you can about the fidget spinners." With not much more direction than that, and Newcleus' *Jam on it* playing in the background, the students engaged in all sorts of curiosities and devised ways of testing their constantly emerging questions about the spinners. They had "contests" for time wondering if the ones with two "arms" spun faster than the traditional ones with three. They were curious about spinning in relation to the different colors and materials (some were metal and others were plastic). They piled them on top of each other to see how that affected the spinning. The cell phones came out and they started timing and videotaping their "experiments" all without being prompted to do so. It was a very open-ended engagement and they used all of the tools at their disposal to learn all they could about these familiar objects. This experience allowed me to glean a few things about creativity and learning environments that foster creativity.

First, the activity was open-ended and collaborative. The students were seated in groups around tables, the spinners were initially placed in the middle and although they each grabbed one spinner, quite soon after they were trading and asking questions about the other spinners before almost seamlessly moving into a collaborative mode. Because it was open-ended, there were no "rules" about any particular sequence to obey; they just followed their inquiries as the emerged.

Second, although they have all had experience with the object, no one was an "expert." They were able to use their own experiences with the spinners (they were experts in their own understandings) and learn from each other's experiences, some of them even pulling out their own spinners from their pockets to add to the testable objects on the table.

Play was essential to their engagement with the spinners. Play, according to Stetsenko (2018), "allows for the use of signs and tools in ways that afford distance from the immediately given context through the power of imagination" thus allowing freedom to become possible (p. 7). The play allowed them to explore different aspects of the spinners and formed the basis of the "experiments" that eventually captured their attention. Through play, they were able to use both the spinners and their cellphones in new ways. The mixing of ideas (along with the constraints and affordances of available tools) allowed new ideas to emerge.

Through the engagements with the fidget spinners, the students produced new creative artifacts that included the use of existing materials and tools in new ways and they investigated the spinning of the objects as new phenomena. Even the ubiquitous cell phones were used in a new way in this context – to collect, analyze and save data.

If we take the fidget spinners to represent a socioscientific problem, we see how this approach allows for the imagination and bricolage of ideas that could lead to new insights, practices and innovations. The open-endedness of the problem presented the challenge of approach and the opportunity of limitless possibilities. Students accessed their individual and collective physical, intellectual and experiential resources to explore the fidget spinner as a "new" object. In just forty-five minutes students together generated data sheets, drawings and videos that they planned on bringing back to the classroom to review, all without being prompted to do so. They tapped into their innate and adaptive creativity as they quickly and effortlessly moved from spinning and playing with the spinners to testing different ideas they generated about the spinners. The problem at hand was of immediate sociocultural importance to them, as middle schoolers. However, if they continued to learn and grow in an environment that fostered creativity in STEM learning, they would have opportunities to engage in sustained collaborative creative practices and contribute to addressing problems of broader importance (Kelly, 2016). Creativity is about new engagements and producing new knowledge, even if that knowledge or insight is new to a learner or group of learners.

The youth in Barbuda engaged in a similar process through the exploration of artistic materials to learn more about their relationship to the environment. Through this process they created a mural with an enduring message of hope for the future that came more quickly than anyone expected. With agentive creativity, Stetsenko (2018) notes,

> The primary emphasis is on people *en-countering, con-fronting and overcoming* the circumstances and conditions that are not so much given as *taken up* by people within the processes of actively grappling with them and thus, realizing and bringing them forth in striving to change and transcend them. (p. n/a)

In order to prepare students to imagine transgressive and transcendent futures, they must be allowed to access and advance their creativity in the present. Engaging students in creative science practices within a Crit-Trans framework would position them to co-produce knowledge towards personal growth and collective well-being.

ACKNOWLEDGEMENT

This work was supported by the National Science Foundation (AISL #1254075).

REFERENCES

Adams, J. (2014). Embodied travel: In search of the Caribbean self in tropical places and spaces. In G. Lean (Ed.), *Travel and imagination* (pp. 25–38). London: Ashgate.

Adams, J., Hefele, N., & Perdikaris, S. (2014, April). *Art/science nexus as pedagogy of imagination and transformative science education.* Symposium: Critical Imagination: Creativity, Curiosity and Critical Pedagogy in Diverse Teaching-Learning Contexts, American Educational Research Association, Philadelphia, PA.

Bai, H. (2015). Peace with the earth: Animism and contemplative ways. *Cultural Studies of Science Education, 10*(1), 135–147. Retrieved from https://doi.org/10.1007/s11422-013-9501-z

Bencze, L., Sperling, E., & Carter, L. (2012). Students' research-informed socio-scientific activism: Re/visions for a sustainable future. *Research in Science Education, 42*(1), 129–148. Retrieved from https://doi.org/10.1007/s11165-011-9260-3

Boger, R., Perdikaris, S., Potter, A. E., & Adams, J. (2016). Sustainable resilience in Barbuda: Learning from the past and developing strategies for the future. *International Journal of Environmental Sustainability, 12*(4), 1–14.

Boy, G. A. (2013, August). From STEM to STEAM: Toward a human-centred education, creativity & learning thinking. In F. Barcellinni & D. Murray (Eds.), *Proceedings of the 31st European conference on cognitive ergonomics* (p. 3). Edinburgh: ACM. Retrieved from https://doi.org/10.1145/2501907.2501934

Brown, D. (2017, September 23). Skerrit to UN: I come to you straight from the frontline of the war on climate change. *Barbados Today.* Retrieved September 23, 2017, from https://www.barbadostoday.bb/2017/09/23/skerrit-to-un-i-come-to-you-straight-from-the-front-line-of-the-war-on-climate-change/

Chakraborty, J., Tobin, G. A., & Montz, B. E. (2005). Population evacuation: Assessing spatial variability in geophysical risk and social vulnerability to natural hazards. *Natural Hazards Review, 6*(1), 23–33. Retrieved from https://doi.org/10.1061/(ASCE)1527-6988(2005)6:1(23)

Clark, B., & Button, C. (2011). Sustainability Transdisciplinary Education Model: Interface of arts, science, and community (STEM). *International Journal of Sustainability in Higher Education, 12*(1), 41–54. Retrieved from https://doi.org/10.1108/14676371111098294

Collins, T. (2006). *Can or should artists attempt to create verifiable change?* Retrieved from https://www.researchgate.net/publication/237772059_Collins_T_2008_Can_or_Should_Artists_Attempt_to_Creative_Verifiable_Change_In_the_publication_from_the_conference_Sense_in_Place_Site-ations_International

Daniel, Y. (2005). *Dancing wisdom: Embodied knowledge in Haitian vodou, Cuban yoruba, and Bahian candomblé*. Champaign, IL: University of Illinois Press.

Fox, N. (2013). Creativity, anti-humanism and the 'new sociology of art, *Journal of Sociology,* 51(3), 522-536. Retrieved from https://doi.org/10.1177/1440783313498947

Glăveanu, V. P. (2010). Paradigms in the study of creativity: Introducing the perspective of cultural psychology. *New Ideas in Psychology, 28*(1), 79–93. Retrieved from https://doi.org/10.1016/j.newideapsych.2009.07.007

Gough, A. (2015). STEM policy and science education: Scientistic curriculum and sociopolitical silences. *Cultural Studies of Science Education, 10*(2), 445–458. Retrieved from https://doi.org/10.1007/s11422-014-9590-3

Kawasaki, J., & Toyofuku, D. (2013). A distributed intelligence approach multidisciplinarity: Encouraging divergent thinking in complex science issues in society. *The STEAM Journal, 1*(1), 1–8. Retrieved June 24, 2015, from https://doi.org/10.5642/steam.201301.10

Kelly, R. (2016). *Creative development: Transforming education through design thinking, innovation, and invention.* Edmonton: Brush Education.

Land, M. H. (2013). Full STEAM ahead: The benefits of integrating the arts into STEM. *Procedia Computer Science, 20*, 547–552. Retrieved from https://doi.org/10.1016/j.procs.2013.09.317

Landivar, L. C. (2013). *Disparities in STEM employment by sex, race, and hispanic origin* (American Community Survey Reports). Washington, DC: U.S. Census Bureau.

Levinson, R. (2018). Creative science teaching: An uncertain, emancipatory and perturbing endeavor. In L. A. Bryan & K. Tobin (Eds.), *Critical issues and bold visions for science education: The road ahead.* Rotterdam, The Netherlands: Sense Publishers.

Livingston, L. (2010). Teaching creativity in higher education. *Arts Education Policy Review, 111*, 59–62. Retrieved from https://doi.org/10.1080/10632910903455884

Lloro-Bidart, T. (in review). "The bees wore fuzzy yellow pants": Feminist intersections of animal and human performativity in an urban community garden. In J. Brady, B. Parker, S. Belyea, & E. Power (Eds.), *Feminist food studies: Exploring intersectionality*. St. Paul, MN: Women's Press.

Madden, M., Baxter, M., Beauchamp, H., Bouchard, K., Habermas, D., Huff, M., Ladd, B., Pearon, J., & Plague, G. (2013). Rethinking STEM education: An interdisciplinary STEAM curriculum. *Procedia Computer Science, 20*, 541–546. Retrieved from https://doi.org/10.1016/j.procs.2013.09.316

Maeda, J. (2013). Stem+ art= steam. *The STEAM Journal, 1*(1), 34. Retrieved from https://doi.org/10.5642/steam.201301.34

McKittrick, K. (Ed.). (2014). *Sylvia Wynter: On being human as praxis*. Durham, NC: Duke University Press.

Medin, D., Lee, C. D., & Bang, M. (2014). Points of view affects how science is done. *Scientific American, 311*(4), 44–45. Retrieved from https://www.scientificamerican.com/article/point-of-view-affects-how-science-is-done/

Mignolo, W. D. (2009). Epistemic disobedience, independent thought and decolonial freedom. *Theory, Culture & Society, 26*(7–8), 159–181. Retrieved from https://doi.org/10.1177/0263276409349275

Mpofu, V., Otulaja, F. S., & Mushayikwa, E. (2014). Towards culturally relevant classroom science: A theoretical framework focusing on traditional plant healing. *Cultural Studies of Science Education, 9*, 221–242. Retrieved from https://doi.org/10.1007/s11422-013-9508-5

Nicolescu, B. (2005). Towards transdisciplinary education. *TD: The Journal for Transdisciplinary Research in Southern Africa, 1*(1), 5–15. Retrieved from https://doi.org/10.4102/td.v1i1.300

Nkrumah, K. (1966). *Neo-colonialism: The last stage of imperialism*. New York, International Publishers. Retrieved from https://www.marxists.org/subject/africa/nkrumah/neo-colonialism/introduction.htm

Parker, K., Roraback, E., & LaMore, R. (2013, October). *A young Picasso or Beethoven could be the next Edison*. Retrieved from http://msutoday.msu.edu/news/2013/a-young-picasso-or-beethoven-could-be-the-next-edison/

Root-Bernstein, R., & Root-Bernstein, M. (2013). The art and craft of science. *Educational Leadership, 70*(5), 16–21.

Rosebery, A., Warren, B., & Bang, M. (2015). *Expansive meaning and makings in artscience*. Paper presented at the annual American Education Research Association, Chicago, IL.

Shavinina, L. (2013). Innovation education: The emergence of a new discipline. In L. V. Shavinina (Ed.), *The Routledge international handbook of innovation education* (pp. 3–14). London: Routledge.

Simonton, D. K. (2003). Scientific creativity as constrained stochastic behavior: The integration of product, person, and process perspectives. *Psychological Bulletin, 129*, 475–494. Retrieved from https://doi.org/10.1037/0033-2909.129.4.475

Sousa, D., & Pilecki, T. (2013). *From STEM to STEAM: Using brain-compatible strategies to integrate the arts*. Thousand Oaks, CA: Corwin Press.

Sternberg, R. (2003). *Wisdom, intelligence and creativity synthesized*. Cambridge: Cambridge University Press.

Stetsenko, A. (2018). Agentive creativity in all of us: An egalitarian perspective from a transformative activist stance. In M. C. Connery, V. John-Steiner, & A. Marjanovic-Shane (Eds.), *Vygotsky and creativity: A cultural-historical approach to play, meaning making, and the arts* (2nd ed., p. n/a). New York, NY: Peter Lang.

Strong, L., Adams, J. D., Bellino, M. E., Pieroni, P., Stoops, J., & Das, A. (2016). Against neoliberal enclosure: Using a critical transdisciplinary approach in science teaching and learning. *Mind, Culture, and Activity, 23*(3), 225–236. Retrieved from https://doi.org/10.1080/10749039.2016.1202982

Wynter, S. (2003). Unsettling the coloniality of being/power/truth/freedom: Towards the human, after man, it's overrepresentation–an argument. *The New Centennial Review,* 3(3), 257–337. Retrieved from https://doi.org/10.1353/ncr.2004.0015

ABOUT THE AUTHOR

Jennifer D. Adams is an associate professor in the Department of Chemistry with a joint appointment in the Werklund School of Education at the University of Calgary, Alberta Canada. Jennifer holds master's degrees in nutrition from Brooklyn College, CUNY and science education from New York University, New York. She also holds a doctoral degree from The Graduate Center, CUNY. Her research is broad and emphasizes sociocultural and critical approaches to understanding science teaching and learning in formal and informal settings. Her current work examines the intersection of creativity and science teaching and learning in postsecondary science contexts.

ARNAU AMAT

9. THE ENGAGEMENT OF COMMUNITY STAKEHOLDERS IN SCHOOL SCIENCE EDUCATION

Tensions and Transformations in a Project Based on Agroecology

ABSTRACT

In this chapter, a case study in which the manner that community stakeholders are engaged in a school science learning project is analyzed. The educational project was framed within the school agroecological approach as the school food system is changed to a more sustainable and fair system using the vegetable garden management. From an event-oriented perspective, four events that allowed the transformation of the school and its community stakeholders are discussed in order to identify the kinds of difficulties faced and how they could be managed. These difficulties are interpreted through a sociocultural lens as a dialectical relationship between agency and structure.

Keywords: school agroecology, community stakeholders, event-oriented inquiry, dialectical relationship, Education for Sustainable Development

It was a warm June afternoon in a small rural school in the middle of Catalonia, relatively close to Barcelona. In the shadow provided by the school porch, two Morocco-Catalan mothers were preparing a mint tea on a small camping stove. They were boiling tea and mint with a lot of sugar in a metal kettle. After a few minutes of waiting while they were explaining to everybody the tea preparation, they started to pour the tea from the metal kettle to the glasses. All the kindergarten and the elementary students were so amazed to taste the tea, and so were all the parents.

Even though in that moment we did not know yet, that was the final chapter of a four-year project, whose main aim had been trying to work in school science activity from an agroecological approach with the community stakeholders. In that spring afternoon, as once a month in the last four years, three workshops were organized to work about the vegetable garden based on agroecological topic. In that case, the topic was the uses and the identification of Mediterranean aromatic plants, such as mint, thyme or rosemary. These workshops were designed and prepared by the teachers and me, but they were conducted by parents. We called these kind of workshops 'shared activities', because they were moments when parents, neighbors shared a

 | DOI:10.1163/9789004389663_009

concrete space and time in the school, and, of course, they shared their knowledge. Aromatic plants had been the main topic of our work in the vegetable garden and had let us approach to a new cultural knowledge.

Not until I finished swallowing all the mint tea, did I really realize what time it was. It was a half past four, just when the afternoon school time arrived at the end. While the teachers were coming back to the classes with the students, I was talking about the tea with the two mothers and I shared with them some of my memories about a trip to Morocco that I had made several years earlier. Their faces and smiles showed how they were so glad to be involved in this shared activity and sharing their cultural knowledge with the children and the other parents.

Actually, the engagement of the Morocco-Catalan mothers in the shared activities had been an exciting challenge, because the language barriers were revealed as obstacles to engage them who hardly spoke Catalan or Spanish in the shared activities. And I suppose for this reason, I felt pride and happiness, because it had been a long journey since 2009 when the project began. During these years, we were challenged by a lot of different difficulties, but at the same time, the school was transformed, as were the people within.

In this chapter, I will describe and interpret from a case study research approach, one of the central issues in science education from my point of view: how to engage community stakeholders in the school science learning process. The literature on this particular issue shows a variety of approaches. From Education for Sustainable Development (ESD) approach, as is suggested by Sφren Breiting, Finn Mogensen and Michela Mayer (2005) schools must be understood as institutions connected to the real world and as active members of the community, which can play a remarkable role in the social transformation. Thus, school can become a social center' with different sources of knowledge. Furthermore, as argued by Mariona Espinet and Aravella Zachariou (2014), these collaborations establish a "new framework of learning through school's opening to the community," where it is produced intergenerational interactions and creates networks amongst various key community actors.

Moreover, different approaches in science education have been taken in recent years also. Third space theory for instance, defended by different authors as Elizabeth Moje (2004), describe how everyday science discourses, which are typically placed at home and the community, are associated with school science discourses. In their approach, teachers should encourage students to bring their everyday discourses, the first space, at school, the second space, in order to build a Third space, where new discourses and new ways of knowledge would emerge. Recently, Cassie Quigley (2011) has proposed that Third space "occurs when the physical spaces of home and school containing these Discourses are blended in a manner that creates a space that is congruent with the physical space and the specialized knowledge sets of the first space (home) and the second space (schools)" (p. 555).

Even though all these studies have indicated the relevant role of school as a physical space and as an active institution where different stakeholders can meet, little attention has been given to the process of the construction of this kind of collaboration. The main concern of the study is to provide some thick descriptions and interpretations about what kind of difficulties that were faced by the school community stakeholders while this project was going on, and how they could manage them. These descriptions are salient products of the research that was conducted in a kindergarten and elementary school in order to involve parents and neighbors in a school science activity, where the vegetable garden became the main tool in fostering collaborative work for more than four years.

THE TENSIONS IN THE FIELD: REPRODUCTION AND TRANSFORMATION FORCES IN SCIENCE EDUCATION

I am aware that this kind of research sometimes looks suspicious to some science education researchers; more than once, my research was qualified as a "non-science education research" by some of my colleagues. However, when I am thinking about the field of science education, I am reminded of Mikhail Bakhtin's (1981) idea that two forces rule any social language and any field of knowledge: on the one hand, a centripetal force which moves all the features of language and thinking to a central point, in order to unify and to create a standard point of view and, on the other hand, a centrifugal force which tries to escape from the standardization and create new forms of language and thinking. The first one could be represented by the mainstream movements in science education, that sometimes constrain and limit the field. By contrast, the second one force could be represented by the science educators and researchers who are looking for the expansion of the field through new theoretical and methodological frameworks, or the involvement of new actors in learning processes.

The equilibrium between these two forces allow the reproduction and the transformation of any field of knowledge, such as science education. Although all science educators and researchers must understand the mutual necessity of these centripetal and centrifugal forces because of their dialectical nature, my research is focused on the exploration of new interventions and new methods of research, to try to expand the field of science education. This conscious decision which I made some years ago, allowed me to move within the porous boundaries of the field, trying new research methodologies and reflecting about what counts as science education.

In 2009, when I started my PhD, I decided to enroll in a new area of research to focus on sustainability, science education, interdisciplinary and community that was carried out by my adviser, Mariona Espinet, and by my colleagues from GRESC @ research group at the Universitat Autònoma de Barcelona. Our studies focused where science education, ESD and agroecology meet. Moreover, we explored different research methodologies, always in the interpretative and qualitative approach, from critical discourse analysis to more ethnographical methodologies. In a few years, our research provided different examples of how to enrich science education through the framework of agroecology and a different set of research strategies to analyze more closely into community practices and participant identities.

In my particular case, my research was carried out in a small school in a rural area, sixty kilometers north from Barcelona. When I started this research, I was working as environmental educator for a local administration that managed the education for sustainable development school program in the region. My job in that moment consisted of visiting all the elementary schools in the region in order to help teachers to promote educational activities and educational projects on sustainability. From a teacher's point of view school should become a space for the articulation of the school and the community, that means neighborhood and families.

First of all, our decision was to use the vegetable garden as a learning context to promote science learning and community interaction from a school agroecology perspective. School vegetable gardens are becoming one of the most popular topics to work ESD approaches in Catalonia, probably because the possibility to act directly in a real context (Espinet, Junyent, Amat, & Castelltort, 2015). However, adopting a school agroecology framework, according to Miren Rekondo, Mariona Espinet, and Germán Llerena (2015), means emphasizing the food school system, suggests the transformation towards a more sustainable, and fair system, where the students and community stakeholders have an important role.

THE TENSIONS IN THIS RESEARCH PROCESS: BEING AN EDUCATOR AND RESEARCHER IN THE SCHOOL

My research in this school is framed from a hermeneutical phenomenology approach. This approach is interpretative, participant-centered, as defined by Konstantinos Alexakos (2016). For four years, I was engaged in this school research project conducting a participant observation study. During this time, data were obtained from my field diaries, which let me describe and reflect on different significant events in the collaboration through the school vegetable garden. Moreover, some semi-structured interviews were carried out with parents who had an important role in the project; teachers who were in charge of the vegetable garden; and neighbors who participated. Finally, some activities that took place among the school and community stakeholders were video recorded.

While I was working as an environmental educator, I was developing my research project. Even though sometimes it was difficult to act as a researcher and educator at the same time, from my perspective the two roles that I played in the project enriched each other. The processes of acting and researching became indistinguishable as described by Alexakos (2016), "We do not wait for the research to be concluded to suggest changes; instead, interventions are implemented and refined while the research is ongoing and become part of that research" (p. 1). I reflected about what I was doing, and this reflection changed my action.

My lenses for the interpretation are built from a sociocultural theory, thus difficulties are conceptualized as dialectical tensions that afford and constrain community agency and afford and constrain the transformation of school science practices. In order to explain what was happening in this collaboration and why it was happening, event-oriented research was adopted. According to William Sewell (2005), "Social life may be conceptualized as being composed of countless happenings or encounters in which persons and groups of persons engage in social action. (…) Events may be defined as that relatively rare subclass of happenings that significantly transforms structures" (p. 100).

Therefore, events may be defined in terms of contradictions, which are the inherent force of transformations in any social activity. Wolff-Michael Roth and Yew Jin Lee (2007) claimed that these contradictions are expressed as tensions or difficulties while the activity is going on and it is important to understand that it is not possible to eliminate tensions because they are inherent in any social system. As is described by Kenneth Tobin and Steve Ritchie (2011), event-oriented research is analogous to using a zoom lens, that allows the researcher to explore more deeply what was going on in the social phenomena that is studied. Furthermore, Sewell (2005), argues that events usually are contingent because "what has happened at an earlier point in time will affect the possible outcomes of a sequence of events occurring at a later point in time" (p. 100).

In this study, events were selected because they are the expressions of tensions and difficulties in the community, which foster school community transformation. From a sociocultural perspective, these tensions and difficulties are going to be interpreted as a dialectical relationship between agency | structure. According to Tobin (2007), without structure, there can be no agency and without agency, there can be no structure because of their dialectical nature. Therefore, it is important to keep in mind that structures promote and, at the same time, constrain the agency of stakeholders who participate in the collaboration.

Thus, in the follow part four salient events are going to be provided with the purpose to understand what kind of tensions community had to deal with. These events are going to be presented from a chronological perspective with the intension to describe what was going on in this four-year project and to enforce the idea of contingency. Moreover, in each event I am going to try to invoke the voice of the participants, in order to keep the complexity of points of view and understandings around the collaboration.

THE TENSION IN GARDENING METHODOLOGY: FROM AN ECOLOGICAL VEGETABLE GARDEN TO A TRADITIONAL FOOD GARDEN, TO BUILD A MORE SUSTAINABLE VEGETABLE GARDEN

From the very beginning, teachers of the school wanted to involve some people from the neighborhood to help them in the gardening management. The school was in the middle of a rural area and in the surroundings of the school were plenty of small vegetable gardens, which were taken care by old and retired people. Therefore, it should not be particularly difficult to find a neighbor who wanted to help us with gardening. In order to find them, Joan, the Mayor of the village was invited to the school for a meeting.

Joan had worked as a peasant all his life and he was raised in a Catalan traditional agricultural environment. He was so amazed at the invitation, because he thought it was a great opportunity to transmit this cultural tradition to the students. In fact, the school already had a beautiful and a new ecological vegetable garden, which had been built in the framework of a course and paid for a local administration just one year before the beginning of the project. In that meeting, teachers, Joan, and I discussed the logistical arrangements in the school vegetable garden. Actually, our main concern was about two things: the calendar of the plantation and the gardening methodology.

Firstly, calendar plantation, which had been followed by teachers, was the typical summer gardening calendar, which is characterized by planting between February and March and gathering all the vegetables and fruits between July and September. Despite the huge diversity of vegetables that can be planted in a summer garden, they are difficult to fit in the school calendar because from July to September students are on their summer holidays. This means that they cannot harvest and enjoy any production of the garden. For this reason, in the course of this meeting Joan recommended teachers start a winter garden calendar, characterized by planting from December to January, and picking all the production in Spring. Even though the number of vegetable species which can bear the frost (onion, garlic, spinach...) are less amazing than the others (tomatoes, eggplants...), students could follow the whole plantation circle within the school year.

Secondly, the ecological methodology, which the vegetable garden had been built, was very hard to follow. The methodology was patented and the measures were pre-designed to optimize space and water, in order to be as efficient as possible. This was meant to measure and calculate any intervention in the physical garden space, such as plant, water, crops rotation and soil supplies. In this kind of garden, for example, a thick layer of mix of compost and sand is placed above the soil, and each vegetable must be planted in a particular separation. Obviously, this ecological gardening methodology was not designed for neighbors, and it was quite evident that was constraining their agency. They were gardening in the same traditional gardening methodology since they were children. Therefore, this ecological methodology was a huge obstacle for their participation.

For this reason, in that meeting it was decided to flatten the predesigned-ecological vegetable garden to build a new one in a traditional gardening methodology. If we wanted to engage some neighbors in the school, ecological methodologies should be obligatorily sacrificed. Even though at the beginning I was against this, soon I concluded we had made the right decision. But how could I justify this decision if I was supposed to work as an environmental educator? In addition, why did I think it was not an unwise decision? I had not been able to answer this question at first. Only some weeks after, did I understand how important it had been to change the methodology.

After the Christmas holidays, Joan introduced me to Miguel, who became our expert gardener for the first school year. He was in his sixties and lived in small house five minutes' walk from the school where a beautiful garden was managed by him and his wife as a hobby. He had come to the village from the south of Spain more than forty years ago, and he had worked as a stonecutter all his life although he was retired then.

The first thing that was arranged by teachers was visiting his vegetable garden with the oldest students of the school. They visited Miguel's garden on a chilly day at the beginning of the spring, but they could already see some vegetables. He was a great host and, with his strong and southern Spanish accent, he kindly explained to them how he and his wife managed their garden. It was a typical vegetable garden of the region, where people plant the vegetables in furrows where the water drained. After the visit to Miguel's home, the students invited him to their vegetable garden. And at the school, while Miguel was describing how to plant to the students, the teacher told me how many things had been learnt about Miguel in a short while. Then, I realized that nothing like this could have happened in a predesigned-ecological garden. Time and space structure the garden itself, but the agency of the participants also. Therefore, managing time and space tensions was revealed as a serious challenge that we had to face.

In my view, this tension fostered the creation of a new kind of vegetable garden. The school vegetable garden was not a pre-designed impersonal ecological garden anymore, but it was not as the gardens of the village either. A new entity was created: with a traditional methodology which the plants grown in typical Catalan furrows; with the clear dominance of winter vegetables species, such as onions, garlics or fava beans, instead of tomatoes, egg plants or zucchini; and without any kind of pesticide or artificial fertilizers.

Furthermore, in my opinion this is exactly what differentiates ecology projects from sustainability projects: the implication not only of ecological tenets, but also of social, political, cultural and economic tenets. As claimed by Breiting et al. (2005), any school sustainability project must consider three main quality dimension criteria: the teaching and learning processes; the school policy and organization and, finally, the school's external relations. Our garden was transformed successfully from an ecological school garden to a sustainable school garden because we could engage the neighborhood in it.

THE TENSIONS IN TEACHING AND LEARNING PROCESS: BEING ENGAGED THROUGH SHARED ACTIVITIES

Apart from neighbors, our main concern was family engagement, because we viewed family's participation as the real keystone in our collaborative project. In spite of our experience, which indicates that parents rarely participate in school science activities, some publications and reports have appeared in last decades documenting how parents' involvement in school activity plays a significant role in the student's educational outcomes. Joyce Epstein and Mavis Sanders (2000), for example, argued how parental involvement in school activity promotes a better understanding between teachers and families. In their study, teachers who collaborate with parents would have a less stereotypical image of families than other teachers. Moreover, parents who are engaged in school activities usually have a better impression of the school methodologies and teacher's work than the other parents.

It was at the end of May in our first year, it was already possible to harvest some vegetables that had been grown in the garden. For this reason, teachers decided to invite parents to a gardening activity. However, teachers and I were quite tired about the usual activities in which parents regard community stakeholders as passive agents. Often, in these kinds of activity parents are viewed as builders, when they are involved to build something in the school such as a garden, a greenhouse or a fence for the school. At other times, parents are like observers and they are invited only to watch some of their children performance in the school, whether it is a scenic performance or it is a schooling performance.

I support Marta Civil, Jill Bratton, and Beatriz Quintos (2005) and Angela Calabrese Barton, Corey Drake, Jose Perez, Kathleen St. Louis, and Magnia George (2004) who claim that parents and community stakeholders must be seen as cultural and intellectual resources for the school and they must be seen as active agents of school activities. Therefore, parent engagement must be promoted from this agential perspective. For this reason, the main question that I had then was why they would like to be involved in such a collaborative project. Maybe if we were able to answer this question, we could find a way to engage them in the project.

For instance, Anna, who was a mother of three pupils at the school, wanted to be more involved in everyday school activities. However, she was quite cautious with the role of parents in school participation. She believed in the vegetable garden as a way to teach something "kids can't usually see in everyday life." In her opinion, agriculture had been separated from the school curriculum although the school was in a rural area, and for this reason the vegetable garden could be seen as a thing that could help the construction of school identity.

This idea of a context to learn a kind of culture that is becoming distant for students was also shared by other parents. Bruno, a father of an eight-year old girl, told me how important it was to preserve this traditional culture: "Actually, the traditional peasant culture is a complex culture, that nowadays is being lost, but it is so rich, because it is related to the knowledge on the land, on the seeds, on the seasons (…) In my opinion, it is fundamental knowledge that is important to bring closer to the students." Bruno had been involved in the collaboration since the first day and from his point of view, the main purpose of the vegetable garden was to "supply a small part of the food that is consumed by children in the lunch time. So, they can see what they eat was grown in a someplace." Thus, the vegetable garden was understood as a teaching space by parents, where it was possible to inquire about not only the food production and consumption, but also the traditional agriculture background of the region.

Furthermore, from the interviews, it can be shown how parents could be engaged in school activity as learners as well, as it is illustrated by Anna's words: "I would like to learn something about how to can vegetables, because I don't know how to do it." Even though she thought it as a learning space for pupils, she also argued that could become a place where community could learn. Anna and other parents assumed the vegetable garden was a good context to learn for both students and all the stakeholders.

Taking their voices into account, we started to understand that maybe the clue was to construct a teaching and learning activity based on the vegetable garden, where parents play an active role. For this reason, teachers and I agreed to start a new kind of activities that we called shared activities, where the vegetable garden were understood as a learning space where parents play both positions: as learners and as educators, too. In our view, the vegetable garden was a learning space where different agroecological issues could be discussed and parents should be engaged in an active role.

However, our first and immediate headache was the management of physical and temporal structures again. One of the main difficulties was the structuration of space in the shared activities. The vegetable garden in the school was so small, that it was completely impossible for it to hold all of the parents, teachers, and pupils. To solve this difficulty, the shared activities were organized in different workshops, which allowed us to work in small groups in a more comfortable and convenient way. While some groups were working in inquiry-scientific based workshops about agroecological topics, other groups were gardening in the vegetable garden. Moreover, at the same time, working in small groups allowed that the leadership of the workshops could be taken by parents.

Thus, each group in shared activities was composed of about six students, from three to twelve years old, and one or two parents who conducted an observational workshop and an experimental workshop. These workshops were designed to inquire about different agroecological issues from a scientific perspective. Therefore, seeds, soils, plagues, plant physiology or vegetable garden biodiversity became our main topics. In order to support parents in these workshops different guides were provided just to structure and organize the workshops and to facilitate parents' involvement. Although sometimes, when the workshop was particularly difficult to carry out, a previous meeting with parents was arranged in order to explain the organization of the workshop, usually parents knew what the workshop was about while they were dealing with the children.

At first, some parents, such as Anna and Bruno, were quite reluctant to see themselves in the position of educators. Anna though they did not have enough knowledge to teach about gardening and even less in scientific approaches. Bruno told that he could hardly imagine parents leading educational schooling activities. This opinion can be shown from teachers' perspective also. For example, principal of the school had many doubts whether or not parents had to participate in the school schedule: "I would like to engage parents. However, I do not know how to involve them. I think that family's implication is a matter that must be decided by all the teachers' team…"

However, this perception changed in a natural way, through the participation in the shared activities. Despite their first reluctance, in the last year Anna told me "Sometimes, parents who look from outside of the school though 'what the hell is it that I am supposed to do in the school, if I don't know anything about gardening. Apart from being on charge of a group, such a responsibility' that it means I am not a teacher and this is precisely the teacher work." She added that when someone participated in the shared activities, realized that it is not such a difficult thing. Another mother, Elisenda, who participated in almost all the shared activities, saw their participation as a learning process: "The more I participate, the more comfortable I feel [...] Actually, I didn't feel uncomfortable, but I am not used to teach to kindergarten and elementary students, still less if they are in the same working group. It is not a question about difficulty, but I am not used to participate in this kind of activity. And even if you don't know anything about the topic, my spirit is let's see what I am going to learn today!"

It is exciting thinking how these shared activities changed the collaboration and our minds as well. Parents who had expressed their inability to be part of a teaching and learning process were able to manage an experimental workshop about seeds germination with children from three to twelve years old. My view to interpret is framed on Sewell's (2005) theory of social transformation. According to this author, social transformation occurred when structures change. From his point of view, structures can be defined as schemas, that are virtual, and resources, that are actual, which are related to each other in a dialectical relationship. It is precisely this dialectical relationship that allows both to change, because the reproduction of schemas depends on the validation by resources, but at the same time, resources can be modified because of mental schemas.

Even though at the beginning it was hard for all of us to imagine this kind of collaboration, teachers, parents and I changed our minds while the project was going on. Our idea of how to promote parents' participation was constrained by actual structures such as space, because there was not enough room for all parents teachers and students in the garden. In order to find a solution to deal with that, our mental schemas changed and we started to consider parents not only as passive participants, but as active agents who can participate as learners and educators at the same time. Only after that, did we take advantage of the resources of the school to build the shared activities.

This equilibrium and mutual influence between actual resources and virtual mental schemas let me explain how shared activities were created and sustained from 2010 to 2013 in that project. All the arrangements that we did in these four years in order to improve the parents' participation can be understood from this point of view. As described by Sewell (2005), "placing the relationship between resources and cultural schemas at the center of a concept of structure makes it possible to show how social change, no less than social stasis, can be generated by the enactment of structures in social life" (p. 142).

THE TENSION BETWEEN DIFFERENT ONTOLOGIES AND AXIOLOGIES: LEARNING FROM TRADITIONAL GARDENING IN ORDER TO TRANSFORM SCIENCE EDUCATION CURRICULA

For personal reasons, Miguel just collaborated with us in the first year. Since the second year, Pablo, a man in his seventies who lived in the building just next to the school, became our expert neighbor who was on hand to assess teachers. Actually, it was curious how I met Pablo. In the second year, kids were plowing the garden while Pablo was looking at them from his balcony, which was just a couple of meters above the school vegetable garden. Even though I had already noticed that he was looking at us in each shared activity, he had never said anything. He was always observing us in his usual angry expression. However, on that particular day it was literally too much for him, because it was quite evident, that he was getting more and more nervous because kids were not using the plow in an appropriate way. Finally, he started yelling them from his balcony explaining how to use the plow. I started to laugh and I said to him that we would be so glad if he could teach us how to do it. In less than a minute, he was in the garden showing us how to use the plow. Pablo was so committed to the collaboration that he became a perfect expert in the teacher's assessment.

Pablo's strong character created some odd situations. However, one of them was actually one of the most meaningful events in the collaboration. It occurred in the spring of our third year, when the teachers and I were preparing the monthly shared activity. We needed Pablo's opinion to decide what could be planted. When Marta, the teacher who was responsible for organizing the shared activities, asked him about this, he appeared to be upset: "It's impossible! Don't you know that it is impossible to plant during the waxing moon"? And he had told her that he did not want to participate in that shared activity if we decided to plant. In that moment, all the families had already been informed and it was impossible to postpone the shared activity. Therefore, we decided to keep the shared activity. It was funny to see the shadow of Pablo behind the curtains of his window looking at us, while parents and kids were together in the vegetable garden.

The moon calendar is a component of an iconic peasant culture that is followed not only by Pablo, but also by a lot of people who own a vegetable garden in Catalonia. Actually, from the very first moment, the school garden plantation calendar was designed according to the phases of the moon because Miguel was a supporter of this ancestral methodology also.

In my view, this is a relevant vignette because it is the collision of two different ontologies about gardening. On the one hand, community stakeholders, such as Miguel and Pablo, felt that the vegetable garden must produce as much food as possible. Their vision was completely comprehensible, because all of their lives they had been managing a garden to promote efficiency of production. Therefore, all the structures, such as the calendar, spatial organization, plant varieties or soil supplies must be designed to achieve this productive outcome. However, teachers and parents thought that the production of food was not our main aim. As Marta said: "Pablo represents the voice of experience, but at the same time he has a very particular way of gardening and sometimes he wasn't able to adapt to our way… Sometimes, he doesn't understand that productivity is not the most important thing, but learning.

Of course, the garden must produce some food, but the main goal was to learn through food production. The collision of different purposes became very evident when plagues, such as snails or plant lice, attacked the vegetable garden. Meanwhile, the first instinct of neighbors was to act in the same way that they used to act with a non-school vegetable garden, in other words, with pesticide. However, the teachers would rather take advantage of the situation to inquire about the plague, in order to learn more about snails and different traditional ways to control them, rather than wipe them out with pesticides.

This event could be understood also as the encounter of different ontologies and axiologies. It is not difficult to infer how the structures of traditional gardening were constraining the agency of school gardening; but, at the same time, how the tension of the structures of school gardening were constraining the agency of traditional gardening. On the other hand, this collision of ontologies not only limits the agency of the stakeholders, but also allows the transformation of social practice.

A few days after that shared activity, when I came back to the school, I noticed that there were some telescopes in one of the classrooms. I asked Marta about that and she answered that after the problem of planting during the waxing moon, she had decided to study the moon and its phases with the kindergarten students. A father, who was an amateur astronomer, was going to show the moon to them through the telescopes. Thus, a topic that was not planned in the course syllabus emerged after the encounter of these two different ontologies. As presented by Sewell (2005), the multiplicity of structures and their overlapping can also be a source of transformation. In that case, the tensions between these two ways of understanding the same vegetable garden modified the learning and teaching processes.

Finally, I would like to highlight how important the neighborhood was for the school. Pablo became a perfect assessor for the teachers. In different moments, Marta expressed her lack of knowledge about gardening: "I had to accept the difficult challenge of the management of the vegetable garden, but I didn't have the faintest idea how to manage it." Despite Pablo's strong character, Marta explained. "Today I was wondering how we have to plant carrots… And no problem, I just have to ask him."

THE TENSION IN CULTURAL DIVERSITY: HOW TO DEAL WITH LANGUAGE BARRIERS

Like most schools in the region, this one was very culturally diverse. In this particular case, the level of cultural diversity was a slightly above the region mean, with families coming mostly from Morocco, but also from South America and other European countries. In fact, how to bring all this diversity to the collaboration was a daunting challenge.

The main problem was their lack of knowledge of not only Catalan, the vehicular language in all the Catalan schools, but also Spanish. Since the creation of the shared activities, the families' engagement was based on their participation in the shared activities, for which it was necessary to be able to read and comprehend Catalan. Only two of the Moroccan-Catalans were able to do that. Now, it was so obvious that shared activity structures were restricting their agency, but while the collaboration was in progress, it was difficult to pay attention to these things. Unconsciously, an invisible wall was erected between the school and these mothers.

In the second year, when the teachers and I recognized this problem, we tried to resolve it by making some changes to the shared activities. One solution was to mix parents who were not fluent in Catalan or Spanish with Catalan or Spanish native speakers in the workshops. Even though it was not a completely successful solution, this adjustment to the structure of the shared activities afforded a greater participation of some mothers in them.

In my opinion, the fact that the shared activities were based on the production of food was another obstacle. From the very beginning, seeds and plants were revealed as a valuable treasure to keep the cultural diversity in the collaboration, but basing the shared activities just on the production of food could be a real problem. In one of the first meetings with parents, some of the Moroccan-Catalan mothers expressed how important it would be for them to plant mint in the garden.

Mint was planted in the first year. However, the power of mint, spices, or vegetable varieties was mostly revealed when they were cooked. Moroccan-Catalan mothers were very happy in the last shared activities not because mint had been planted in the vegetable garden, but because they could show part of their native culture to other people through the typical Moroccan cuisine. This would suggest that in order to bring the cultural diversity to the school from an agroecological point of view, maybe it is important to combine not only food production, but also food transformation and most importantly, food consumption.

Final considerations and the collapse of the community collaboration

I would like to highlight that after four years, not only had the school changed a lot, but we had also. It was difficult for all of us to maintain the project for longer. From my point of view, the end of the collaboration was the result of a change in the personal lives of some of the stakeholders, in addition to macrostructures, which had a huge impact.

In my case, in the fourth year I took a new position as a professor at the Universitat de Vic, and I needed to dedicate myself to writing my dissertation full-time. It was, therefore, impossible to continue the project. In fact, the main problem was the continuity of the teachers; only two out of the five teachers who started the collaboration were in the school in the fourth year: the principal and Marta. The erratic educational policy of Catalan and Spanish governments does not allow the fostering of long term projects, because of the incomprehensible difficulty in maintaining teachers' places in the school for a long time. Therefore, only Marta had enough energy to keep the project going, but in the fifth year she was going on maternity leave. If it had been possible to pass the baton to the new teachers, perhaps it would have been possible to continue the collaboration for a few more years. However, it was difficult to engage the new teachers in the collaboration, because it was not a project that they had participated in since the beginning and because the new teachers probably would not stay in the same school for more than two years.

From our research, it can be suggested that it is important to select a topic that could be shared with different stakeholders. In fact, other studies had already claimed the opportunity of the food system as a tool to engage community, such as Sumi Hagiwara, Angela Calabrese Barton and Isobel Contento's (2007) study, which concluded that "by using food as an instructional tool, culturally relevant connections can be made between science taught in both school and the home" (p. 496). In our case, working on agroecological approach allowed any stakeholder to be engaged with the project, because everyone has experience of the food system, whether it is food production, preparation or consumption. However, it would probably be a good idea not just to work on the food production, but also to combine different parts of the food system, in order to promote the inclusion of more participants.

As claimed by Breiting et al. (2005), a key aspect of partnership is regular and systemic development and exchange of information. In fact, the originality of our solution lies in the creation of a systematic way of engaging families and neighbors through the shared activities. The systematization of this community encounters can be viewed as the creation of new structures that allowed the engagement of the community stakeholders. Probably without this kind of activities, we would not have this mid-term success in the collaboration.

In this chapter, the focus of the attention was on showing the agential role of the stakeholders in a school science collaboration. Parents and the neighborhood are understood as valuable resources that from their own way of understanding the world can enrich the science teaching and learning process. Of course, the implication of different ontologies and axiologies can cause different tensions and difficulties, but they must be understood as expressions of the transformation of school structures.

ACKNOWLEDGEMENTS

The author would like to thank all the school community stakeholders (parents, teachers, neighbors and students) who participated in this collaboration during four years for their valuable and unconditional support. The author's thanks are also extended to Mariona Espinet and my GRESC@ colleagues for their insights and recommendations while the collaboration was going on. Finally, I appreciate the efforts and assistance of Milford Edge for his support and help in completing the chapter.

REFERENCES

Alexakos, K. (2015). *Being a teacher | researcher: A primer on doing authentic inquiry research on teaching and learning*. Rotterdam, The Netherlands: Sense Publishers.

Bakhtin, M. M. (1981). *The dialogic imagination: Four essays*. Austin, TX: University of Texas Press.

Breiting, S., Mayer, M., & Mogensen, F. (2005). *Quality criteria for ESD-schools: Guidelines to enhance the quality of education for sustainable development.* Vienna: BMBWK.

Calabrese Barton, A., Drake, C., Perez, J. G., Louis, K. S., & George, M. (2004). Ecologies of parental engagement in urban education. *Educational Researcher, 33*(4), 3–12. Retrieved from https://doi.org/10.3102/0013189x033004003

Civil, M., Bratton, J., & Quintos, B. (2005). Parents and mathematics education in a Latino community: Redefining parental participation. *Multicultural Education, 13*(2), 60–64.

Epstein, J. L., & Sanders, M. G. (2000). Connecting home, school, and community: New directions for social research. In M. T. Hallinan (Ed.), *Handbook of the sociology of education* (pp. 285–306). New York, NY: Kluwer Academic/Plenum Publishers.

Espinet, M., Junyent, M., Amat, A., & Castelltort, A. (2015). Moving schools towards ESD in Catalonia, Spain: The tensions of a change. In R. Jucker & R. Mathar (Eds.), *Schooling for sustainable development in Europe: Concepts, policies and educational experiences at the end of the UN decade of education for sustainable development* (pp.177–199). Dordrecht: Springer. Retrieved from https://doi.org/10.1007/978-3-319-09549-3_11

Espinet, M., & Zachariou, A. (Eds.). (2014). *Key stones on school-community collaboration for sustainable development.* Retrieved from http://ensi.org/global/downloads/Publications/369/CoDeS-Key%20Stones.pdf

Hagiwara, S., Calabrese Barton, A., & Contento, I. (2007). Culture, food, and language: Perspectives from immigrant mothers in school science. *Cultural Studies of Science Education, 2*(2), 475–499. Retrieved from https://doi.org/10.1007/s11422-007-9063-z

Moje, E. B. (2004). Working toward third space in content area literacy: An examination of everyday funds of knowledge and discourse. *Reading Research Quarterly, 39*(1), 38–70. Retrieved from https://doi.org/10.1598/rrq.39.1.4

Quigley, C. F. (2011). Pushing the boundaries of cultural congruence pedagogy in science education towards a third space. *Cultural Studies of Science Education, 6*, 549–557. Retrieved from https://doi.org/10.1007/s11422-011-9335-5

Rekondo, M., Espinet, M., & Llerena, G. (2015). La construcción discursiva de la competencia eco-ciudadana en la escuela: La realización de un diseño tencológico colaborativo en agroecología escolar. *Investigación en la escuela, 86*, 7–19.

Roth, W. M., & Lee, Y. J. (2006). Contradictions in theorizing and implementing communities in education. *Educational Research Review, 1*(1), 27–40. Retrieved from https://doi.org/10.1016/j.edurev.2006.01.002

Sewell, W. H. (2005). *Logics of history: Social theory and social transformation.* Chicago, IL: The University of Chicago Press.

Tobin, K. (2007). Collaborating with students to produce success in science. *Journal of Science and Mathematics Education in South-East Asia, 30*(2), 1–43.

Tobin, K., & Ritchie, S. M. (2011). Multi-method, multi-theoretical, multi-level research in the learning science. *The Asia-Pacific Educational Researcher, 20*(3), 117–129.

ABOUT THE AUTHOR

Arnau Amat is a professor at the Universitat de Vic – Universitat Central de Catalunya in Spain. His research focuses on sociocultural approaches in three different areas: involving the community in the school through science education, education for sustainable development and science teaching education. Prior to that he worked as environmental educator for various private companies and education authorities and as a science teacher in high school.

GEETA VERMA, ANTON PUVIRAJAH AND HELEN DOUGLASS

10. EXAMINING THE MEDIATION OF POWER IN INFORMAL ENVIRONMENTS

Considerations and Constraints

ABSTRACT

The purpose of this chapter is to examine two out-of-school learning contexts to gain insights into the mediation and manifestation of power among students. We investigated and report herein the participation of mainstream high school students in a robotics competition and elementary-aged underrepresented minority students in a supplementary and optional take home project (SOTHP). Informal learning contexts (including out-of-school contexts) provide alternative sites for knowledge production in the sense that students' participation is mostly voluntary, free-choice, self-paced, and non-sequential. We argue that the language and culture of science, cultural products of the scientific community, are profoundly important as they have a unique system of resources for creating meaning and scientific knowledge. Examining students' experiences in informal social settings allows for studying the manifestation and mediation of power and privilege (Puvirajah, Verma, & Webb, 2012). We contend that power is a phenomenon that is played out in everyday social interactions and therefore is not a distant construct. Michel Foucault (1975) argues that power is omnipresent in all human social endeavours. Artifacts such as language and social interactions reveal, create, reflect, obscure, and depoliticize power (Ng & Bradac, 1993). That is, these artifacts can show the extent to which power is exerted, used, negotiated, realized, abused, accepted, and challenged. As such, it is important to examine how these artifacts are used in doing science and especially how language is used in learning and teaching science. This examination allows us to create experiences for students to engage in and understand the language and culture of science so that they can move toward authentic membership in the scientific community. The chapter focuses on the following conceptual question: How do social contexts and discourses of two informal learning environments reveal the mediation of power among students?

Keywords: informal science learning, afterschool STEM, STEM discourse, high school robotics, mediation of power

 DOI:10.1163/9789004389663_010

INTRODUCTION

Michel Foucault (1975) argues that power is omnipresent in all human social endeavours. Thus, by extension, we say that power is present in everyday school and classroom experiences and social interactions. However, oftentimes in these experiences and social interactions the distribution and enactment of power is skewed or asymmetrical disfavouring the student to advantage teachers and all who represent the school as an institution. In this chapter we first briefly discuss the asymmetrical nature of power in the school in general, and more specifically in the science classroom to advance our position that informal science learning can offer experiences and social interactions where power is more evenly distributed among all participants. We then consider the role of context (e.g. formal vs. informal) and Discourses (Gee, 2003) in the way power and oppressive acts are manifested. Using power as a lens and based on our previous and ongoing work, the chapter subsequently presents an examination of social contexts and discourses of two informal learning environments. We conclude the chapter by reasserting that learning experiences need to be examined through the lens of power if the intent is for students to take control and ownership of their learning and to enable students to create communities practices where they can engage with adults, contexts, and time in ways that create interdimensional spaces that go beyond the traditional science classroom. Essentially, the chapter attempts to answer the following question: How do social contexts and discourses of two informal learning environments reveal the mediation of power among students?

Oppression and Schools

As we discuss the nature of power and its relationship to science learning, we preface our discussion by presenting a brief and somewhat critical view of schooling in the United States. We base our argument on the premise that schools in general are inherently oppressive and amplify the hegemonic realities of the larger society within which they are situated. Life conditions primarily arising from poverty and low socioeconomic status and ontological factors such as race and ethnicity contribute to these oppressive and hegemonic acts enacted toward students. The oppressive and hegemonic activities of schools are especially troublesome when they are directed toward underrepresented and underserved Latino/a and Black students. This is because, more often than not, the same students are also subjected to various forms of oppressive conditions outside the bounds of the school, at home and in their community. That is, for these students, there is no respite from the oppressive acts of various institutions and individuals around them. For example, lack of (denial of) adequate housing, safe and healthy neighbourhoods, lead-free drinking water, nourishing meals, parental and familial support (emotional, spiritual, physical), and materialistic things are oppressive conditions under which a large number of underrepresented minority students live their daily lives. Added to this are the oftentimes-oppressive conditions of the schools. Cesar Rossatto, Cecilia Rivas, Daniel Heiman, and Juanita Esparza (2015) have stated that students "are violated in schools by oppressive systems that practice authoritarian abuse, place needless academic restrictions on them, labels, tracks, and promotes a cultural infringement on others that fail to fit into their educational, and personal, ideologies" (p. 6) and work to domesticate students within the educational institution.

Anecdotally, as researchers and educators, we have seen oppressive practices in the schools being played out on a daily and routine basis from the moment the student walks into the school building. Dishearteningly, these practices often resemble routines observed in the U.S. prison systems and seem to prepare students for the prisons rather than for transforming the world in which they live (Wadhwa, 2015). For example, in the middle school that we work with, students enter the school building through a doorframe metal detector guarded by a school resource officer, a sworn law enforcement officer. Once in the school building students are subject to dress code inspection, reprimanded for gathering in groups, and moved along quickly out of the hallway and into the classroom. The talk and the interactions of adults during these events are demeaning and authoritarian in tone, largely without regard to the dignity of the student. Additionally, it is routine practice in the schools that we work in for students to walk in single-file on the right side of the hallway along the colored taped line on the floor, with one hand behind their backs and other hand's index finger on their lips. The teacher, who acts more like a prison warden, monitors the students' actions. These and other (restroom breaks, lunchroom routine) institutionalized practices that focus on control, management, and compliance strip students of their identity and agency and have a deeply demoralizing effect.

An important aspect of the oppressive practices in the school is the type of instruction that is carried out by the teachers and the degree of authentic learning that occurs. When teachers center their attention to controlling and managing their students over providing authentic learning experiences, students reciprocate by either complying or resisting. That is, students adapt (a form of learning) their actions and interactions within the school in response to the oppressive actions of the institution. Again, we stress what is learned by the student is more suitable for the prison pipeline than for the world beyond it. Thus, a general observation of the school as an institution compels us to think of schools as places where the oppressed continue to be oppressed and the oppressors act in a way to keep power and maintain the status quo and prevent the oppressed from challenging their power. This status quo in power differential is maintained by the powerful through the control and regulation of the school, where students are proselytized into being in the subordinate class and accepting the asymmetrical nature of power. Morton Deutsch (2005) states that there is a form of self-fulfilling prophecy in that the conduct and actions of the oppressed resulting from the oppressive conditions are used by the oppressors to defend and rationalize their oppressive actions. Additionally, as Paulo Freire (1970) points out, there exists a dichotomy of sorts in the way the oppressed view freedom; they desire it on one hand, but on the other, they fear it because of the internalization of the oppressive conditions. Thus, the oppressive conditions of the schools make it difficult for the students move from their subordinate class while they are at the school and during their adult life. This results in depressed social and intellectual mobility for underserved and underrepresented students.

Oppression in Science Classrooms

Earlier we alluded to the fact that instructional practices of the teacher can be oppressive to the students in that these practices may not result in students having social and intellectual mobility, but in fact may subjugate students to be subservient to the more powerful. This is primarily because teachers feel a need to control, manage, and garner compliance from the students using effective power to support their instructional objectives and efficacy. In a sense, we say that for teachers and schools as an institution, instructional efficacy and the degree of control a teacher has over the students are highly intertwined. Consequently, the prevailing practices of the classroom, including the science classroom, are disproportionately teacher driven, with little or no input from students. This presents a classroom that is asymmetrical in power and limits the ways in which students experience, gain, and use knowledge.

School science can mitigate or exacerbate the inherently oppressive and hegemonic reality of schooling. Formal science learning contexts (i.e., the science classroom) continue to be the primary site for learning opportunities and social activity for students. Students in schools frequently experience science as a list of facts or conclusions, often referred to as final form science (Roth & Lucas, 1997). More often students experience science as decontextualized list of facts or conclusions to be memorized and regurgitated for tests and exams. As such, students are rarely afforded opportunities to use dialogic language to collaborate, mutually construct knowledge, and generally engage in science as a social activity (Lemke, 1990; Warren & Rosebery, 1996). Rather, students memorize rote inconsequential facts, have limited engagement with science as a practice, and find school science to be dismissive of their life and career goals (Verma, Puvirajah, & Webb, 2015). Michael Ford and Ellice Forman (2006) argue that students in the science classroom should actively participate in productive disciplinary engagement where they co-construct, critique, and reconstruct scientific knowledge. Thus, in science learning experiences (including science classrooms) where power is symmetrical, students have opportunities to ask critical science questions that are more relevant to their lived experiences or participate in discourses that develop epistemic understanding and practices associated with science. These epistemic practices include "articulation and evaluation of knowledge, coordination of theory and evidence, making sense of patterns in data, and holding claims accountable to evidence" (Erduran, Ozdem, & Park, 2015, p. 3).

In addition, many science educators are bound by national, state, and/or district science curricula, which primarily aim to prepare students for the science pipeline

(Aikenhead, 2006). Research indicates that school science fails to engage the majority of science students, especially those who belong to underrepresented cultural and socioeconomic groups (Calabrese Barton, 1998). Further, there has been a major decline in student enrollment in science that has been attributed to students' disenchantment with science (Aikenhead, 2006; Fensham, 2004) and/or the conflict between students' cultural self-identities and school science (Brickhouse & Potter, 2001; Carlone, 2004). Glen Aikenhead (2006) reported that less than half of 7th graders in the United States and even fewer 11th graders thought science would be important, useful, or helpful in their adult lives. Research into students' views of traditional school science shows that students find school science to be useless, meaningless, "socially sterile, impersonal, frustrating, intellectually boring, and/or dismissive of [their] life worlds and career goals" (Aikenhead, 2006, p. 26).

The science education/research community is well aware of this predominant instructional situation as evidenced by the glut of research literature associated with teacher-centered instructional practices and lack of student engagement in science. While our past (Puvirajah, Verma, & Webb, 2012; Verma, Puvirajah, & Webb, 2015) and current work add to the chorus calling for more engaging spaces of learning for all students, we also emphasize examining power relations between adults (including teachers) and students through the lens of critical pedagogy. More specifically we feel, and as alluded to earlier, it is important for students to have science learning experiences that develop scientific practices, attitudes, knowledge, and perceptions (world views) that they can use in the short and long term to recognize and mitigate their own and others' oppressive conditions. We by no means have notions of advocating for an activist oriented science learning (although we see nothing wrong with it); but want teachers to recognize their practices as potentially being driven by control acts that attempt to exacerbate the oppressive conditions students (especially from underrepresented and underserved groups) find themselves in and outside of the school. When this recognition is made, it potentially opens doors for teachers and other adults to facilitate science-learning experiences that are less oppressive and allow students to use evidence-based reasoning to critically examine their (and others) life conditions. So, it must be understood that depending on the nature of the science learning experience, it can either work to dominate, indoctrinate, and oppress students toward domestication or work to liberate and develop their 'essence of consciousness' so they are able to critically reflect and act to transform themselves and their community (Freire, 1993).

The science education standards and reform documents such as the *Next Generation Science Standards* (NGSS Lead States, 2013) and the *Framework for K-12 Science Education* (Schweingruber, Keller, & Quinn, 2012) offer guidance for science teachers to mitigate their oppressive science teaching practices. This may enable teachers to move toward a more discourse oriented and authentic science teaching and learning experience where the teacher and the student authenticate each other's thinking. What we mean is that when the effective power between the teacher and student becomes more symmetrical, authentic student discourse begins to occur among all in the science classroom, and the deep chasm between the role of the teacher and the student begins to close. Thus in a classroom where there is little power differential between the teacher and the students, as Freire (2008) states "the teacher is no longer merely the one who teaches, but one who is himself taught in dialogue with the students, who in turn while being taught also teach" (p. 249). However, we know that even with guidance from reform documents, science teacher professional development, and in-service teacher graduate coursework that espouse a less oppressive science classroom, there are institutional, societal, and teacher barriers. Despite these barriers, we are also familiar with many efforts currently taking place to mitigate the oppressive nature of school science. Unfortunately, these efforts are rarely successful or sustainable for longer periods of time because of the institutionalized nature, culture, and expectations of schools.

Thus while we await a shift from a tradition of oppression to a tradition of authentic dialogue where power among students and the teacher is shared and symmetrical, we must also examine environments and contexts outside of formal schooling for opportunities for students to learn science through authentic discourses that foster equal sharing of power with all participants. For example, our research group is involved in examining the influence of informal learning contexts and discourses (primarily language) in the enactment and distribution of social power. Based on our work, we argue that informal science learning contexts minimize the hegemonic activities associated with formal schooling experience and offer students opportunities to draw on more sites of data gathering and knowledge production, and allow for a contemporary, collaborative, and transdisciplinary sciences that is socioculturally mediated (Falk & Dierking, 2010; Verma, Puvirajah, & Webb, 2015). Additionally, such informal science learning experiences extend school science to alternative spaces where there is increased likelihood for the learning to be "…self-motivated, voluntary, and guided by the learner's needs and interests" (Dierking, Falk, Rennie, Anderson, & Ellenbogen, 2003, p. 109). Hegemonic and oppressive nature of formal schooling are diminished in informal learning environments as students come to have a voice and are able to decide on their own the type and extent of their science experiences. This autonomy gained from participating in informal science learning makes their experiences authentic and meaningful.

Discourses, Power, and Context

Both an explicit and implicit manifestation in the science classroom is the exercise and exertion of power by all occupants. This manifestation of power occurs through various [D]iscourses (Foucault, 1975) and these discourses may be liberating, oppressive, or a combination of both (Candela, 1999). Conversely, it is also appropriate to say that inherent in the nature of human social interactions (i.e. [D]iscourses) is the presence of power. With respect to our work presented here, we broadly define power as human actions or activities of control through [D]iscourses. We also make use of James Gee's (2003) definition of [D]iscourse.

In our discussion, [D]iscourse means the situated or contextualized use of language (words and deeds). So in terms of [D]iscourse, contexts not only represent "the words, deeds, and things that surround our words or deeds, but also our purposes, values, and intended course of action and interaction" (Gee, 2003, p. 717). Discourses include situated social interactions through symbol systems and representations (written and spoken language, mannerisms, actions, designed spaces, creative works, body language etc.). In a sense, [D]iscourses communicate. For example, the presence of a uniformed on-duty police officer at a vehicle accident scene communicates something different than the same police officer at a middle school entrance next to a doorframe metal detector at the beginning of the school day. It can be understood in the above example that context and Discourse influence each other and collectively influence the nature of perceived and real power the police officer has over the situation and the involved individuals. Thus moving forward, one of our premises is that discourse and context reflexively influence one another and that together provide meaning to social interactions (which are steeped in power transactions). Discourses used in specific context can reveal how power is exerted, used, negotiated, realized, abused, accepted, and challenged. This is because power is not some vague construct, but is omnipresent (Foucault, 1975) in all human social interactions. Thus, integral to any examination of the manifestation of (social) power is the inclusion and account of discourses and contexts that reveal the nature of power. Our research team's focus on the interplay of discourse, context, and power is to examine the broader connectedness to issues associated with social justice, equality, equity, and democracy within science education.

Research indicates that science learning experiences that are of interest and of consequence to students (and their communities) enable students, including those from underrepresented minorities, to engage with school science. Therefore, researchers have become more interested in alternative spaces of learning to create relevant and inclusive opportunities for students. These alternative spaces may include informal contexts (e.g., museums, botanical gardens, zoos) and out-of-school contexts (e.g., gardening clubs, robotics clubs) (Dierking, Falk, Rennie, Anderson, & Ellenbogen, 2003; Puvirajah, Verma, & Webb, 2012). Informal as well as out-of-school contexts have been broadly defined as science programs and experiences guided by the intentional nature of student participation that occur outside the classroom (Falk & Dierking, 2002) and are "free choice, non-sequential, self-paced and voluntary" (Falk, Donovan, & Woods, 2001, p. 7).

Learning in informal contexts can take place in many environments (e.g., science museums, natural history parks, zoos, botanical gardens, industry, geological sites, & nature centers); whereas out-of-school learning can take place in non-curriculum based, learner-led, and learner-centered opportunities. This is supported by the notion that the cumulative nature of science learning is strongly socioculturally mediated and takes place in various learning contexts beyond the traditional science classroom. Avi Hofstein and Sherman Rosenfeld (1996) propose that informal and out-of-school learning experiences exist on a continuum, from compulsory to free-choice, and may include the following activities: (1) school-based field trips, (2) student projects, (3) community-based youth science programs, (4) casual visits to museums, (5) after school clubs; (6) summer camps; (7) supplementary and optional take-home projects; and (8) the press and electronic media. Some ways in which students participate voluntarily in these experiences is by taking part in out-of-school experiences, such as science clubs (including robotics clubs), summer camps, and supplementary and optional take-home projects that are less constrained by school bells and lesson times, and provide alternative learning spaces for students.

We have argued earlier that students in a largely adult-dominated and oppressive culture of schools and school-like places are limited in the way they experience and gain knowledge and have fewer opportunities to talk, discuss, and experience the practices of the scientific community and connect those to students' lives (Puvirajah, Verma, & Webb, 2012). Many studies indicate that students' experiences in formal environments could be bridged with out-of-school settings (e.g., robotics clubs, gardening clubs, take-home projects) to create inclusive and engaging spaces of learning (Contento, Koch, Lee, & Calabrese Barton, 2010; Mallaya, Mensah, Contento, Koch, & Calabrese Barton, 2012). In the next section, we share two out-of-school learning contexts to explicate the symmetry in power among students and adults. The two learning contexts we discuss are robotics club (and competition); and supplementary and optional take home projects (SOTHP) (Puvirajah, Verma, & Webb, 2012; Verma, Puvirajah, & Webb, 2015; Douglass, Verma, & Bourdeaux, under review).

Informal, Formal, and In-Between Spaces

We argue that power may be mitigated in these two contexts (robotics club and competition and SOTHP) due to the: (a) voluntarily nature of participation; (b) opportunities to participate in a collaborative community; and (c) participation in authentic science experiences to co-construct and critique mutual knowledge claims (Ford & Forman, 2006; Verma, Puvirajah, & Webb, 2015). The learning contexts presented opportunities for students to: (a) authentically represent the disciplinary practices of science while examining problems/questions that had consequences and held important meaning for students; and (b) co-construct and critique each other's knowledge claims. We will expound on these ideas as we discuss the two out-of-school learning contexts in detail.

Robotics club and competition. This project focused on a group of high-school students and their mentors (teacher and engineers). The high-school students were enrolled in two different suburban schools in the Southeast United States (US) and voluntarily joined the yearlong robotics club activities and participated in a regional robotics competition. The group (students and mentors) met in the evenings (after school) on a regular basis to build their robot for the competition. We share this particular example to demonstrate the mediation of power among project participants using context and language. The teacher mentors took the lead in organizing the students to accomplish various tasks associated with building a successful and functioning robot. These tasks included creating various teams such as (1) the design team – to produce the mechanical design; (2) the build team – to produce the computer interface for the robot; (3) the promotions team – to create the graphic design for doing the public relations activities; and (4) the logistics team – to coordinate the activities of other teams. During these organizational meetings, club members discussed ideas for building a robot, studied the rules proposed by the competition organizers, elected captains of each team, and proposed a production timeline for building the robot. The teacher and engineering mentors met with the team captains to discuss next steps and help appraise the progress of each group. The captains in turn worked closely with their team members to move the robotics club toward successfully building the robot.

Soon after, the club members engaged in the build phase, a 6-week period where the team had to design and build their robot. The teacher and engineering mentors worked closely with the students. Club members used an empty classroom to organize their robot building activities. Due to space limitation, members of the design team took turns to work on the robot. In addition, other obligations (e.g., homework, family obligations) influenced team members' participation in the robotics club activities. Therefore, members of various teams met for 2–3 hours for three-days a week. Since participation in the robotics club activities was voluntary, students who participated actively in the build phase were selected to participate in the regional robotics competition. This included student participants from the build, design, logistics, and promotions team. An important thing to note is the required logistics; team members were responsible for ordering the robot building materials, keeping the team accounts, paying bills, arranging for team's lodging and boarding (during the competition), transportation, and shipping the robot to the competition venue. This is important to note since the diversified roles enabled many students with different interests to participate in a community that supported the robotics club and competition activities.

During the 3-day competition, several new positions were created to operate the robot. These positions did not exist during the build-phase. Three of these positions included robot driver (a student to drive the robot), a coach (a student to assist the driver with supplemental functions such as lifting arms and drive train sensors), and safety captain (to ensure safety of team members in the pit area). The safety captain worked closely with the team members and mentors to help them comply with safety regulations (e.g., wear eye protection, secure electric cords, keeping working areas clear of obstructions).

Additional two positions that were created included scouts and pit crew. The role of scouts was extremely important as they went around the competition area and observed/studied other teams' robots and judged them using the rating sheet provided by the competition organizers. The team assigned six full time scouts during the competition as engaging in peer-review of other teams' robots was found to be very useful by the team members. The pit area was the place where they worked actively with the robot. The pit-crew comprised of various team members depending on what needed to be done with the robot. As an example, during the robot testing, the driver, coach, and the engineering mentor comprised the pit crew. When heavy lifting or other repairs were required, other members joined the pit-crew.

Our research focused on the 3-day robotics competition, as we wanted to delimit the research space to the context of the robotics competition. We examined students' linguistic and social interactions during the robotics competition and wanted to understand the interaction among students and adults in this out-of-school learning context. We proposed a framework called "Acts of Authentication" to make sense of students' experiences. We described Acts of Authentication as, "practices that legitimize an individual or a group's participation in an activity as recognized by the participants of the activity and/or by the learned observer" (Verma, Puvirajah, & Webb, 2015, p. 270). We draw upon three interrelated ideas and activities to conceptualize the notion of Acts of Authentication: (1) student talk (including everyday talk); (2) productive disciplinary engagement; and (3) community of practice (Verma, Puvirajah, & Webb, 2015).

We leverage this framework to explicate the power balance between adults and student participants in the robotics club and competition activities. We argue that this out-of-school context created a space for students to participate in a community of practice to create (and participate) in discursive language (including symbol systems). In fact, students' participation in institutionalized ways of communicating were tangential outcomes that emerged out of their interactions with the expert facilitators (engineering and teacher mentors) as well as their peers. Enacting the acts of authentication created a "third space" (Gutierrez, Rymes, & Larson, 1995, p. 445) where students' everyday talk intersected with productive disciplinary engagement (Ford & Forman, 2006) and blurred the boundaries between mentors' and students roles as all participants focused on creating a functioning robot. The element of uncertainty and solving an authentic problem contributed to students enacting the acts of authentication while balancing the power equation among students and mentors.

Supplemental and optional take home project (SOTHP). SOTHP emerged out of a professional development collaboration with a 5th grade science teacher and his students. The collaborative work took place in an urban school district in Southeast US. The school was classified as a Title I school (Title I funds provides financial assistance to local educational agencies [LEA]) and schools with high numbers or high percentages of children from low-income families to help ensure that all children meet challenging state academic standards). At the time of the study, the school had 90% Latino/a students while Asian, Caucasian, and African-American students comprised 10% of the student population. The Latino/a students came mostly from Mexico and other South American countries whereas the Asian students mostly came from Vietnam and Cambodia.

In this project, we share one such space that bridged formal science experiences with an out-of-school project to possibly create inclusive learning opportunities. In the collaborative professional development, we worked closely with Charlie, one of the 5th grade teachers, over a period of three years. During the first year, we collaborated on revising the curriculum unit on weather, electricity and magnetism, and earth processes. In the first year, most of our work took place in the classroom. During the second year, we collaborated on revising the curriculum unit on various biology topics (e.g., cell, classification) in addition to the topics covered in the first year. Since Charlie taught in a departmentalized setting, science instruction took place every day, four-class periods per day. This allowed us to create robust collaborative opportunities that included researchers' participation in class activities (e.g., co-teaching), bi-weekly meetings, and ongoing email communications.

Our close collaboration with Charlie's students revealed a deep interest from students to participate in supplemental and optional activities outside the classroom (related to the topics covered in the classroom). We proposed the idea of integrating out-of-school learning opportunities for students to supplement the work being done in the science classroom. Specifically, we proposed and designed optional and supplemental take home projects (SOTHP) after each unit to support students' learning and engagement. The two topics for SOTHP included cells (plants and animal) and classification (plants and animals). We will discuss the cell SOTHP in this paper (Douglass, Verma, & Bourdeaux, under review).

The participating teacher, Charlie is a Caucasian teacher and worked primarily with culturally and linguistically diverse students. He taught a total of 26 fifth graders out of which 21 were classified as ESOL by the school district. We collected additional data to give voice to students' self-identified linguistic and cultural identities. We found the students linguistic identities created a nuanced and a richer picture compared to the ones shared by the school district. As an example, 79% of the participating students identified themselves as non-native English speakers, with the modal first language identified as Spanish. Similarly, 59% of the students identified fluency in more than one language. We share this information to help us understand that there may be discrepancies between institutional and students' recognition of linguistic and cultural identities.

After students had finished studying the unit on the cell, we invited students to participate in the cell SOTHP. This entailed students creating an edible model of a plant or an animal cell. This was an optional, supplementary, and voluntary activity and students were not obligated to take part in it. We shared a few parameters and guidelines to create the edible cell (Douglass, Verma, & Bourdeaux, under review). We encouraged students to work with their families and caregivers to create a model of the edible cell. Students were encouraged to integrate their linguistic and cultural backgrounds by using foods they eat at home, working closely with their families, and creating a model of an edible cell. We recognized that this aspect of cell SOTHP was extremely important for the success of this voluntary activity. Many students' and their families' food choices and preferences are marginalized by the mainstream public schools due to various reasons. Some reasons include: these food are not being served in the cafeteria, the teachers may not be familiar with the food, and students do not have a place in the school or the science classroom to share such an important aspect of their backgrounds. The cell SOTHP created a third space (Gutierrez, Rymes, & Larsno, 1995, p. 445) where students' linguistic and cultural backgrounds intersected with the disciplinary knowledge. In addition, the students were able to work with their parents and caregivers and create a community of practice that privileged their funds of knowledge (Moll, Amanti, Neff, & Gonzalez, 2005).

We highlight the importance of the emergence of a third space to identify the power differential that occurs between students and the teacher. We make this argument because students have an opportunity to work closely their families and caregivers and the teacher in a science-related project that is voluntary and optional. The teacher's role in this project was relegated to that of a project facilitator. Therefore, this third space enables students to engage with science learning experiences that are of their choice and relevant and meaningful to them. This space is also inclusive of their funds of knowledge and ways of being.

This conversation becomes extremely important when we work with culturally and linguistically diverse students since the classroom may be oppressive and marginalizing for many students. Therefore, in this project we were interested in finding out about linguistic and cultural diversity of the students' families. We were not able to find this data from the school district; therefore we collected this information using a questionnaire. We found that only 21% students identified English as the primary language spoken at home. Thirty-three percent of the students shared that at least one parent spoke fluent English, and 35% identified that at least one parent can write English. In terms of cultural backgrounds, 60% percent of the students identified themselves as natives to the United States of America, with the modal non-native response being Mexico. However, only 20% of the students reported their parents as natives to the United States, with the modal country of origin being Mexico.

As mentioned earlier, we noticed a discrepancy between personal and institutional ways of documenting linguistic and cultural diversity. Recognizing and rectifying these discrepancies is important since they may have consequences for student learning. School districts may be cognizant to institutional notions (data) of diversity and may use these to organize learning opportunities. However, these learning opportunities may not align with the nuanced linguistic and cultural backgrounds of students. This is understandable (though not acceptable) since the budget and resources available to them limit schools and school districts. We know that the impediments for students learning are profound for underrepresented minorities. Student population is increasingly becoming diverse across the classrooms in the United States (and around the word). How do we then create learning opportunities to encompass the increasing linguistic and cultural diversity of students in the science classroom?

Since the SOTHP project was voluntary and optional, the unexpected 95% return rate became overwhelming for the teacher and researchers to manage the submissions.

As indicated earlier, the intent of this SOTHP was to create science-related project in this out-of-school contexts. Most students (with the help of their parents and caregivers) created the edible cell project and not all students had an opportunity to present their completed projects in class due to lack to time. We encouraged students to showcase their final product and do a gallery-walk during the lunch period. This led to a discussion about future SOTHP projects and students came to an agreement that presentations would take place in small groups and sample projects will be presented to the whole class. The high participation rate created an opportunity for us to explicitly connect school science with the SOTHP experience. We observed that students included various edible food items in their cells and students were able to articulate their choice of food items to represent cell-organelles in the cell model.

We discussed the cell SOTHP in this chapter and situate it in the context of out-of-school learning opportunities and conversations related to power. The SOTHP project reinforced to us that there is not only a need for students to learn science and connect it to their lives, but also have spaces where these boundaries can be fused. The learning experiences become meaningful and relevant to students if they are rooted in students' backgrounds and their communities while allowing students to engage with productive disciplinary engagement (Ford & Forman, 2006).

What Does it All Mean: Creating an Interdimensional Space

A few key elements emerge in this chapter. In the robotics club and competition, students chose to participate in the club (and competition activities) because of their inherent interest in the activity that led to their participation in the club's community of practice. This was also visible in the SOTHP project as students chose to complete the project as they were interested in supplementing their science classroom activities with SOTHP that integrated their funds of knowledge. It is important to note that students in both projects had different intentions (and backgrounds) to create and participate in an *interdimensional space*. This interdimensionality is created at the intersection of context, time, and contribution of various adults in these projects. In the robotics club and competition, the interdimensionality was created due to high school students' interests in supplementing their academic preparation with knowledge of building a robot – this project took place beyond bell-times and was facilitated by the engineering mentor who also became co-constructor and critiquer of knowledge claims. Similarly, in the SOTHP project, students participated in an interdimensional space due to their interest in integrating their funds of knowledge and cultural ways of being in a science-related project at home. The final SOTHP products were co-created and critiqued with the help of community of practice created by students with their families and caregivers. It is to be noted that both projects were voluntary and optional for students and therefore the creation of interdimensional space was fluid as it related to contexts, time, and participation of adults. In other words, the interdimensional spaces get created by students' and adults' conscious and concerted efforts and could exist in spaces that are in-between school and life beyond school. These interdimensional spaces could become more accessible, meaningful and relevant for students' equitable participation in science learning.

In both out-of-learning contexts, students taking control and ownership of their learning seem to be an important aspect to diffuse the power differential. The control and ownership happens at various levels such as: (1) posing and solving authentic questions/problems; (2) participating in a community of practice (including peers, expert facilitators, and families/caregivers); (3) enabling interdimensional space intersection of everyday and academic talk as well having access to expert facilitators (mentors and parents/caregivers).

We do recognize that not all out-of-school learning contexts may be of interests to students. Therefore, it is very important to continue to examine various out-of-school contexts to identity elements that gets students interested in science learning opportunities. These elements could include engaging students in authentic and productive disciplinary engagement, linking students learning experiences to community funds of knowledge, creating a community of practice beyond the science classroom and/or a combination of all.

Our argument about power and balancing power differential in science learning environments is an important one. Various reports and policy documents highlight the underachievement and underemployment of students belonging to various racial, ethnic, and socioeconomic groups. Various reform efforts situated in the science classroom examine students' learning experiences vis-a-vis conceptual understandings and other parameters guided by state/districts standards. While these reform efforts (and their outcomes) may somewhat help students bridge the gap in their academic achievement (or not) – the underperformance and underemployment of various student groups continue to persist. Examining these learning experiences through the lens of power may unfold contexts where students engage with adults, contexts, and time in ways that create interdimensional spaces that go beyond the traditional science classrooms. These places enable students to create communities of practices around topics of common interest/s and draw upon the expertise of facilitators. To further this conversation, we strongly argue that researchers examine the power relations embedded in all learning environments including informal learning and out-of-school learning contexts.

REFERENCES

Aikenhead, G. (2006). *Science education for everyday life.* New York, NY: Teachers College Press.

Brickhouse, N. W., & Potter, J. T. (2001). Young women's scientific identity formation in an urban context. *Journal of Research in Science Teaching, 38*, 965–980. Retrieved from https://doi.org/10.1002/tea.1041

Calabrese-Barton, A. (1998). *Feminist science education.* New York, NY: Teachers College Press.

Candela, A. (1999). Students' power in classroom discourse. *Linguistics and Education, 10*, 139–163.

Carlone, H. B. (2004). The cultural production of science in reform-based physics: Girls' access, participation, and resistance. *Journal of Research in Science Teaching, 41*, 392–414. Retrieved from https://doi.org/10.1002/tea.20006

Contento, I. R., Koch, P. A., Lee, H., & Calabrese Barton, A. (2010). Adolescents demonstrate improvement in obesity risk behaviors after completion of choice, control & change, a curriculum addressing personal agency and autonomous motivation. *Journal of the American Dietetic Association, 110*, 1830–1839. Retrieved from https://doi.org/10.1016/j.jada.2010.09.015

Deutsch, M. (2005). "Maintaining oppression." Beyond intractability. In G. Burgess & H. Burgess (Eds.), *Conflict information consortium*. Boulder, CO: University of Colorado. Retrieved from http://www.beyondintractability.org/essay/maintaining-oppression

Dierking, L. D., Falk, J. H., Rennie, L., Anderson, D., & Ellenbogen, K. (2003). Policy statement of the "informal science education" ad hoc committee. *Journal of Research in Science Teaching, 40*, 108–111. Retrieved from https://doi.org/10.1002/tea.10066

Douglass, H., Verma, G., & Bourdeaux, B. (under review).

Erduran, S., Ozdem, Y., & Park, J. Y. (2015). Research trends on argumentation in science education: A journal content analysis from 1998–2014. *International Journal of STEM Education, 2*(1), 1–12. Retrieved from https://doi.org/10.1186/s40594-015-0020-1

Falk, J. H., & Dierking, L. D. (2002). *Lessons without limit: How free-choice learning is transforming education*. Walnut Creek, CA: AltaMira Press.

Falk, J. H., & Dierking, L. D. (2010). The 95 percent solution school is not where most Americans learn most of their science. *American Scientist, 98*, 486–493.

Falk, J. H., Donovan, E., & Woods, R. (2001). *Free-choice science education: How we learn science outside of school*. New York, NY: Teachers College Press.

Fensham, P. J. (2004). *Defining an identity: The evolution of science education as a field of research* (Vol. 20). Berlin: Springer Science & Business Media.

Ford, M. J., & Forman, E. A. (2006). Redefining disciplinary learning in classroom contexts. *Review of Research in Education, 30*(1), 1–32. Retrieved from https://doi.org/10.3102/0091732x030001001

Foucault, M. (1972). *The archaeology of knowledge*. London: Routledge.

Foucault, M. (1975). *Discipline and punish: The birth of the prison*. New York, NY: Vintage Books.

Freire, P. (1970). *Pedagogy of the oppressed*. New York, NY: Herder and Herder.

Freire, P. (1993). *Pedagogy of the oppressed*. New York, NY: Continuum Books.

Freire, P. (2008). The banking concept of education. In D. Bartholomae & A. Petrosky (Eds.), *Ways of reading* (pp. 242–254). New York, NY: Bedford/St. Martin's.

Gee, J. P. (2003). Opportunity to learn: A language-based perspective on assessment. *Assessment in Education: Principles, Policy & Practice, 10*(1), 27–46. Retrieved from https://doi.org/10.1080/09695940301696

Gutierrez, K., Rymes, B., & Larson, J. (1995). Script, counterscript, and underlife in the classroom: James Brown versus Brown v. Board of Education. *Harvard Educational Review, 65*, 445–471. Retrieved from https://doi.org/10.17763/haer.65.3.r16146n25h4mh384

Hofstein, A., & Rosenfeld, S. (1996). Bridging the gap between formal and informal science learning. *Studies in Science Education, 28*, 87–112. Retrieved from https://doi.org/10.1080/03057269608560085

Lemke, J. L. (1990). *Talking science: Language, learning and values*. Norwood, NJ: Ablex Publishing.

Mallya, A., Mensah, F. M., Contento, I. R., Koch, P. A., & Calabrese Barton, A. (2012). Extending science beyond the classroom door: Learning from students' experiences with the choice, control and change (C3) curriculum. *Journal of Research in Science Teaching, 49*, 244–269. Retrieved from https://doi.org/10.1002/tea.21006

Moll, L., Amanti, C., Neff, D., & Gonzalez, N. (2005). *Funds of knowledge for teaching: Using a qualitative approach to connect homes and classrooms*. Mahwah, NJ: Lawrence Erlbaum Associates.

NGSS Lead States. (2013). *Next generation science standards: For states, by states*. Washington, DC: The National Academies Press.

Ng, S. H., & Bradac, J. J. (1993). *Power in language: Verbal communication and social influence*. Thousand Oaks, CA: Sage Publications.

Puvirajah, A., Verma, G., & Webb, P. (2012). Examining the mediation of power in a collaborative community: Engaging in informal science as authentic practice. *Cultural Studies of Science Education, 7*, 375–408. Retrieved from https://doi.org/10.1007/s11422-012-9394-2

Rossatto, C. A., Rivas, C. E., Heiman, D. B., & Esparza, J. (2015). "Troublemaking," "making trouble," and "making it" through institutionalized schooling: Critical pedagogy as a transformational exodus.

Critical Education, 6(5), 1–33. Retrieved from http://ices.library.ubc.ca/index.php/criticaled/article/view/184884

Roth, W.-M., & Lucas, K. B. (1997). From "truth" to "invented reality": A discourse analysis of high school physics students' talk about scientific knowledge. *Journal of Research in Science Teaching, 34*, 145–179. Retrieved from https://doi.org/10.1002/(sici)1098-2736(199702)34:2%3C145::aid-tea4%3E3.0.co;2-t

Schweingruber, H., Keller, T., & Quinn, H. (Eds.). (2012). *A framework for K-12 science education: Practices, crosscutting concepts, and core ideas*. Washington, DC: National Academies Press.

Verma, G., Puvirajah, A., & Webb, H. (2015). Enacting acts of authentication in a robotics competition: An interpretivist study. *Journal of Research in Science Teaching, 52*, 268–295. Retrieved from https://doi.org/10.1002/tea.21195

Wadhwa, A. (2015). *Restorative justice in urban schools: Disrupting the school-to-prison pipeline*. New York, NY: Routledge.

Warren, B., & Rosebery, A. S. (1996). "This question is just too, too easy!": Students' perspectives on accountability in science. In L. Schauble & R. Glaser (Eds.), *Innovations in learning new environments for education* (pp. 97–125). Mahwah, NJ: Lawrence Erlbaum Associates.

ABOUT THE AUTHORS

Geeta Verma is an associate professor of STEM education at the University of Colorado Denver's School of Education & Human. Described as a critical, intercultural, and international scholar, Geeta focuses on creating equitable, imaginative, and innovative Science, Technology, Engineering, and Mathematics (STEM) formal and informal learning environments.

Anton Puvirajah is an Assistant Professor of STEM education at the University of Western Ontario in Canada. He has an active and well-developed research and outreach programs in teacher education and STEM learning that align with his interest in examining how teachers and science learners use Discourses to explore, negotiate, and develop their respective roles and identities within a community of practice. Specifically, within the STEM learning context, often using a critical lens, he studies Discourses that are enhanced through the use of technological tools and occurring in informal settings.

Helen Douglass is an assistant professor of STEM Education at The University of Tulsa's Harvey Kendall College of Arts and Sciences. Her research includes examining contextual critical junctures women identify related to their STEM journeys via visual images, and how to make k-12 STEM teaching and learning more inclusive. Her current projects include creating and using maker spaces in both formal and informal environments to increase equitable and inclusive teaching and learning practices for students and teachers.

MICHAEL TAN

11. INNOVATION TO WHAT END? MAKERSPACES AS SITES FOR SCIENCE EDUCATION

ABSTRACT

Makerspaces have risen in prominence in the recent five years as sites for the acquisition of Science, Technology, Engineering, and Mathematics (STEM) competencies. Often spoken in the same breath is the supposed ability of STEM and associated practices to 'change the world', in deliberately disruptive ways, and to euphemistically 'drive the economy' while at it. In this chapter, I argue for a closer examination of the teleological aims of science education, specifically in the light of recurrent movements since at least the 1960s to closely tie education to specific societal goals. Such approaches rely heavily on technocratic rationalities which fundamentally misunderstand the nature of innovation, learning, and human (and non-human) agency. I use considerations of creative problem solving in design, the nature of abductive reasoning practices in science, and Andrew Pickering's *Mangle of Practice* as lenses to develop a curriculum argument for science education generally speaking, and in makerspace activity in particular. In essence, I advocate a transdisciplinary approach to instruction, and a greater embrace of the literal and metaphorical messiness of learning and becoming.

Keywords: makerspaces, creativity, abductive reasoning, tacit knowledge, mangle of practice, STEM

SCIENCE IS NOT ENOUGH

In recent years, talk about the maker movement as a form of disruptive answer to the question of how schools ought to educate have emanated from many quarters within the academy and in the mass media, especially those outlets associated with technological innovation. Making and makerspaces have been proposed as solutions to perceived gaps in preparing students for contemporary challenges, be it as participating in economies as entrepreneurs, engineers, or as active agents possessing what have been widely touted as twenty first century competencies. While it is important to recognise how some of the affordances of contemporary technologies can enable some of the more progressive educational goals previously deemed inaccessible, it is likewise important to not go overboard into a form of techno-utopianism that supposes education to be a solvable problem. In this chapter,

DOI:10.1163/9789004389663_011

I intend to provoke a curriculum argument into the aims of science education, specifically through the pedagogical technology of making in contemporary makerspaces. In essence, I advocate a position that may be construed as being conservative – that the essential problem of *education* has hardly changed even though its particular formulation has shifted with the times. This is not to claim an easy rhetorical victory, but rather to serve as a reminder of the hard work that remains to be done. As educators, we need, on the one hand, to be more self-critical of the ability of science to bring us the better life, and to take a position that avoids a triumphalism that can be associated with the technocratic perspective. On the other hand, the cynical view needs a thorough disavowal, especially for citizens who find themselves increasingly locked-in to technological forms of control and modes of living. Science education needs to work between these tensions, to prepare students who understand the ontological and epistemological bases of science, who understand science's place in providing truth claims of a particular form, and who, importantly, understand the interplay of the role of science and the humanities in providing answers for the significant problems that we encounter (see, e.g., Duschl & Grandy, 2012; Waddington & Feinstein, 2016; Zeidler, Sadler, Simmons, & Howes, 2005). In this regard, the activity of making, in the context of makerspaces, present much potential for educators to work through these issues. As it is, however, popular depictions and implementations of making activity, especially in public makerspaces where participation times are often limited, have it that these activities are merely means to 'engage' learners in learning STEM concepts, with the complicated ideas implicit at best. Such a position ought to be seen as problematic: to be sure, few people can reject the notion that we ought to celebrate the cultural achievement that is our understanding and partial, limited control of the natural world. Yet, not dealing with these issues presents an overly one-sided view of science that is intolerable.

THE MAKER MOVEMENT

It is not hard to understand the excitement generated around the so-called maker movement. Especially in the United States, where large commercial organisations have taken it upon themselves to promote versions of making compatible to its financial interests, popular sentiment about the place of making in STEM learning have reached such widespread status that even the office of the President has given its explicit support (Obama White House, 2016). As imagined, the maker movement has arisen in large part due to the recent ubiquity of particular technologies. Specifically, computing technologies have become such commodity items that it has become feasible in some cases to deploy nearly disposable computers commonly. Manufacturing technologies have concomitantly kept pace, with lower cost computer driven machines – digital fabrication devices – seemingly lowering thresholds for high quality making. The influence of the global internet and the era of self-publishing to wide audiences have worked together with open standards to further lower thresholds to self-directed learning and participating in virtual (and arguably, real) communities of practice. As imagined, flexibly programmable electronics and computing machinery can be deployed according to, design schematics and programming code freely shared and modified by users all over the world. With stories of 'kick started' (internet campaign funded) hardware designs upsetting the traditional funding models and emulating the wild successes of the internet boom around the turn of the century, it is perhaps inevitable that talk of an 'invention gap' to mirror that of the 'digital gap' is beginning to take place in some circles (and concern for its correlation with family SES) (Bell *et al.*, 2016). Makerspaces are perceived as solutions to preparing students for participation in future economies, where all routinisable jobs will be undertaken by robots and algorithms, and the only productive jobs left will be those in the creative industries, telling the machines what to do, while an underclass will be told by computers what to do.

On the other hand, motivated by such ideas as helping people 'scratch their own itch' (Banzi, 2011), the maker movement can also be seen as a reaction to manufacturers' closing off of technologies to users and increasingly deploying planned obsolescence as a means to increase their profits. With the advent of robotic assembly and manufacturing, especially for highly integrated electronics products, end users are increasingly being confronted by boxes full of what Issac Asimov would have termed as 'magic' – sufficiently developed technologies that non-experts have no way of understanding its function, let alone modify, repair, or repurpose devices for their own needs. Another part of the maker movement then, is a justifiable response to this perceived problem – that of mitigating the excesses of human consumption and its concomitant effects on the natural environment. Besides the environmental perspective, understanding one's tools and machines can be considered as a form of personal empowerment, in that the practice of learning to uncover the magic of technology allows us control and agency over the ends to which technologies may be applied to. Indeed, many of the roots of the current 'DIY movement' can be traced to publications like the *Whole Earth Catalog*, born in the countercultural 1960s, at a time where apocalyptic visions of nuclear assisted mutual assured destruction suggested to people that it was important to learn how to be self-sufficient when support structures do not exist anymore. In some ways, such a situation actively exists in places like Zimbabwe (Kabayadondo, 2015) and rural parts of India (Birtchnell, 2011; Singh, Gupta, & Mondal, 2012), where local values of *kikuya-kuya* and *jugaad* have arisen to repurpose, refashion, and extend the use of technologies imported into the country without a supporting infrastructure to maintain their serviceability. While much of these efforts have successfully dealt with mechanical engineering challenges, it remains to be seen how effective these approaches can be in the microelectronic realm as more products become complex "all or nothing" devices. These 'rural' efforts of repurposing technologies are not to be underestimated; they share a lot in common with the more cosmopolitan hacker ethic (Levy, 2010), not that of nefarious computer access and activity, but the original sense associated with the 'nerds' at MIT's Tech Model Railroad Club, and the original innovators that brought us the computing revolution (Isaacson, 2014).

There appear at least two teleological visions for supporting making as a curriculum proposal for mainstream schooling: innovation, from an economistic viewpoint of supporting 'high value job creation' (e.g., Obama White House, 2016); and self-reliance and personal ownership in demystifying technologies used daily (e.g., Blikstein, 2013). While these visions do not necessarily contradict one another, it is my purpose to problematise the former, and to support the latter more strongly. In essence, the argument is fairly simple: the central role for education has been, and should continue to be, the expansion of the realm of the possible for our learners. By prescribing a goal ("Teach science for X") we are placing bets on the future state of the world, and the history of our collective abilities to foresee trends, especially for rapidly changing fields of technological development and their applications, have demonstrated that we are not particularly prescient. Instead, we should teach science so that students understand the intrinsically creative and innovative nature of the scientific venture, and provide more opportunities for deliberate transdisciplinary interchange of ideas as the basis for seeing and using interconnections between science and other ways of knowing. In this regard, my modest proposal for science education is for educators to more strongly position science in relation to other forms of knowledge, not in any way to lower its epistemological status to become 'but one of several equivalent ways of knowing', but as a means to become aware of the terrain of the disciplinary boundaries separating the natural, social, and *artificial* sciences (for the sciences of the artificial, see Simon, 1996/1969), and how these boundaries can be productively used in deliberate boundary crossings to generate new knowledge and understanding.

Significantly, because educators have had difficulty communicating the practices of science, much of the tacit knowledge of actually *doing* science is lost, and instead, a major focus of science education has been on the rather synthetic explicit knowledge. In planning scientific investigations, we often design these experiences to ensure an effective communication of scientific principles; educators are not as well equipped to deal with circumstances where the 'real world' refuses to cooperate. Yet, it is precisely in these episodes where we have to deal with what Andrew Pickering (Pickering, 1995, 2008) refers to as a non-human agency that we get closer to insightful knowledge of science. Rather than the established demonstrations and lab activities (of oscillating pendulums, standardised titrations, microscope work), students should get experiences dealing with things that may not work exactly as they thought they would, struggle through the mangle of practice (*ibid.*) and emerge at the other end understanding something of science that can be beyond the conventional representations of our language and cultural practices. Making and makerspaces have the *potential* to give students different experiences of science, to educate them in the practices of science, and, importantly, to provide them with the tools to change the world. My support derives, broadly, from (i) the contrast between the nature of scientific knowledge, and the nature of scientific practice; (ii) the role of tacit knowledges and embodied learning process in science; and (iii) the nature of problems that are worthy of our collective attention and problem solving will. I expand on these in the following sections.

THE INNOVATIVE NATURE OF SCIENCE

While science stands as an impressive body of knowledge representing the grand cultural achievement of all humanity, there is a temptation to treat it, metaphorically, as a grand edifice much like a classical museum containing obscure secrets that the hoi polloi need to be in reverence towards. In this regard, it becomes possible and very tempting, to treat science instruction as an activity of initiation into a sacred order to maintain and expand the museum's collection. In some way, this perspective is not wrong; after all, there is much knowledge that can be represented in linguistic and other culturally agreed upon means, and of which we rightly ought to consider an intellectual responsibility to transmit to future generations. It is this perspective that has brought sociologists of education such as Michael Young (2008) to argue for the significance of powerful knowledges, as opposed to knowledges of the powerful. In ongoing curriculum debates about the nature of knowledge, powerful knowledges refer to the kinds of knowledge, the possession of which enable users a more central participation in social deliberations. Such knowledge transcends their sites of generation, and can connect individuals across space and time, as when classical mathematical problems become solved in contemporary times, and then applied to cryptographic problems that could not have been foreseen. On the other hand, knowledge of the powerful refers to knowledge claims that are made on the imprimatur of the powerful. Although it had been a great Enlightenment advance to have been able to become justifiably sceptical of truth claims based on whether "the Chief, the Party, or the Pope says so" (Moore & Muller, 1999, p. 193), it appears that the 'post-truth' times that we find ourselves in may have made some reversals on these gains. Possibly contributory to this state of affairs has to be a form of poststructuralist thought. In the academy, the emphasis on analysis of discourse has seen, in science studies, sociological examination into the processes by which scientific knowledge is generated. While illuminating, Maxim Waldstein (2008) points out that an emphasis on linguistic analysis has left out an analysis of the performativity of scientific practice: "The performative idiom, on the contrary [to the linguistic], embeds these textual, semiotic, or representational chains in observable, palpable practices of coping with material agency, constructing machines, making experiments, and engaging in other risky enterprises "in the thick of things." In short, the performative idiom portrays science as something that "does things" rather than simply represents or stands for them" (p. 226).

Because of a curriculum focus that privileges knowledge as textual inscriptions, as concepts that can be communicated through structured experiences, it should not be surprising that entire forms of knowledge related to practicing science can be completely neglected from typical learning experiences in science. To be careful, this is not to say that conventional science curricula are made up completely of propositional knowledge claims. Rather, after Michael Polanyi, a vital difference lies in the distinction between knowing that, and knowing how (2009). It is important that students acquire knowledge of the form that may be reportable ultimately as quantitative differences in international comparative assessments; nonetheless, as Polanyi (*ibid.*) points out, while it can be quite easy for individuals to acquire knowledge about science, gaining the competence to generate new knowledge can be a completely different challenge. Key to understanding this challenge is the process of conceiving a new idea – of creative discovery – is fundamentally different from its logical examination (Walton, 2005). While the latter relies on logic, and can be expressed or represented in terms that allow for interpersonal interrogation of truth claims, the former is highly personal, idiosyncratic, and psychological to the point of being ineffable. How is it that August Kekulé's famed discovery of the structure of benzene could be based upon a dream of the Ouroboros, or that a falling apple may inspire Newton to consider its implications to gravitation? Such leaps of innovation, while somewhat understandable after the fact, do not appear to follow any patterned logic, and certainly does not seem to be something that can be foreseen. As educators, the challenge becomes especially acute: taking as a given that any future task that requires routinisable skill will be automated, surely there is something approaching a moral responsibility for educators to prepare learners for this and other kinds of uniquely human skills?

Perhaps the question ought to be asked: what is known about the nature of innovation in science? How does scientific knowledge progress; not at the sociological, scientific-revolution level, but at the level of the individual? Or, perhaps in more accessible terms, what is the significance of scientific practice on scientific knowledge? For starters, it appears that much of the progress in scientific understanding progresses not through the stereotypical image of a gradual accretion of evidence in an inductive manner, but rather through a process that is better described as having an abductive nature. While inductive reasoning is a process where general principles are produced from particular observations, and while deductive reasoning can be thought of as the 'opposite' process where details of a particular are inferred from general principles, abductive reasoning has been referred to as a process of inference to the best explanation. For instance, given an instance (a white object), and knowledge of a general rule (magnesium carbonate is white), one could plausibly conclude that the object could be magnesium carbonate. This form of reasoning, of course, has to be recognised as very tentative, and of course, tests need to be done to rule out the possibility that the selected hypothesis is wrong; but the strategy is nonetheless central to any form of innovation, and practical reasoning. Abductive reasoning is in use when teachers sum up the classroom situation and decide on the optimal deployment of instructional strategy among the routines that she is familiar with; it is in use when doctors are presented a symptom, and they have to make a diagnosis, with or without further tests to rule out complications; abductive reasoning is also in use when designers decide that the optimal solution for what they perceive to be a client's problem should be of a certain form.

In science, for instance, Albert Einstein's famous thought experiments were certainly not in any way accumulation of discrepant phenomena (near light speed motion); rather, these were instances in which Einstein made inspired suppositions, and worked through their logical implications to derive new knowledge. Similarly, discovery of the planet Neptune was made based on first observing anomalous motion of Uranus, and then speculating that a planet ought to exist in Neptune's orbit, before the planet was actually observed. In more recent times, James Watson and Francis Crick's method for elucidation of the double helix structure of DNA triumphed over other competitors, most notably Rosalind Franklin. While Franklin worked methodically from primary data derived from x-ray crystallography, Watson and Crick developed physical models, again as with the abductive method, almost 'jumping to conclusions' and then testing models for fit, dismissing mistakes, and reiterating until success. As Anton Lawson (2010) describes it, this approach to scientific argumentation is distinct from the established approach of the academy; I would add that its popularity is certainly not on the same scale. A recent search on Google Scholar reveals about 16,800 results for 'science education' and 'argumentation', whereas 'science education' and 'abductive'/'abduction' only garnered 1960/1740 results.

Contributing to the unpopularity of the abductive reasoning approach has to be the relative difficulty of communicating these ideas; in schools infatuated with a 'meritocratic' desire to rank sort their students, abductive reasoning does not offer neat and tidy, 'right' and 'wrong' answers. Because there are no explicit rules as to how to assess an abductive conclusion to be better than another, teachers are faced with the challenge of trying to express tacit rules by metaphors or analogy, neither of which are completely adequate for the task. This is not to say that we are completely bereft of methods; Lawson (2010) points to the Predict-Observe-Explain protocol from the early 1990s (White & Gunstone, 1992) as an effective method, if instructors can avoid certain pitfalls. Notably, instructors were often unaware of the distinction between hypotheses (candidate explanations) and predictions, often conflating the two (Oehrtman & Lawson, 2008, in Lawson, 2010). According to Lawson, the only other significant development in teaching students how to reason scientifically (i.e. abductively) has been a framework proposed by Katherine McNeill and Joseph Krajcik (2007), a difference of fifteen years. With such little research attention focussed on this form of reasoning, it is probably safe to adduce that the general state of students' abilities should be fairly under-developed.

Making activities in makerspaces, if appropriately conceived as creative problem-solving activities, can be of benefit to science education. Makerspaces can focus on design as a method to speculatively identify and solve problems. Here, it is useful to note that the design method bears amazing resemblance to the abductive approach used by scientists as they confront discrepant evidence and build theories and predictions to either subsume phenomena into the current explanatory framework, or (re)build another theoretical edifice to house the discrepancy. For instance, just as Watson and Crick supposed that the DNA structure *ought to be* a double helix, because genetic reproduction requires a pair of parents, designers make use of general design principles such as humans desire expansive views to suppose that a living room *ought to be* filled with large windows. Of course, the challenge persists for educators – how could this principle be used to communicate the nature of abductive reasoning? Given the likely context sensitivity of abduction, the problem of transfer across contexts becomes an issue that has yet to be studied. More importantly, it is clear that the role of instructors in these learning processes needs to be one of skilful craftsperson, rather than an industrialised production line operator; given the increasing pressures internationally for school systems to measure up against rationalised international comparisons such as TIMSS or PISA (Labaree, 2010), there is concern that educators may experience increasing pressures to conform to standardised processes, and pass the normalising tendencies on to their students.

TACIT KNOWLEDGE

Compounding the effect of a general lack of research attention and instructional methods, the nature of this form of reasoning makes its skilful acquisition hard. The rules for abductive reasoning are based on practice and are difficult to explicate. While there may be some general principles, exceptions abound, and these exceptions make all the difference between a successful abductive conclusion or otherwise. Taking a familiar example, if, in laboratory investigations designed to verify the rate of acceleration due to gravity, a student produced a result that was in significant deviation from the 'standard' 9.81 ms^{-2}, how should she conclude? That her experiment was in error (and what could be the cause of it), or could it be that the lab was located directly above a large oil deposit? What rules determine the selection of an appropriate conclusion in this case? As above, the decision-making rules are tacit, or, as succinctly put by Polanyi (2009/1966), "we can know more than we can tell." While glib, such statements still leave us without some empirical basis for thinking about the nature of these forms of knowledge and how they can be developed in learners. For this, researchers have turned to theories of embodied cognition, the essential premise of which is that the mind consists of more than just the brain.

As argued by philosophers such as Daniel Dennett (2009) and Andy Clark (2008), the classical problem of mind-body dualism, and the intolerable nature of arguments for the unnatural nature of 'mind stuff', constituting consciousness and our apparently superior cognitive abilities have been progressively resolved with the recognition that apparently fantastical feats of computation do not require fantastical mechanisms for its realisation. The computing revolution, originating from devices such as Charles Babbage's difference engine, showed that appropriately deployed mechanisms could be used to make complex logical operations; the current computing devices we use are only different in scale and speed, but not in principle. While we have been able to exorcise Descartes' ghost, Clark argues that there should be no reason that we should privilege the brain as the sole locus of cognition; just as the heart remains a vital organ of the circulatory system but is useless without the blood vessels, the brain is a central organ of cognition, but requires the rest of the body, and the external environment as elements of the cognitive system.

These insights have led to a reinterpretation of how we learn; for instance, one interpretation of the embodied cognition hypothesis considers abstractions to be metaphorical extensions grounded in bodily experiences which are culturally common. For instance, George Lakoff and Rafael Núñez (2000) propose that, in combination with some basic innate human abilities, even sophisticated and apparently abstract mathematical relationships such as imaginary numbers can ultimately be traced back to familiar bodily experiences. If this is so, it would stand to reason that educators can engineer particular bodily experiences to aid comprehension and retention of particular abstract concepts. Indeed, such findings have demonstrated, for example: the benefit of teacher gestures in elementary grades mathematics instruction (Novack *et al.*, 2014); the role of a computer assisted feedback device to improve intuitive grasp of ratio relationships (Abrahamson & Lindgren, 2014); a computer vision based system to monitor a class full of children in movement to simulate molecular motion in different phases of matter (Lindgren & Johnson-Glenberg, 2013), and others.

The foregoing has to be seen in the context of the much wider problem of the epistemological status of knowing-that versus knowing-how. The essential problematic is between the prior status of propositional knowing-that, or of the experiential knowing-how. To be fair, as with the resolution of Meno's paradox, it is possible for theoreticians to create false dichotomies; we can have intermediate amounts of knowledge; these forms of knowledge are not all-or-nothing but can mutually inform each other. However, for instructional design, more principled direction is desired. For this, at least a recent review and position paper by Mitchell Nathan (2012) has been motivated to support a position contrary to the formalism-first perspective. As described, the formalism-first perspective still holds great influence, at least as informal wisdom regarding instructional sequencing. Significant to matters of curriculum and instructional sequencing, is the conflation of the disciplinary structure and developmental trajectory suggested for learners:

> The FF [formalism first] view uses the disciplinary structure as its developmental roadmap: What is foundational to the discipline is also deemed developmentally primary; what constitutes secondary and peripheral topics to the field then follow in the learning experience; and applications of disciplinary knowledge to practical problems comes last in the scientific process, and therefore are expected to occur later developmentally. (p. 135)

Especially for a hierarchically structured form of knowledge as that of the natural sciences (Bernstein, 2000), it is easy to assume a standard sequence of instruction to develop through a combination of inductive generalisation from physical phenomena to theoretical representations, or (as above) deductive application of theoretical representations to contexts apparently increasing in its similarity to the original context where knowledge was acquired. Similar to the arguments presented here, however, Nathan also argues for a progressive-formalisation perspective with an emphasis on the primacy of ground up experience of physical phenomena.

To be sure, curriculum and pedagogical arguments have been made from somewhat more conservative positions too, even considering the principles of tacit knowing-how: Christopher Winch (2013) proposes that the distinction between propositional and tacit knowledges to not be as clear as the perhaps stereotypical exemplar (after Polanyi) of knowledge of bicycle riding versus ability to do so. Winch argues that "an examination of propositional knowledge yields the requirement that knowledge of individual propositions implies some understanding of the concepts that such propositions express, and that this consideration implies that knowledge requires an indeterminate but nevertheless significant ability to understand and to make inferences within the subject. It implies, therefore, a degree of KH [knowing how] even among novices" (p. 135). Winch argues that the kinds of knowledge needed for students is distinct from the forms required by practitioners at the forefront of knowledge production; there exists forms of knowledge which, in some cases, are virtually impossible to inductively generate over particular experiences – for these, he advocates pedagogical forms such as direct instruction. In similar fashion, Elizabeth Rata (2016) argues for the transcendental nature of particular forms of knowledge, and the tension that exists between "the concreteness of experience and the abstractness of thought" (p. 169). Rata suggests that conceptual knowledge may be acquired without necessarily having to pass through the requisite experiences, and that such knowledge be used to guide the development of students' conceptual understanding, much as a means to direct student noticing of particular pertinent aspects of reality.

It is most likely the case that no firm resolution of the experiences versus formalisms debate will be within reach of investigators. It is certainly probable that there are some formalisms that cannot be readily programmed into a series of practical learning experiences. On the other hand, as instructional designers, it is important that we find as many ways as possible to recruit the social, affective and bodily-cognitive systems to maximise student learning of these concepts, especially if these are not straightforward. We ought not to forget that as intelligent species, we have merely co-opted and jury-rigged various bodily, social, and cognitive systems in order to conceive and process the abstractions that we now take for granted. Evolutionarily speaking, our brains and innate biological structures have not had sufficient time to develop adaptations specific to understanding such unnatural concepts as frictionless surfaces, platonic objects, electronic orbitals or gene editing. Kim Sterelny's (2003) thesis suggests that, as with the co-evolution of species such as otters and their dams, humans have been able to co-opt and extend our human artefacts to provide for evolutionary advantages over time. It should be perhaps not surprising how much of conceptual change literature of the past century dealt with the distinction between innate models of the workings of reality (diSessa, 1993; Pinker, 1997), and more accurate scientific understandings.

In science education, these ideas should not be too foreign; after all, much of the advancement in contemporary sciences derives from how we have made use of sensory amplification to derive more accurate measurements of phenomena. While experts are able to trust such devices and translate readings into theoretical interpretations, the ability of novices to convert device readings into abstractions requires the development of a certain sense of professional vision (Goodwin, 1994) that is complicated in proportion to the complexity of the device. With increasing miniaturisation, it could be argued that these black boxes are increasing the distance between the phenomena and its theoretical interpretation. While schools do not often have overly sophisticated laboratory equipment, computer supported simulations perhaps need slightly more critical scrutiny in the manner in which they may close off the inner workings of the simulation and subtly coerce students into orthodox, received views of science. While it may be argued that such simulations allow for the examination of phenomena at physically impossible scales of time and space, one wonders at the costs associated with losing touch, metaphorically and literally, with a physical sense of the world in which we live. In this regard, a possible makerspace goal of learning to take things apart, to reverse engineer and understand complexity, may be used as a means to educate students in ways to trust their instruments and to be familiar in just what these mechanisms do. With expanded ranges of bodily based experiences to create metaphorical extensions of, increasing exposure to physical experiences in makerspaces may in fact stimulate metaphorical insights into theoretical problems. As recounted by (Miller, 1981, in Nathan, 2012), Einstein's insights into relativity were likely a result of his more down-to-earth experiences as a patent examiner considering the problem of time synchronisation in railway systems, a recurrent problem of his time. While it may be true that not every student may have the intellect to extend such experiences in the manner that Einstein did, it may nonetheless still be appropriate here to advocate increased opportunities for learners to extend bodily experiences in theoretically informed ways.

MANGLE OF PRACTICE

If we become concerned with the ways in which the bodily senses are amplified by machines and the ways in which theoretical advances (public or private) may be derived in empirical investigations, it is pertinent to consider what Pickering (1995) has termed the "Mangle of Practice". Developed in consideration of his research into high energy physics, the essential problematic is that of the role played by non-human agents in the practice of science. Instead of thinking of objects as passive recipients and channels of human agency, and focussing as with the sociologists of scientific knowledge have done on human activities and discourse, Pickering suggested that non-human artefacts and materials can possess particular agentic abilities. This is not to ascribe conscious agency to non-human objects, but to consider that these objects do not necessarily behave as we believe (or design) that they will. This is also not to suppose that there is a non-determinism to the behaviour of objects; our natural sciences most certainly give us a good control and predictive power over the natural world, but there exist limits to our knowledge. Our science works in limited, special conditions which often turn out to work only as approximations to the universe in which we exist. In a rather dramatic example, despite our extensive knowledge of hydrology and geology, no amount of modelling, virtual or physical, could account for and control the movement of the Mississippi river as it winds its way through its drainage basin towards the Gulf of Mexico (Pickering, 2008). The system of levees and relief drainage systems designed to keep New Orleans safe from the river have not been particularly successful:

> The Army Corps of Engineers actually speaks of "stopp[ing] time in terms of the distribution of flows". But this project has always itself been embedded in a decentred and open-ended becoming of the human and the non-human, a "dance of agency," as I would call it, between the engineers and the river. The human agents, the engineers, try something – raising the levees, say – and the nonhuman agent takes its turn by rising still higher and flooding New Orleans. In response the humans do something else – building the weir between the Mississippi and the Atchafalaya – to which the river does something else – ripping and tearing away at it. And so on, forever […] The Army Corps of Engineers have not acted blindly in its struggles with the river. Plenty of *science* has entered its strategies. The Corps of Engineers would like to get out of the thick of things by finding the timeless hidden essence of the river and hence dominate it, first conceptually and then materially […] but […] it failed to work […] The moral of this part of the story is twofold. First, *science itself is caught up in the flow of becoming*; the contents of science change emergently. And, second […] scientific knowledge helps to conceal becoming from us […] *science itself thus appears as a veil*, clouding our perception of how things actually are. As Heidegger put it, science is at best in the domain of the "correct" rather than the "true." (pp. 7–8, emphases in original)

The fundamental human desire for control over nature has resulted in the development of the kind of natural sciences that we come to take for granted, and cause us to suppose that as educators, the power that these knowledges afford its possessors is a significant reason why students ought to learn science. Yet, the perspective of the mangle informs us that, while these theories 'work', the interface between theory and practice is decidedly not clean and consistent, but highly complex. Even designed artefacts do not necessarily work as they are intended to, with designers' visions of its use case colliding with the inevitability of material decay and the simple rule that there are more ways for a system to be in disorder than there are ordered states. It is not uncommon to explain in lessons the distinction between the theoretical, ideal case, and the practical limitations that prevent the achievement of ideal conditions; the question arises as to the ultimate practical utility of these ideas, and what is lost as we come to understand only a partial, veiled, clouded, perception of the way things are. We certainly do not wish our students to only become expert symbolic manipulators, but to be able to have practical influence on the world; it is here that the trouble lies.

The purpose of science education, or education in general, ought to be such that learners have an expanded sense of possibility in their lives. Given the time that we find ourselves in, widely dubbed the post-truth era, it is increasingly important that students not only understand the claims of science, but also the ways in which we are able to conclude in the ways that we have. Besides the basis in rational argumentation that has been the research program in science education in the past several decades and into contemporary times, it is also important to consider means of extracting conclusions from practical experiences, especially in recognition of the ways in which practical evidence needs to be coaxed into compliance with theoretical predictions, or vice versa. This coaxing, also referred to as 'tuning' by Pickering (1995) refers especially to large or complicated mechanisms used in research teams to carry out scientific investigations, but could equally well be applied to such simple mechanisms as the pendulum mentioned previously. Very often dismissed as 'experimental error', or non-ideal conditions, we may be doing our students a disservice if we do not embrace the mangle and acknowledge a different ontological basis for science – not quite so much as an eternal truth that has been revealed for all time, allowing for control and mastery over nature; but rather as partial knowledge that depends on delicately *working together* with materials and artefacts that may or may not behave exactly as we expect them to.

The distinction, as with Heidegger, is in the careful difference between being correct and being true. In science education, I do not believe that there have been enough approaches that address this significant challenge to this delicate ontological difference. In making artefacts, a similar understanding is expressed by Tim Ingold (2013), where careful study of art, architecture, archaeology, and anthropology has shown that recurrently, and across these fields, humans have never been able to achieve the hylomorphic ideal: that of being able to create artefacts by impressing abstract forms onto passive matter. Instead, humans have only been able to participate in the act of mutual becoming, in subtly altering the flow of materials as they move from one state to another, while humans are similarly changed by these modifications to matter. The example of architecture, where it may appear that carefully laid out plans are subsequently built, does not stand up to scrutiny as incessant forces of weathering, human use and abuse within and beyond the architect's vision, put waste to careful schemes for control. In this regard, Ingold is likely to agree that any control is but transient and perhaps illusory. As human beings capable of manipulating matter, and being able to co-evolve with matter, it is important that students not only learn to be consummate users, but also makers, in the broadest sense of the term possible. Making and makerspace then, ought not be merely considered as a context to introduce STEM concepts, but as sites where this vital learning of how to dance with nature can occur.

HOW DO WE CHANGE THE WORLD?

It is understatement to say that knowledge of science is necessary in our contemporary times. A line of research (see, e.g., Jenkins, 1994; Longbottom & Butler, 1999; Roth & Desautels, 2002; Roth & Calabrese Barton, 2004) has argued forcefully for the centrality of educating students in science, not merely for academic excellence, but for participation in a strong, socially relevant purpose of changing the world for the better. While the question of what 'better' constitutes is contested, to say the least, the nature of problems that confront societies and deserve scientific attention require some examination. Generally speaking at least, the argument could be made that this class of problems fall into the category of being what can be termed wicked. So called, not for its evil nature, but for its intractability, wicked problems resist solution, are ill defined, do not have end points, and have solutions that often become further problems (Rittel, 1973; Rith & Dubberly, 2007). While issues like the provision of public health care and education are obvious for its scale, essentially any design problem falls into the category of the wicked problem (Nelson & Stolterman, 2003). The design of seating for humans, for instance, presents so many opportunities for solutions, all of which will inevitably cause problems from the different postures that their designers assume will be optimal.

In considering social issues for which science ought to play a part in solution, it is important that we do not portray these issues as tame problems that are completely resolvable. Such a depiction mischaracterizes wicked problems that they may eventually face as young adults, leading to the inevitable disillusionment as they wonder why the methods of science do not seem to work. Perhaps what makes this situation worse is the interplay between the natural sciences, and the humanities in arriving at solutions – if we understand Hume's Guillotine accurately, whereby there is no logical necessity connecting descriptions of the current state with normative states, we come to understand science as only providing half the solution. Science provides an excellent means of description, of describing and explaining why the world is the way it is; however, science has little to say when it comes to telling us what the world *ought to be* – for that we rely on the human sciences, our philosophical understanding of our ethico-moral place in the world.

If properly deployed to attend to these issues, making needs to provide learners with experiences of making propositions of 'oughtness' – we need to equip learners with opportunities to develop well thought out solutions, originating not from a position of scientific hubris providing technological solutions-looking-for-problems, but rather conscientiously and humbly reasoning through our understanding of the human condition and what improvement actually looks like.

MORE QUESTIONS THAN ANSWERS

Making and makerspaces present themselves as interesting candidate solutions to what may be long standing problems of education. Aided by the computing and communications technologies of the last several decades, it is not hyperbolic to say that the kinds of making available today are significantly different are present affordances not previously accessible to users of manufacturing technologies. While this may be the case, it is nonetheless important to pay attention to the nature of educational problems to ensure that appropriate solutions are generated. In this chapter, I have reviewed what I believe to be significant problems that confront science education: the difficulty of nurturing students to produce new knowledge; the challenge to learning posed by contemporary understandings of tacit and embodied knowledge; a different ontology marked by the mangle; and an expanded view of how problems ought to be approached by a deliberate transdisciplinary approach. To say that these problems are solved would certainly conceit of the highest order. There is much work to be done – how, exactly, ought to inform the development of abductive reasoning skills, and how would abductive reasoning transfer across contexts are currently open questions. What would a pedagogy that recognises the mangle in science look like? How would learning happen in such circumstances? What kind of curriculum animal is a transdisciplinary course? Who would teach it? How would we assess it? These questions, and more, await much patient work.

REFERENCES

Abrahamson, D., & Lindgren, R. (2014). Embodiment and embodied design. In R. K. Sawye (Ed.), *The Cambridge handbook of the learning sciences* (2nd ed., pp. 358–376). Cambridge: Cambridge University Press.

Banzi, M. (2011). *Getting started with arduino* (2nd ed.). Sebastopol, CA: O'Reilly Media.

Bell, A., Chetty, R., Jaravel, X., Petkova, N., & Van Reenen, J. (2016). *The lifecycle of inventors* (Working Paper). Retrieved from https://ssrn.com/abstract=2838018

Bernstein, B. (2000). *Pedagogy, symbolic control and identity* (Rev. ed.). Lanham, MD: Rowman & Littlefield.

Birtchnell, T. (2011). Jugaad as systemic risk and disruptive innovation in India. *Contemporary South Asia, 19*, 357–372. Retrieved from https://doi.org/10.1080/09584935.2011.569702

Blikstein, P. (2013). Digital fabrication and 'making' in education: The democratization of invention. In J. Walter-Herrmann & C. Büching (Eds.), *FabLabs: Of machines, makers and inventors*. Bielefeld: Transcript Publishers.

Clark, A. (2008). *Supersizing the mind: Embodiment, action, and cognitive extension*. Oxford: Oxford University Press.

Dennett, D. (2009). Darwin's "strange inversion of reasoning". *Proceedings of the National Academy of Sciences, 106*, 10061–10065. Retrieved from https://doi.org/10.1073/pnas.0904433106

diSessa, A. (1993). Toward an epistemology of physics. *Cognition and Instruction, 10*(2–3), 105–225. Retrieved from https://doi.org/10.1080/07370008.1985.9649008

Duschl, R. A., & Grandy, R. (2013). Two views about explicitly teaching nature of science. *Science & Education, 22*, 2109–2139. Retrieved from https://doi.org/10.1007/s11191-012-9539-4

Goodwin, C. (1994). Professional vision. *American Anthropologist, 96*, 606–633. Retrieved from https://doi.org/10.1525/aa.1994.96.3.02a00100

Ingold, T. (2013). *Making: Anthropology, archaeology, art and architecture*. Abingdon: Routledge.

Isaacson, W. (2014). *The innovators: How a group of hackers, geniuses, and geeks, created the digital revolution*. New York, NY: Simon & Schuster.

Jenkins, E. W. (1994). Public understanding of science and science education for action. *Journal of Curriculum Studies, 26*, 601–611. Retrieved from https://doi.org/10.1080/0022027940260602

Kabayadondo, Z. (2015). The prototyping mind: Rethinking perception, affordances, and the mediation of cultural artifacts. *Mind, Culture, and Activity, 23*, 154–174. Retrieved from https://doi.org/10.1080/10749039.2015.1087570

Labaree, D. (2014). Let's measure what no one teaches: PISA, NCLB, and the shrinking aims of education. *Teachers College Record, 116*(9), 1–14.

Lakoff, G., & Núñez, R. E. (2000). *Where mathematics comes from: How the embodied mind brings mathematics into being*. New York, NY: Basic Books.

Lawson, A. (2010). Basic inferences of scientific reasoning, argumentation, and discovery. *Science Education, 94*, 336–364. Retrieved from https://doi.org/10.1002/sce.20357

Levy, S. (2010). *Hackers: Heroes of the computer revolution*. Sebastopol, CA: O'Reilly Media.

Lindgren, R., & Johnson-Glenberg, M. (2013). Emboldened by embodiment: Six precepts for research on embodied learning and mixed reality. *Educational Researcher, 42*, 445–452. Retrieved from https://doi.org/10.3102/0013189x13511661

Longbottom, J. E., & Butler, P. H. (1999). Why teach science? Setting rational goals for science education. *Science Education, 83*, 473–492. Retrieved from https://doi.org/10.1002/(sici)1098-237x(199907)83:4%3C473::aid-sce5%3E3.0.co;2-z

McNeill, K. L., & Krajcik, J. (2007). Middle school students' use of appropriate and inappropriate evidence in writing scientific explanations. In M. C. Lovett & P. Shah (Eds.), *Thinking with data: The proceedings of the 33rd Carnegie symposium on cognition* (pp. 233–265). Mahwah, NJ: Erlbaum.

Miller, A. I. (1981). *Albert Einstein's special theory of relativity: Emergence (1905) and early interpretation (1905–1911)*. Reading, MA: Addison-Wesley.

Moore, R., & Muller, J. (1999). The discourse of 'voice' and the problem of knowledge and identity in the sociology of education. *British Journal of Sociology of Education, 20*(2), 189–206. Retrieved from https://doi.org/10.1080/01425699995407

Nathan, M. J. (2012). Rethinking formalisms in formal education. *Educational Psychologist, 47*(2), 125–148. Retrieved from https://doi.org/10.1080/00461520.2012.667063

Nelson, H. G., & Stolterman, E. (2003). *The design way: Intentional change in an unpredictable world: Foundations and fundamentals of design competence*. Englewood Cliffs, NJ: Educational Technology.

Novack, M. A., Congdon, E. L., Hemani-Lopez, N., & Goldin-Meadow, S. (2014). From action to abstraction: Using the hands to learn math. *Psychological Science, 25*, 903–910. Retrieved from https://doi.org/10.1177/0956797613518351

Obama White House. (2016). *Nation of makers*. Retrieved from https://obamawhitehouse.archives.gov/nation-of-makers

Oehrtman, M., & Lawson, A. E. (2008). Connecting science and mathematics: The nature of proof and disproof in science and mathematics. *International Journal of Science and Mathematics Education, 6*, 377–403. Retrieved from https://doi.org/10.1007/s10763-007-9111-x

Pickering, A. (1995). *The mangle of practice: Time, agency, and science*. Chicago, IL: University of Chicago Press.

Pickering, A., & Guzik, K. (Eds.). (2008). *The mangle in practice: Science, society, and becoming*. Durham, NC: Duke University Press.

Pinker, S. (1997). *How the mind works*. New York, NY: Penguin.

Polanyi, M. (2009). *The tacit dimension*. Chicago, IL: University of Chicago Press. (Original work published in 1966)

Rata, E. (2016). A pedagogy of conceptual progression and the case for academic knowledge. *British Educational Research Journal, 42*(1), 168–184. Retrieved from https://doi.org/10.1002/berj.3195

Rith, C., & Dubberly, H. (2007). Why Horst W.J. Rittel matters. *Design Issues, 23*(1), 72–74. Retrieved from https://doi.org/10.1162/desi.2007.23.1.72

Rittel, H. W. J., & Webber, M. M. (1973). Dilemmas in a general theory of planning. *Policy Sciences, 4*(2), 155–169. Retrieved from https://doi.org/10.1007/bf01405730

Roth, W.-M., & Calabrese Barton, A. (2004). *Rethinking scientific literacy*. New York, NY: RoutledgeFalmer.

Roth, W.-M., & Désautels, J. (2002). *Science education as/for sociopolitical action*. New York, NY: Peter Lang.

Simon, H. A. (1996). *The sciences of the artificial* (3rd ed.). Cambridge, MA: MIT Press. (Original work published in 1969)

Singh, R., Gupta, V., & Mondal, A. (2012). Jugaad – From 'making do 'and 'quick fix 'to an innovative, sustainable and low-cost survival strategy at the bottom of the pyramid. *International Journal of Rural Management, 8*(1–2), 87–105. Retrieved from https://doi.org/10.1177/0973005212461995

Sterelny, K. (2003). *Thought in a hostile world: The evolution of human cognition.* Malden, MA: Blackwell.

Waddington, D. I., & Feinstein, N. W. (2016). Beyond the search for truth: Dewey's humble and humanistic vision of science education. *Educational Theory, 66*(1–2), 111–126. Retrieved from https://doi.org/10.1111/edth.12157

Waldstein, M. (2008). Mangle of practice or empire of signs? Toward a dialogue between science studies and soviet semiotics. In A. Pickering & K. Guzik (Eds.), *The mangle in practice: Science, society, and becoming* (pp. 221–242). Durham, NC: Duke University Press.

Walton, D. (2005). *Abductive reasoning.* Tuscaloosa, AL: University of Alabama Press.

White, R., & Gunstone, R. (1992). *Probing understanding.* London: Falmer Press.

Winch, C. (2013). Curriculum design and epistemic ascent. *Journal of Philosophy of Education, 47*(1), 128–146. Retrieved from https://doi.org/10.1111/1467-9752.12006

Young, M. F. D. (2008). *Bringing knowledge back in: From social constructivism to social realism in the sociology of education.* Abigndon: Routledge.

Zeidler, D. L., Sadler, T. D., Simmons, M. L., & Howes, E. V. (2005). Beyond STS: A research-based framework for socioscientific issues education. *Science Education, 89*, 357–377. Retrieved from https://doi.org/10.1002/sce.20048

ABOUT THE AUTHOR

Michael Tan is currently a research scientist at the Learning Sciences Lab of the National Institute of Education, Singapore. His research interests have brought him on a meandering path through studies in the sociology of curriculum through to his current interest in embodied cognition and the learning sciences, all in a bid to understand and propose solutions to the multiply layered complexity that is education. He now studies middle schoolers making 'random things' in makerspaces, and hopes this effort will bring advances to our knowledge of learning.

CHANTAL POULIOT

12. BEING A SCIENCE EDUCATION RESEARCHER AND A CONCERNED CITIZEN AGAINST EPISTEMOLOGICAL ANESTHESIA

ABSTRACT

In Quebec City (in the province of Quebec, Canada), a highly publicized environmental controversy emerged in late 2012. Concerning the presence of several heavy metals in the air, this situation involves the bulk transshipment activities of the Port of Québec, located just a few kilometres from the downtown area of Quebec City. In this chapter, I present the main lines of the citizen engagement in the controversy, describe various aspects of my participation in sociopolitical discussions and address the engagement of researchers.

Keywords: environmental issues, citizen engagement, controversies, sociopolitical discussions, dust, epistemological anesthesia

> There's a slow train coming
> It's movin' on down the line
> Steel wheels on iron rails
> Tonight I'm fixin' to die
> Woo, I hope you don't mind pretty mama
> Woo-hoo, hope you don't mind if I go
> 'Cause when the steam from the slow train rises
> I ain't gonna see you anymore (…)
>
> Joe Bonamassa (*Slow Train*, 2011)

The year 2013 in Quebec was marked by a major spillover effect (in the sense of Callon, Lascoumes, & Barthe, 2001) of the expansion of oil production in Western Canada and the transport of oil by rail. Images of the downtown core of Lac-Mégantic engulfed in flames quickly made their way around the world (Transportation Safety Board of Canada, 2016). The population of Lac-Mégantic and the citizens of Quebec are still coming to grips with this disaster and its social, environmental and economic effects, including the death of 47 people. To summarize events, on July 6, at 12:56 a.m., a train carrying 72 tank cars of light crude oil, parked in the town of Nantes, Quebec with no crew members on board,

DOI:10.1163/9789004389663_012

slowly started rolling down a descending grade of 1.2% towards the town of Lac-Mégantic. Gathering speed as it approached, it derailed near the centre of town, exploding and causing a major fire that destroyed more than 35 buildings across two square kilometres. The fire was so intense that it was detected by a NASA weather satellite passing overhead at 2:21 a.m. In addition to the human tragedy, this incident also had devastating environmental and economic consequences. It is estimated that between 100,000 and 200,000 litres of crude oil spilled into the Chaudière River (Côté, 2013).

The Musi-café bar was destroyed by the explosions that night. Almost half the people who died were in that bar. For the people of Quebec, the Musi-café became a symbol of the anguish experienced by the residents of Lac-Mégantic, but also a powerful symbol of resilience. (Since music played an integral role in this tragedy right from the start, I chose to highlight the words of the song "Slow Train," written by Blues singer and guitarist Joe Bonamassa with Kevin Shirley [2011, track 1]).

In Quebec, 2013 and the years that followed were also marked by public demonstrations against pipeline projects in this province and elsewhere in Canada and plans to expand oil sands production. A major project to reverse the flow of a 38-year-old pipeline and increase the amount of light and heavy oil passing through it made news headlines. Many citizens from different quarters expressed anger over how the government was handling this file and stressed their desire to participate in discussions aimed at both defining the issues and deciding on the actions to be taken. Environmental issues (air, water, earth, biodiversity) as well as ethical and economic issues were raised. The risks and uncertainties related to this project (e.g., ruptures in the pipeline due to cracking) continue to be the subject of intense debate in the media. Recently, a coalition of environmental groups filed a motion against TransCanada aimed at forcing the company to subject the pipeline project to Quebec's environmental regulations. The Calgary-based pipeline giant claims that it is not under the obligation to do so, asserting that its pipeline is under the sole responsibility of the federal government (Shields, 2016). In brief, since 2013, the future of the oil industry and the risks associated with oil transport have been the subject of heated sociopolitical discussions in Quebec.

Moreover, another highly publicized environmental controversy also emerged in late 2012. Concerning the presence of several heavy metals in the air, this situation involves the bulk transshipment activities of the Port of Québec, located just a few kilometres from the downtown area of Quebec City. The alarm was sounded by a resident of one of the affected neighbourhoods (Véronique Lalande) and her partner (Louis Duschene) in October 2012. Since that time, the controversy has continued to unfold, raising a number of ethical, social, environmental, economic, legal and legislative concerns. This controversy serves as the backdrop for this chapter, in which I will address the participation of science education researchers in sociopolitical discussions.

The chapter is divided into four sections. The first section recounts my experience as a researcher interested in the teaching of socio-scientific controversies. The second section presents the main lines of the controversy surrounding the presence of metal dust in Quebec City (in the province of Quebec, Canada). The third section describes various aspects of my participation in sociopolitical discussions. The concluding part addresses the engagement of researchers, raising more questions than it answers.

BEING A RESEARCHER IN SCIENCE EDUCATION…

During my doctoral studies, I examined the way science students appropriated the controversy surrounding the health effects of cell phone use (Pouliot, 2007). More specifically, I investigated these students' relationships with science experts and how they viewed the participation of various social actors in the development and management of this controversy (Pouliot, 2011).

Over the years, I have developed a research program that aims, among other things, to shed light on the way students and future science teachers view the development of socioscientific controversies and attribute roles and capacities to citizens, governments, industries and scientists in the management of these controversies (Pouliot, 2014). Seeking to identify cases where citizens have been able to document situations and influence the course of events, I explored the fields of science and technology studies (STS) and the public understanding of science (PUS). Research and reflection on STS and PUS helped me to identify useful models for understanding and reporting on interactions between citizens and scientists (Callon, 1999; Callon, Lascoumes, & Barthe, 2001) or between citizens and the organizers of public consultations (Rowe & Frewer, 2000, 2005) and for interpreting the ways citizens can participate in the co-production of scientific knowledge (Pouliot & Godbout, 2014).

As a science education researcher, I address with my students (future science teachers at the secondary and junior college levels) the relevance of studying socio-scientific controversies in the classroom (Pouliot & Groleau, 2011). The educational aims of appropriating such controversies can vary, from learning the art of argumentation to developing a value system, can include mastering the key notions of a given discipline (Sadler, 2004) or the promotion of activism (Bencze & Alsop, 2014). From the perspective of what researchers such as Lawrence Bencze, Ralph Levinson, Matthew Weinstein, Lyn Carter and Isabel Martins refer to as *politicized science education*, addressing these thorny issues can, for example, allow students to consider how political and economic actors are entwined (Bencze & Pouliot, in review; Levinson, Bencze, Carter, Martins, Pouliot, & Weinstein, in preparation) but also to consider themselves as legitimate participants in sociopolitical decision-making processes, able to question the discourse of experts mobilized by the existing political and scientific authorities (Hodson, 2011; Larochelle & Désautels, 2006) and produce original knowledge (Lalande & Duchesne, 2013a, 2013b, 2013c; Pouliot, 2015).

… AND A CITIZEN EXPOSED TO METAL DUST

I live in a neighbourhood (called La Cité-Limoilou, referred to simply as "Limoilou") located in the downtown area of Quebec City. Laid out like New York City, with it streets and avenues at right angles, Limoilou is bordered by a paper mill and the city's incinerator. It is located approximately 2 km from the Port of Quebec, a hub for international trade in the mining, agricultural, energy and iron and steel sectors (33 million tons of cargo passed through the Port in 2012). In the last ten years, many families have come to live in the neighbourhood. Parks, cafés, bakeries, community gardens and a mixed population (different ethnic and socio-economic groups) make it a welcoming place to live (Saltzstein, 2015).

On October 26, 2012, a controversy began that is still playing out as I write this chapter. This is what took place that day and during the weeks that followed. Véronique Lalande, a resident of Limoilou noticed that there was red dust on the tires of her son's stroller and all around her house. She notified the City, which sent an environmental technician to assess the complaint. The file was then forwarded to Urgence-Environnement, Quebec's environmental emergency service. With her partner, Louis Duchesne, Lalande then took samples of the dust (from balconies and cars) and sent them to an independent laboratory. A few days later, St. Lawrence Stevedoring (SLS), which undertakes the transshipment of dry bulk in the Port of Quebec, confirmed that iron oxide dust had settled on the Limoilou area.

In early November, Lalande and Duchesne got the results of the analysis of their dust samples from AGAT Laboratories. In addition to iron oxide, their samples were found to contain higher concentrations of heavy metals such as nickel, zinc and copper than those found in major cities like New York, New Delhi or Hong Kong (Lalande & Duchesne, 2013).

On November 19, 2012, Lalande attended the City Council meeting and asked the mayor what the City intended to do to ensure that the Port of Québec would carry out its operations in a way that was safe for the residents of the central neighbourhoods of Quebec City. On November 21, SLS received a notice of non-compliance from Quebec's Ministry of Environment for infractions to the *Environment Quality Act*. The same day, the neighbourhood council for Vieux-Limoilou adopted a resolution to support Lalande's efforts in the metal dust file. During the same period, it came to light that the DRSP (regional public health department) had been concerned about the dust coming from the Port of Québec since 2009, when residents of the borough of Beauport brought the problem of dust to their attention. Then, on November 23, 2012, residents of the Cap Blanc neighbourhood complained that there were volatile substances in the air. It subsequently emerged that these residents had gathered 200 signatures on a petition in September 2012 to express their displeasure with this situation. The same day (November 23), a resident of Lévis contacted the media to say that, the previous summer, clouds of dust had regularly been seen rising from the facilities used for the transshipment of dry bulk in the Port of Quebec. The same day again, the mayor of Quebec City announced that he was referring this file to the

Ministry of Environment, citing the worrisome levels of some metals in the samples that Lalande and Duschesne had had analyzed.

On December 10, 2012, on the website of the Initiative Citoyenne de Vigilance du Port of Québec (ICVPQ, a citizens' initiative aimed at monitoring Port of Québec activities; see vigilanceportdequebec.com), Lalande and Duchesne published the results of an analysis of other dust samples collected in the Limoilou, Maizerets, Saint-Jean-Baptiste and Baie de Beauport neighbourhoods in November 2012. The aim was to show that the dust episode of October 26 was not a singular incident and that dust regularly settled on the central neighbourhoods of Quebec City. On December 18, 2012, the results of this analysis were presented at a press conference at the *Fistons* restaurant in Limoilou. On December 23, 2012, noticing that an unusually large amount of dust had settled on their snow-covered doorstep, Lalande and Duchesne took yet another sample of the dust. This would not be the last. Dust was collected on several other occasions, at different times and from different locations.

Thus, these citizens "stirred up the dust," bringing the issue of metal dust into the sociopolitical sphere. Citizens continued to document the situation. After the analyses of dust samples mentioned above (those collected in Limoilou on October 26, 2012, and at various locations in the central neighbourhoods of Quebec City in November, 2012), the ICVPQ set out to establish a link between the dust on the ground and the dust in the air. To this end, Lalande and Duchesne submitted an access to information request and obtained all the data collected by the Ministry of Environment for the entire Greater Quebec City area. They analyzed the data from the monitoring stations in the air quality monitoring network. The results suggested that heavy metal dust was also present in the air (and not just on the ground) and showed that the average concentration of nickel for the three monitoring stations between March 2010 and April 2012 was 52 ng/$m^{3.}$ (As a point of reference, over the period of 2003–2009, the average concentration of nickel in Canada was 0.9 ng/$m^{3)}$. Based on an examination of the nickel/cobalt ratio, Lalande and Duchesne concluded that, regardless of place and date, the settled dust and airborne dust came from a single common source. The data were also interpreted in relation to the wind, leading the authors to conclude that the dust was coming from the Port of Québec (a later report by the Ministry of Environment supported this argument, Ministère du Développement durable, de l'Environnement, de la Faune et des Parcs [MDDEFP], 2013).

PARTICIPATING IN SOCIOPOLITICAL DISCUSSIONS

Metal dust settles on the porches of my house almost every day, as well as on my car or any toys that my two sons leave outdoors. The metal dust situation concerns me as a researcher, a neighbourhood resident and a mother (Zahran, Mielke, Weiler, Hempel, Berry, & Gonzales, 2012). Since this controversy began, I have written letters and opinion pieces for the general press (Pouliot & Roux, 2013) and participated on two occasions in a public affairs show. I have also written a book addressed to citizens concerned by and interested in this issue (Pouliot, 2015).

Since it came out, the book has been read by people from different quarters: elected representatives, teachers at the primary, secondary and junior college levels, lawyers, and researchers in the social and natural sciences. I have given interviews for radio and newspapers (Mouterde, 2015). Moreover, I have attended numerous citizen meetings and as many representations before the courts as possible related to the two authorized class-action lawsuits against the Port of Québec and the iron ore transshipment company. As an academic, I have made presentations locally and at the international level describing how citizen engagement has contributed to the development of this controversy.

With regard to my teaching activities, I invited Véronique Lalande and Louis Duchesne into my class for future science teachers to present the steps involved in their investigation and production of knowledge. Following this encounter, some students wrote a letter that was subsequently published in a Quebec City newspaper (Pouliot et al., 2015). I have also had conversations with professors of sociology, nursing and social medicine who have covered this controversy in their courses.

Several factors have led me not only to highlight the contributions and challenges of citizen investigation from an academic perspective but also to participate in the sociopolitical discussions that have taken place since late 2012. Three appear to be particularly relevant for the aims of this chapter. First, in the pubic discussions that have marked this controversy, the idea that citizens are incapable of understanding complex issues and producing knowledge on them has been presented by the authorities as a given. Citizens have been infantalized (being called "sorcerer's apprentices") and the suggestion that the knowledge they produced was derived from playing with children's toys was put forward on at least two occasions, including once during a City Council meeting at which the Port Authority was defending its expansion project. Furthermore, over the last three years, the tone of the mayor's comments has alternated between being aggressive and benevolent, with the express aim of putting citizens *in their place*. The use of the "knowledge deficit" model (Callon, 1999) by the existing authorities forces citizens to act within a discursive space that is hostile to them and to react to discriminatory and prejudicial comments. The second factor concerns the authorities' public discourse regarding the decisions they made to control the dust emissions and identify their source. Although these decisions were made in response to the political actions taken by citizens and on the basis of knowledge produced by the ICVPQ, they were regularly presented in public statements as spontaneous actions. This appropriation of the work of citizens mainly by the representatives of the City and Port was reinforced by repeated efforts to depoliticize the issue, rejecting the criticisms made by citizens and dismissing the concerns of the residents who are exposed to the dust. A study to be conducted by the DRSP (regional public health department) over the next two years is a prime example of the way the citizens' efforts to problematize the issue have been appropriated. This study was negotiated by three of the four groups of actors concerned: the Port of Québec, the City, and the DRSP. The fourth group, the ICVPQ, was not invited to the discussions (and asserts that the study ignores its concerns). And yet, it could be argued that Louis Duchesne's input, as a researcher in ecosystems and the environment, would have been just as relevant as that of the Port or City representatives. The third factor that motivates me to participate in these sociopolitical discussions is that, in their many public statements, the authorities have disregarded a fundamental aspect of this controversy, namely, that the gains made through these citizens' struggle have benefitted society as a whole. Indeed, *everyone* wins if the residents of the central neighbourhoods of Quebec City are able to live in a healthy environment, not least because these neighbourhoods are also visited by the residents of the suburbs and by many international tourists. However, rather than recognizing the relevance of these citizens' efforts and the time, financial resources and energy invested by them to bring this situation to light, the mayor and the Port Authority CEO have repeatedly suggested that their demands for a healthier living environment go too far, to the point of undermining the economic development of the City and the Port.

TAKING ACTION AGAINST EPISTEMOLOGICAL ANESTHESIA: THE IMPERATIVE FOR SCIENCE EDUCATION RESEARCHERS TO ENGAGE

> Who is speaking? Who can speak? And who is not speaking when others speak? But also, what would those people who aren't speaking say if they could? (Didier Eribon, 2015)

In today's society, many actors in positions of authority show epistemological arrogance, imposing their points of view on their fellow citizens by virtue of the power that is socially attributed to them (Oreskes & Conway, 2010; Pignarre & Stengers, 2007). This violent behaviour does not go unchallenged: some people under their rule are well aware of the value of the knowledge they hold or co-produce (Lalande & Duchesne, 2016). This knowledge is often referred to as lay knowledge and the ontologies they help to develop are said to be contingent, subjective and intuitive. Epistemological anesthesia is one of the strategies used by citizens to lessen the pain or anxiety caused by these asymmetrical knowledge and power relationships (Darré, 2006). It refers to putting aside one's own worldview and adopting a different view, one that is considered to be more objective and better informed.

Intellectuals from varying backgrounds agree that the greatest challenges facing the 21st century are water, soil and air quality as well as social inequality (Atkinson, 2015; Klein, 2014) and researchers are being called upon to participate in public sociopolitical discussions (Vose, 2015). While some observers see the participation of experts in these discussions as a mean of educating citizens who lack knowledge or have negative attitudes towards techno-scientific developments (Quirion, 2015), others see it as a way to mobilize points of views that could add insight to a collective understanding and collective decision-making concerning the future of the planet (Seguin, 2016).

Problems related to the conservation and sustainability of ecosystems, the future of our species, and the geopolitical sharing of responsibility through economic and political mechanisms (Bonneuil & Pestre, 2015) involve contentious issues that science education endeavours to introduce in the classroom, in particular, by examining the nature of science and by teaching about controversies. It appears to me that, as science education researchers, we can play a significant role with regard to the way citizens structure their relationships with public authorities (Hodson, 2011; Larochelle & Déseautels, 2006; Levinson, 2010). Our sociopolitical engagement can aim to tackle citizens' epistemological anesthesia and support individuals and collectives in the process of empowerment. Among other possible actions, we can lead people to conceive of knowledge relationships (Charlot, 1997) that involve experts, corporations and citizens (users, consumers, and producers of lay knowledge) (Bencze & Alsop, 2014; Pierce, 2013). The decision to strive towards this form of critique can be driven by thinking about the consequences of our inaction (for example, a continuation of inequality and the trivialization of citizen struggles related to health and environmental issues) as opposed to the consequences of our action. For example, one of the identifiable consequences of writing my own book for citizens is that representatives of the municipal, provincial government, Port, and industrial authorities have learned that citizens are tenacious and courageous and have the capacity to produce legitimate (and scientific) knowledge even in a difficult context involving the asymmetrical distribution of power and resources.

I am well aware of the danger of oversimplifying this issue and realize that I could be accused of being particularly sensitive to the metal dust situation affecting the residents of the central neighbourhoods of my city. Far from being the culmination of my reflection on the imperative for science education researchers to venture into the area of public involvement, I would like this chapter to be understood as an invitation to deepen our reflection and envision greater interaction between our research field and the sociopolitical spheres. I firmly believe that the engagement of researchers, understood from the perspective of the specific intellectual as defined by Michel Foucault (2001), the collective intellectual as defined by Pierre Bourdieu (2001), or the generic intellectual as defined by Diane Lamoureux (2013), should be more fully addressed in our field of research.

Many questions could emerge from a reflection on the potential sociopolitical action of science education researchers. For example, what types of involvement in public discussions should be promoted? How can we support and encourage one another, considering that such involvement requires a great deal of energy and entails a set of rules other than academic rules? What unique contributions can we bring to sociopolitical discussions through our individual and collective expertise? Who should our contributions target and to what aims? What challenges might emerge along the way? How can we prepare ourselves to overcome them? What discussions can we begin right away that might open up new paths in this regard? These are far-reaching but necessary questions for those who believe that science education is for everyday life (Aikenhead, 2006) and that it must enable people to think for themselves, take a stand and accomplish together what no one person can accomplish alone (Pignarre & Stengers, 2007, p. 176, Trans.). In short, both intellectual and political will is needed to lead this collective reflection, the will to provide those who, to varying degrees, are subject to symbolic violence with the theoretical means and tools to combat it. I will conclude by quoting an inspirational comment made by Anthony Atkinson (2015) in his recent book on inequality: "The world faces great problems, but collectively we are not helpless in the face of forces outside our control. The future is very much in our hands" (p. 13).

REFERENCES

Aikenhead, G. (2006). *Science education for everyday life*. New York, NY: Teachers College Press.

Atkinson, A. B. (2015). *Inequality: What can be done?* Cambridge, MA: Harvard University Press.

Bencze, L., & Alsop, S. (2014). *Activist science & technology education*. Dordrecht: Springer.

Bencze, L., & Pouliot, C. (in review). Battle of the bands: Toxic dust, active citizenship and science education. *Cultural Studies of Science Education*.

Bonamassa, J., & Shirely, K. (2011). Slow train [Recorded by J. Bonamassa]. On *Dust Bowl* [studio album]. Netherlands: Provogue Records.

Bonneuil, C., & Pestre, D. (2015). *Histoire des sciences et des savoirs: Le siècle des technosciences*. Paris: Le Seuil.

Bourdieu, P. (2001). *Contre-Feux 2*. Paris: Raisons d'agir.

Callon, M. (1999). The role of lay people in the production and dissemination of scientific knowledge. *Science Technology & Society, 4*, 81–94. Retrieved from https://doi.org/10.1177/097172189900400106

Callon, M., Lascoumes, P., & Barthe, Y. (2001). *Agir dans un monde incertain: Essai sur la démocratie technique*. Paris: Éditions du Seuil.

Charlot, B. (1997). *Du rapport au savoir*. Paris: Anthropos.

Côté, C. (2013, July 9). 100 000 litres de pétrole sur la rivière Chaudière. *La Presse*.

Darré, J.-P. (2006). *La production de la connaissance pour l'action; arguments contre le racisme de l'intelligence*. Paris: Éditions de la Maison des sciences de l'homme.

Eribon, D. (2013). *Pierre Bourdieu l'insoumission en héritage*. Paris: Presses Universitaires de France.

Foucault, M. (2001). *Dits et écrits II, 1976–1988*. Paris: Gallimard.

Hodson, D. (2011). *Looking to the future: Building a curriculum for social activism*. Rotterdam, The Netherlands: Sense Publishers.

Klein, N. (2014). *This changes everything: Capitalism vs. the climate*. Toronto: Simon & Schuster.

Lalande, V., & Duchesne, L. (2013a). *Analyse des concentrations de métaux dans l'air ambiant du territoire de Limoilou. Initiative citoyenne de vigilance du Port de Québec*. Retrieved from http://www.vigilanceportdequebec.com

Lalande, V., & Duchesne, L. (2013b). *Analyse de la provenance du nickel dans l'air ambiant du territoire de Limoilou en fonction de la direction des vents. Initiative citoyenne de vigilance du Port de Québec*. Retrieved from http://www.vigilanceportdequebec.com

Lalande, V., & Duchesne, L. (2013c). *Analyse des concentrations de nickel dans l'air ambiant des quartiers Limoilou et Vanier* (d'avril 2012 à mai 2013). Retrieved from http://www.vigilanceportdequebec.com

Lalande, V., & Duchesne, L. (2016). *Collection de témoignages de citoyens victimes d'empoussièrement dans les quartiers centraux de la ville de Québec. Initiative citoyenne de vigilance du Port de Québec*. Retrieved from http://www.vigilanceportdequebec.com

Lamoureux, D. (2013). *Le trésor perdu de la politique*. Montréal: Écosociété.

Larochelle, M., & Désautels, J. (2006). L'éducation aux sciences et le croisement des expertises. In A. Legardez & L. Simonneaux (Eds.), *L'école à l'épreuve de l'actualité* (pp. 61–78). Issy-les-Moulineaux: Entreprise Sante Formation.

Levinson, R. (2010). Science education and democratic participation: An uneasy congruence? *Studies in Science Education, 46*(1), 69–119. Retrieved from https://doi.org/10.1080/03057260903562433

Levinson, R., Bencze, L., Carter, L., Martins, I., Pouliot, C., & Weinstein, M. (in preparation). *Politicized science and technology education*. Dordrecht: Springer.

Larochelle, M., & Désautels, J. (2006). L'éducation aux sciences et le croisement des expertises. In A. Legardez & L. Simonneaux (Eds.), *L'école à l'épreuve de l'actualité* (pp. 61–78). Issy-les-Moulineaux: Entreprise Sante Formation.

Ministère du Développement durable, de l'Environnement, de la Faune et des Parcs (MDDEFP). (2013). *Origine des concentrations élevées de nickel dans l'air ambiant*. Limoilou.

Mouterde, P. (2015). Contamination aux poussières métalliques: Les luttes citoyennes face aux pouvoirs. *Droit de parole, 42*(7), 3.

Oreskes, N., & Conway, E. (2010). *Merchants of doubt*. London: Bloomsbury Press.

Pierce, C. (2013). *Education in the age of biocapitalism: Optimizing educational life for a flat world*. New York, NY: Palgrave Macmillan.

Pignarre, P., & Stengers, I. (2007). *La sorcellerie capitaliste*. Paris: La découverte.

Pouliot, C. (2007). *Appréhension d'une controverse sociotechnique et rapport aux experts scientifiques: Une étude de cas* (Doctoral dissertation). Université Laval, Québec.

Pouliot, C. (2011). Post-secondary students' relationship to people they consider to be scientific experts. *Research in Science Education, 41*, 225–243. Retrieved from https://doi.org/10.1007/s11165-009-9162-9

Pouliot, C. (2014). Citizens as concerned but knowledge-poor watchdogs: Asymmetrical attributions of legitimacy to social actors in the management of biotechnology issues. In L. Bencze & S. Alsop (Eds.), *Activist science and technology education* (pp. 511–530). Dordrecht: Springer.

Pouliot, C. (2015). *Quand les citoyen.ne.s soulèvent la poussière* [When citizens raise dust]. Montréal: Carte Blanche.

Pouliot, C., Carignan, M., Arseneau, I., Lavallée, S. B., Breton, Bergeron, A. N., … Tremblay, S. (2015, May 26). Question politiquement vive. *Le Soleil*.

Pouliot, C., & Godbout, J. (2014). Scientists could achieve more fulfilled professional lives by embracing the skills needed for effective interactions with the public. *Embo Reports, 15*, 833–835. Retrieved from https://doi.org/10.15252/embr.201438590

Pouliot, C., & Groleau, A. (2011). L'approche des îlots de rationalité interdisciplinaires, éducation aux sciences et à la citoyenneté: illustrations en enseignement collégial. *Pédagogie collégiale, 25*(1), 9–14.

Pouliot, C., & Roux, S. (2013, Novembre 9). Soulever la poussière. *Le Soleil*.

Quirion, R. (2015). *Messages à la communauté: une année de transition* (Fonds de recherche du Québec). Retrieved from http://www.frq.gouv.qc.ca/messages/une-annee-de-transition-budgetaire

Rowe, G., & Frewer, L. (2000). Public participation methods: A framework for evaluation. *Science, Technology, & Human Values, 25*(1), 3–29.

Rowe, G., & Frewer, L. (2005). A typology of public engagement mechanisms. *Science Technology, & Human Values, 30*, 251–290.

Sadler, T. D. (2004). Informal reasoning regarding socioscientific issues: A critical review of research. *Journal of Research in Science Teaching, 41*(5), 513–536.

Saltzstein, D. (2015, August 11). In Quebec, creativity fuels neighborhood revivals. *The New York Times*.

Seguin, E. (2016). Rapports science et société: Le monde de la recherche en difficulté d'apprentissage! *Découvrir*. Retrieved from http://www.acfas.ca/publications/decouvrir/2016/02/rapports-science-societe-monde-recherche-en-difficulte-apprentissage

Shields, A. (2016, February 18). Des groupes écologistes déposent un recours contre TransCanada. *Le Devoir*.

Transportation Safety Board of Canada. (2016). *Railway investigation R13D0054*. Retrieved from http://www.bst-tsb.gc.ca/eng/enquetes-investigations/rail/2013/R13D0054/R13D0054.asp

Vose, R. (2015). The imperative to engage. *CAUT Bulletin, 62*(10). Retrieved from https://bulletin-archives.caut.ca/bulletin/articles/2015/12/the-imperative-to-engage

Zahran, S., Mielke, H. W., Weiler, S., Hempel, L., Berry, K. J., & Gonzales, C. R. (2012). Associations between standardized school performance tests and mixtures of Pb, Zn, Cd, Ni, Mn, Cu, Cr, Co, and V in community soils of New Orleans. *Environmental Pollution, 169*, 128–135. Retrieved from https://doi.org/10.1016/j.envpol.2012.05.019

ABOUT THE AUTHOR

Chantal Pouliot is a Full Professor of science education at the Faculty of Education at Université Laval (Québec, Canada) and a researcher at the Center for Research and Intervention on Student Success (CRIRES). Her research interests include both documentation and teaching of current environmental and health controversies. She wrote *Quand les citoyens soulèvent la poussière* (2015, Carte blanche) for people exposed to airborne metal dust in the central neighborhoods of Quebec City.

COLIN HENNESSY ELLIOTT, SHRUTI KRISHNAMOORTHY,
CATHERINE MILNE AND KATHRYN SCANTLEBURY

13. SCIENCE EDUCATION AS A MATERIAL ISSUE?

Exploring the Role of Materiality in Science Education through the Lens of Baradian Theory

ABSTRACT

In what ways can human-matter intra-action as a teaching lens frame science-teaching learning in the classroom?

Applying Karen Barad's post humanist theory, we explore how techno-scientific practices allow us to investigate the rich context of education in the making. Such perspectives lead us to argue for a focus on how material-discursive practice emerge through the intra-activity between human and matter agents as a source of locally intelligible phenomena. Traditional humanist models of teaching and learning locate agential responsibility in a specific human agent, which serves to separate the world of education into dichotomies: the teacher from the taught, the knower from the known, the self from other, and the object from the subject. Baradian theory challenges us to focus instead on intra-actions, which do not assume pre-existing object agents but accepts that it is through specific intra-actions that phenomena in all their complexity emerge. We must work from the perspective that agential responsibility, which constitutes responsibly embodied engagement in a material world, and recognize that there are consequences, possibilities, and responsibilities for intra-actions. Drawing on examples from co-teaching, self-assessment and emergent classroom practice, we use Barad's theory, specifically her notions of agential realism, diffraction and entanglement, to propose models of practice that make intra-actions the focus with the goal of making learning more ethical, affective and inclusive.

Keywords: new materialism, agential realism, classroom implications, diffraction, feminist theory, mattering, post-humanism

If you were to say to science teachers that material in the forms of matter and instruments were important to the teaching and learning of science, you could probably be met with a collective response, "Of course!" However, an examination of the resources developed to support science teaching and learning does not reflect

 | DOI:10.1163/9789004389663_013

this collective endorsement. For example, a word search through the Next Generation Science Standards (NGSS) has scant mention of instruments or materials. Statements usually take a form similar to the following:

> There is a strong connection to engineering through the instruments and technologies that have allowed us to explore the objects in our solar system and obtain the data that supports the theories that explain the formation and evolution of the universe. (NGSS Lead States, 2013, p. 51)

In this statement, we can assume that the "us" applies to humans and only humans are presented as active as objects in the solar system are explored. Note how this statement depicts instruments and technologies as passive objects under the control of humans, and not directly part of knowledge making in science that is theoretical. However, such presentation stands in stark contrast to how research scientists often talk about the matter or instruments they work with. Barbara McClintock is an example of a scientist who appreciated the active nature of the materials with which she worked when she said, reflecting on her award of a Nobel Prize, "It might seem unfair, however, to reward a person for having so much pleasure, over the years asking the maize plant to solve specific problems and then watching its responses" (Bioscience, 1984, p. 131). McClintock claimed that her deep engagement with corn chromosomes helped her to construct phenomena including "jumping genes", a process now called transposition or "crossing over" (Fox Keller, 1983) but her comment makes one wonder if these phenomena were a co-construction involving her and maize plants.

However, science education policy documents, such as the NGSS, project a sense of objects as being just there, gaining their significance based on how they are used symbolically by humans in language or text-based interactions. As Andreas Reckwitz (2002) notes, from this perspective, "the material world exists only in so far as it becomes an object of interpretation within collective meaning structures" (p. 202) of social culture, as the material is considered only to be important in its passive contributions as symbolic objects providing meaning and value. However, throughout time, science has struggled with such a perspective because, from a naive realist or scientific realist perspective, the material elements of the world constitute elements that are real and need explanation.

This symbolic placement of material objects and explanatory unobservable objects on a continuum views objects as only having meaning once they are assigned a symbolic label. However, more recently scholars, such as Karen Barad, a feminist philosopher and physicist, argue that language has too much power and have proposed an agentic role for objects in her ontology. Barad (2007, p. 132) asks, "How did language come to be more trustworthy than matter?" She wonders how the belief that representations are needed to mediate the interaction between knower and known came to be an accepted ideology supporting the untenable social constructivist view of the world by giving words the power to mirror existing phenomena. Barad argues that "a performative understanding of discursive practices challenges the representation list belief in the power of words to represent pre-existing things" (p. 133) and that matter deserves its due as "an active participant in the world's becoming" (Barad, 2003, p. 803). She presents a philosophy of *agential realism* thereby removing the human from the center of cultural interactions and noting that matter has agency. The resultant material-discursive practices between human and non-human are "intra-actions" in which reality and knowledge are constructed through the intra-actions between material object and human.

Barad's thinking is influenced in part by the work of philosophers, such as Ian Hacking (1982) who argues that manipulating something to experiment on something else is a criterion for identifying what is real. According to Barad, intervention is a key element of realism toward entities versus realism toward theory, which tends to be a normative perspective in science education. Regardless of NGSS's efforts to give a higher profile to practice in schools, science continues to be presented as a theoretically based conceptually mired discipline. Let us look at an example to illustrate what we mean. Consider a relationship between light and an instrument like the camera obscura (vaulted room [camera] dark [obscura]) (see Figure 13.1).

The first camera obscura was built with a pinhole as the aperture. Looking to see an inverted image through a camera obscura is not a passive action but embodied. You need to learn to "see" through the camera obscura. However, even before you can do that, you need to build the camera obscura. Then you have to learn how to hold it so that light is able to enter through the aperture and form an image on the screen and then be convinced that you see an image. Seeing that others also capture images with their camera obscuras provides a context that convinces us that we have done the same and thereby produced similar images or convinces us that we need to refine or modify our camera obscura so that we can achieve similar results to those achieved by others. Note the variety of practices that are involved in "seeing" images or representations of the world in which we are using this apparatus. Barad argues

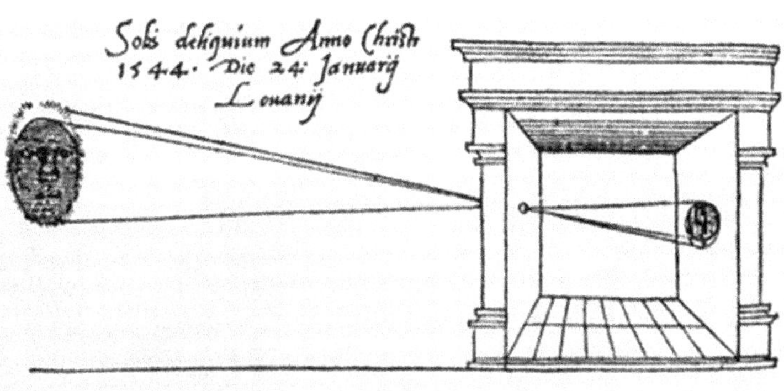

Figure 13.1. An image of a large camera obscura used by Reinerus Gemma-Frisius in 1545 to observe an eclipse of the sun. Note the pinhole aperture

that we should take into account the practices through which representations are produced. However, in accounts of science and in science education, often the reverse happens and the practices are "blacked out" (2007, p. 53) and the representations are what is valued. However, perhaps science would have greater meaning for learners if representations were understood, not as depictions of what we should observe, but as concentrations or traces of the varied "practices of engagement" (2007, p. 53).

Barad makes the argument that theories are also practice-based, entangled with apparatus of culture, globalization, values, and that materials, including objects such as lenses, mirrors, and chemicals, all contribute to the production of scientific theories. Theorizing, like experimenting, are material practices and if one accepts that these practices are entangled, then these domains of theorizing and experimenting are entwined. She wants us to take the material nature of practices seriously proposing the notion of agential realism in which her ontology does not take things or words for granted and a way of knowing or epistemology that is predicated on correspondence between these two elements.

A cultural perspective leads us to accept the constructed nature of science and science education. However, we do not want to be left arguing that science mirrors culture as a remedy to discounting that science mirrors nature, especially if culture is viewed as social and human with languages as the only basis for the construction of knowledge. Much constructivism focuses on conceptual understanding and formation of individual constructions as the basis for human knowing. Even social constructivism struggles to come away from the humanism, which places humans at the center. Building on theories from Judith Butler (1990) and Niels Bohr, Barad (2007, p. 51) argues that practices are constitutive and in the enactment of "knowledge-discourse-power practices" play significant roles in producing phenomena, so that knowing what is excluded is as important as identifying the elements of the phenomenon.

AGENTIAL REALISM AND THE CONSTRUCTION OF PHENOMENA

> We do not obtain knowledge by standing outside the world; we know because we are of the world. We are part of the world in its differential becoming. (Barad, 2007, p. 185)

When we think of knowledge construction or production, we might see ourselves as outside observers from an objective point of view, or maybe as co-constructors of our world where nature and culture interact. Whether from this performative, constructivist or sociocultural lens our understanding of reality assumes innate boundaries between human beings, nature and culture. We see our physical and social environments and ourselves as pre-existing entities that we come upon to study and manipulate. Barad's post-humanist performative framework challenges this innate segregation between entities in the world arguing that boundaries come into existence through the enactment of apparatus. Through her agential realist lens, matter does not pre-exist and is in a process of continual 'becoming' through manipulations in the universe. So instead of seeing matter, apparatus, nature and culture as fixed entities in the world, agential realism views matter's agency through the performance of apparatus where phenomena are produced.

Agential realism challenges pre-defined static dichotomies that separate the object from subject, the knower from known and as a result, the hegemony of dominant ideas or worldviews as objective truth. We commit ourselves to specific arrangements in the universe by assuming pre-existing boundaries and therefore by extension, the concepts and measurements we derive from our observations. However, instead of assuming these boundaries and attributing a passivity to matter, our experiments involve matter in its becoming valuing matter as actively contributing to ontology and epistemology. An agential realist approach examines the effect of boundary making within the context of experimental observation and the production of phenomena. It reassesses the cuts made in defining boundaries around entities defined as the object and those considered outside the process of phenomena. Barad addresses the implicit assumption that the effect of measurement on objects by the apparatus is differentiable from the phenomena being measured and proposes apparatus as these specific arrangements where boundaries continually (re)create an "exteriority within" (Barad, 2007, p. 135).

Through the performance of apparatus, what we see is the 'intra-action' of matter in its state of becoming, as boundaries are determined and phenomena produced. Intra-action acknowledges that matter is not a fixed pre-existing entity and comes into being through the unfolding of material configurations. Furthermore, phenomena form the basic ontological unit of an agential realist approach and are characterized as dynamic articulations of the world. Therefore, our meaning making and understanding of reality is through intra-actions of matter in its state of continual becoming and the material-discursive practices. When conducting experiments, the measurements we make are partly entangled with the phenomena, and cannot be separated from instruments or the human. Measurements in this sense are marks that appear on objects within this entangled phenomenon. Cause and effect are therefore "causal entanglements" (p. 176). Intra-action is causal in revealing the marks of measurements.

An agential realist approach acknowledges matter as continual configurations of materiality and attributes agency to this becoming through intra-action. The 'mattering of matter' is entangled with the production of phenomena and challenges the passive role we attribute to non-human entities in science. Instruments are not fixed objects that we come upon to manipulate, nor are humans independent of the measurement process, rather their cultural practices are entangled with the engagement of using instruments. In this approach, we are challenged to see boundary making as an ongoing process that involves the decisions by all evolving entities within the phenomenon. In a classroom, instruments are not separate from the students who use them, from the object they are 'measuring'. Instead, their entanglement in mattering and boundary making creates material-discursive arrangements, which include theoretical concepts. Barad challenges educators to rethink how teachers frame the role of instruments and material in knowledge construction in classrooms and encourages us to problematize this relationship. So rather than focusing exclusively on using instruments or matter correctly, the classroom should be a site where teachers and students explore how instruments and measurement construct the phenomena that we observe.

DIFFRACTING CLASSROOM SCIENCE

Diffraction or Reflection

> This is the part of the creativity and difficulty of doing science: getting the instrumentation to work in a particular way for a particular purpose (which is open to the possibility of being changed during the experiment as different insights are gained). Furthermore, any particular apparatus is always in the process of intra-acting with other apparatuses, and the enfolding of phenomena (which may be traded across space, time, and subcultures only to find themselves differently materializing) into subsequent iterations of particular situated practices constitutes important shifts in the particular apparatus in question, and therefore in the nature of the intra-actions that result in the production of new phenomena, and so on. (Barad, 2007, p. 203)

We, as teachers, are cajoled to reflect; reflect on our practice, on a specific lesson, on the day, on our goals, on our vision of the school, and much, much more. However, Barad (2007) argues the physical act of reflecting merely shines back what is being reflected upon. In that sense, when we reflect on lessons or lab activities, or classroom communities, we are merely seeing – although maybe seeing for the first time – what it is we aim to reflect like a mirror. Yet, science classrooms are places of community, full of different perspectives. Valuing all perspectives requires an embracement of polysemy, or the understanding that there are multiple ways of knowing – through seeking polyphonic, the acknowledgement of multiple voices.

In that sense, reflection is a solitary, even reductionist, practice. Instead of reflecting, it might be important to take the methodological stance that comes with diffracting, just as waves diffract, different understandings, or insights, through each other (Barad, 2007); this does not have to be just an individual process, much like reflection seems to be. Instead, diffraction can encompass insights from different members of a classroom community to create new meaning from what has been experienced from the viewpoint of many. Diffraction generates differences.

Barad (2003) describes diffraction as a metaphor for her methodology. Borrowing from Donna Haraway (1997), Barad describes "reading insights through one another in attending to and responding to the details and specificities of relations of difference and how they matter" (2003, p. 71). Throughout her writing, she diffracts insights from vastly different disciplinary and cultural spaces in order to generate new meanings, new ways of seeing the social and physical world. As teachers and learners or, teacher learners – in addition to teacher researchers (Alexakos, 2015) – taking account of different viewpoints into how we see science teaching and learning is important to create inclusive classrooms for all. Science teachers who embrace diffraction as a methodological – and pedagogical – approach to informing their classroom practices will view laboratory practices, classroom discussions, and the influence of the material resources on student learning in new and important ways helping teachers and students to be comfortable with difference.

Informing Classroom Practice

Diffraction can also inform a tool of classroom improvement where all participants of a classroom community – students and teachers – are given equal space to participate. Ken Tobin and his colleagues (Roth & Tobin, 2002; Roth & Tobin, 2006) have theorized and implemented a classroom apparatus called cogenerative dialogues; these are a methodological apparatus for the whole community, which can be reconstituted through the lens of diffraction. Cogenerative dialogues are a space where students – from just two students to the whole class – and teacher(s) bring their viewpoints on the teaching/learning practices and cultures within a specific classroom community, with the goal of improvement for all within that classroom context (Roth & Tobin, 2002). These cogenerative dialogues must be framed, or at least attempted to be framed, outside of the classroom power structure to allow all voices to be heard equally. Through a Baradian lens, we can view this as a diffraction space, where different insights from different levels of the traditional classroom power hierarchy are given equal weight as a way to collectively understand and to collaboratively generate improvement in practice (Lenz Taguchi, 2010).

Embracing diffraction to inform classroom practice can foster agency and responsibility upon each learner – both teachers and students – to express their specific point of view to counter oppressive narratives that overtake classrooms and can create negative spaces, both emotionally and academically. Barad's agential realism provides a different way to think about what it means to know, where "knowing is a matter of differential responsiveness (as performativity articulated and accountable) to what matters" (Barad, 2007, p. 149). Therefore, as teachers, students and material "intra-act" in the process of knowing and understanding the classroom through diffraction, they are generating meaning. Without participation from all, meanings are produced which can result in boundaries/practices/understandings that exclude and alienate participants and matter within the classroom community.

More specifically in the science classroom, knowledge and meaning making about science as a discipline to be learned is an iterative process of collaboration and diffraction. As evidenced by the quote above, Barad sees apparatuses as the important part of knowing/seeing different phenomena. Phenomena is the fundamental unit of science (instead of particle, energy, measurement, etc.).

Therefore, the intra-actions of apparatuses with humans and nonhumans play important roles in the shaping of phenomena. This strongly informs science-teaching practices; in laboratory tasks, each student will intra-act with the tools and materials given to them with slight difference, leading to outcomes of slightly different phenomena. Therefore, science teachers should not provide laboratory assignments with the purpose of confirming a scientific phenomenon that has been discussed completely previously. When we do, we are forcing many parts of apparatus – material, human and social – to a predetermined set up, suppressing the possible understandings students may bring from their experiences. Instead, laboratory exercises must make use of the differing meanings student make by facilitating diffraction. If students, whether the whole class or in small groups, bring their insights about a laboratory experience to a conversation framed through diffraction, then their differing opinions, worked through each other, will come to new meanings about their experiences, and about science. This is both an inclusive and authentic scientific experience.

Focusing on phenomena as the "intra-action" between material and person means that each student, being a part of their own apparatuses will see phenomena differently; bringing the scientist/learner/student into the apparatus used to measure the world pushes us as science teachers to rethink how we see the science we are teaching. This requires community/cultural based pedagogy of science, where the student walks into each classroom as a shaped piece of the apparatus they will use to see the world. They are part of the world, part of making the phenomena that science teachers hope they can interrogate. Therefore, science learning is incredibly student/community/culture/history nuanced. In this sense, science is not something to be learned, but something to participate in. In this way, we must bring the complex interactions between the social and material of a science classroom to the forefront of how we think about teaching and learning.

Baradian Approach to Classroom Experiments

In the following section, Shruti Krishnamoorthy describes a specific experiment she conducted in a Liberal Arts class: I used Barad's agential realist framework in approaching an undergraduate course, focused on developing discussions around citizen science and science in the community for non-science majors. The particular section I focused on had the goal of supporting students to relate rising carbon dioxide levels in the atmosphere to ocean acidification, and the impact that more hydrogen ions in the ocean could have on marine life. As part of the lesson intentions, I hoped to explore ocean acidification through a practical experiment in the classroom that saw the degradation of shells submerged in various concentrations of acid and ocean water. From my perspective, my goal was to have students in the classroom gain some appreciation of the impact of human activity on marine life while building connections between different ecosystems. To approach this class using a Baradian theoretical lens, I re-framed my own lens in deciding where to make agential cuts in defining the boundaries of the apparatus while considering diffractive gratings that I would use to intra-actively produce phenomena in the classroom.

Barad describes apparatus as "specific material configurations, or rather, dynamic (re)configurations of the world through which bodies are intra-actively materialized" (p. 170). In this sense, Barad sees apparatus as dynamically formed and continually configured. Using her view of apparatus as "open ended structures" (p. 170), my approach to presenting the experiment was to first create intentional dynamic boundaries around the defined experimental apparatus such that the cuts made were done so with consideration for the possible entangled entities of inclusion and exclusion. Given Barad argues that matter is not an entity on its own but produced through intra-action, my decision around boundary cuts for inclusion and exclusion were not cemented, but shifted and developed through diffraction as the lesson continued. That is to say, the apparatus for the experiment did not merely consist of beakers, solutions, shells and indicators, but included the students, their laptops, notebooks, cellphones, the desks, chairs, projector, slides, the classroom and myself. I also considered 'outside' elements such as people walking by the room, the university context we were situated in and so on. However, the first cut I decided on was to limit the physicality of the apparatus to the classroom. In a sense, although the students bring elements from outside the room with them, this boundary constrained our context for mattering to those elements present within the physical space. We therefore defined our apparatus to the classroom and everything contained within it.

Next, students focused on a discussion to determine their experimental procedure. We engaged in discovering what was important for us to investigate, why it was important and elements in an experimental design we thought would help in materializing objective phenomena. I sought to employ a diffractive approach in the intra-active discussion between the students and me. This involved using each students' suggestions arranged with, against and through each other has to highlight differences in their ideas and possibilities that could ensue. Simply put, I allowed students to suggest apparatus and protocols they would use to test the effect of ocean acidity on shells and using a combination of layering their suggestions upon each other, combining their suggestions and talking through them, looked for highlighted areas of difference and superposition. In this process, the apparatus for experimentation included (at various times) slides of an experimental set-up I placed in front of them, their own *laptops used as resources*, materials in the classroom and their physical arrangement within the classroom space to help materialize their suggestions and ideas. In this sense, apparatus boundaries for the experiment were in flux, creating different phenomena at each moment.

At the end of the discussion, through their diffractive intra-action, students had decided on areas of similarities and differences in the next stage of their experiment. Areas of superposition highlighted their agreement on using ocean water for one beaker, vinegar for another and a mixture of vinegar and ocean water for the third. Nodal points in their diffractive discussion revealed differences in their approach to the mixed beaker where they felt the need for disparate concentrations to understand the effect of ocean acidification. While some students felt that it was important to mimic pH levels that most closely resembled the ocean today, other students felt the need to create samples that measured pH levels much lower than found in the oceans. Effectively, the students were able to create space for their own agency in knowledge creation and through the diffractive process, found ways for similarity and difference in their exploration of the phenomena.

In the next stage, as students collected their samples and observed the effect of their experimental set-up, the mattering of knowledge for each student and group of students was different since each apparatus was different. Considering knowledge creation as dynamic, their understanding of the particular objective phenomena they created was different. So once again through diffraction, the classroom discussion involved each of their observations layered on top of, through and with the rest of the class to facilitate the mattering of phenomena by the group as a whole. In this sense, the dynamic nature in the apparatus follows a dialectical movement from the student, her/his own knowledge creation through the experiment set-up, to the classroom as the apparatus, and the knowledge created as a product of that diffractive discussion. In this particular instance, highlighted differences focused on the differential pH levels and the ways in which increasing hydrogen ions can create differential effects on the shells that were part of the study.

An issue I struggled with when engaging in the diffractive discussion was navigating differences that emerged through diffraction. As the teacher, I found myself aware of the way in which I intra-acted with students' explanatory theories and worked to avoid privileging my views over the students' when creating an understanding of ocean acidification with them. I felt the urge to 'summarize' our findings and wrap the discussion up using my own interpretation of the experiment that did not include the areas of difference that came up in our diffractive discussion. To practice a Baradian approach, I was pushed to be comfortable with difference and embrace no uniformity in students' experience and understanding of phenomena.

Matter as Emergent through Dynamic Intra-Active Discursive Practice

Barad suggests a move away from the nature versus culture dichotomy and proposes instead, mattering of matter created through intra-action such that nothing exists before the creation of the phenomena. Therefore, the creation of phenomena is particular to a specific space-time and apparatus configuration, ever changing in its production and existence. The dynamic nature of her proposed lens creates room for a non-static approach to knowledge creation and understanding of nature. Here, knowledge is constantly created and re-created such that each iteration is dependent on the entangled entities that are included and excluded from measurement. In this sense, Barad's work pushes to re-analyze her understanding of apparatus as applied to scientific principles.

Science Education as a Material Issue

In this chapter, we focused on science education as a material issue from the role of matter in the form of instruments, what can be learned from using instruments and how as instruments are refined, they can extend and expand human knowledge. The second part of the chapter considers matter and the material from a post human, Baradian perspective to diffractively examine our teaching of science when matter has agency. Barad's theory describes the relationship between material and discursive as entangled through the dynamics of intra-activity. The challenge is to resist privileging the human and assuming that matter/material is passive. The camera obscura influences the learner's intra-actions as s/he develops the practices to effectively utilize the instrument. How did the camera obscura influence the learner's material discursive practices? What representations are produced through using the camera obscura? Because of having access to the camera obscura, what new knowledge did the learners generate?

Barad's theory also challenges us to re-examine knowledge from a diffractive rather than reflective stance. Polyphonia proposes that we understand there are multiple ways of knowing but we need to push this interpretation further to ask questions about the material discursive forces that are produced through the intra-actions between learner and the material. How did the learner engage with the material? How does their science identity intra-act with their engagement with the material? What emergent material-discursive practices are produced through building and using a camera obscrua? Are there positive emotions from successful intra-actions? How do cultural values and norms, such as who is capable in science affect learner's engagement in science and a teacher's acceptance that the students' have come to know? Shruti struggled with her intra-action with the students, framing her role as teacher, noting how she privileged her science knowledge over that of her students.

Yet if we are to embrace a Baradian perspective in teaching and learning, the challenge is to constantly and consistently problematize the focus on the human and engage with the material. Shruti struggled with engaging in the discussion but there is also the struggle with recognizing matter's agency. Shruti also needs to examine what role did laptops, ocean water, vinegar, beakers and eggshells contribute to the material-discursive practices. We become engaged with what Hillevi Lenz Taguchi has called intra-active pedagogy (Lenz Taguchi, 2010). In intra-active pedagogy, material is a performative agent, because there are no boundaries between materials and discursive, rather they are co-constitutive, inseparable and interconnected.

These performative agents are critical in the becoming of entities of learners, who are 'being-*of*-the-world', not being-*in*-the-world (Barad, 2007, p. 160). As science educators who embrace this perspective of materiality is to engage with Barad's onto-epistemological theory, that is the theory of knowing and being are not distinct, the implications of this approach is to view teaching and learning as active, with the material having agency in knowledge generation (Lenz Taguchi, 2010). Moreover, we recognize that our *being-of* the knowledge produced in our teaching, our responsibility to critically examine the material-discursive practices that emerge in the pedagogical spaces where we engage in the teaching and learning of science, taking matter seriously and respecting its agency.

REFERENCES

Alexakos, K. (2015). *Being a teacher|researcher. A primer on doing authentic inquiry research on teaching and learning.* Rotterdam, The Netherlands: Sense Publishers.

Barad, K. (2003). Posthumanist performativity: Toward an understanding of how matter comes to matter. *Signs, 28*(3), 801–831. Retrieved from https://doi.org/10.1086/345321

Barad, K. (2007). *Meeting the universe halfway: Quantum physics and the entanglement of matter and meaning.* Durham, NC: Duke University Press. Retrieved from https://doi.org/10.1215/9780822388128

Bioscience. (1984). People and places. *Bioscience, 34*, 130–131. Retrieved from https://doi.org/10.1093/bioscience/34.2.130

Butler, J. (1990). *Gender trouble: Feminism and the subversion of identity (Subversive bodily acts, IV bodily inscriptions, performative subversions).* New York, NY: Routledge.

Fox Keller, E. (1983). *A feeling for the organism.* New York, NY: W. H. Freeman.

Hacking, I. (1982). Experimentation and scientific realism. *Philosophical Topics, 13*, 71–87. Retrieved from https://doi.org/10.5840/philtopics19821314

Haraway, D. J. (1997). *Modest−Witness@Second−Millennium. FemaleMan−Meets−OncoMouse: Feminism and Technoscience.* New York, NY: Psychology Press.

Lenz Taguchi, H. (2010). *Going beyond the theory/practice divide in early childhood education: Introducing an intra-active pedagogy.* New York, NY: Routledge.

NGSS Lead States. (2013). *Next Generation Science Standards: For states, by states.* Washington, DC: The National Academies Press.

Reckwitz, A. (2002). The status of the "material" in theories of culture: From "social structure" to "artefacts". *Journal for the Theory of Social Behavior, 32*, 195–217. Retrieved from https://doi.org/10.1111/1468-5914.00183

Roth, W.-M., & Tobin, K. (2002). *At the elbow of another: Learning to teach by coteaching.* New York, NY: Peter Lang.

Roth, W.-M., & Tobin, K. (Eds.). (2006). *Teaching together, learning together.* New York, NY: Peter Lang.

ABOUT THE AUTHORS

Colin Hennessy Elliott is a Ph.D. student in science education at New York University. Originally, from Easthampton Massachusetts, he received a B.S. in Physics and an M.S. in Science and Technology Studies from Rensselaer Polytechnic Institute. Before beginning his current program, Colin was a high school physics teacher in Newark, New Jersey. His research interests include student science standpoints, student agency, urban science education, culturally relevant/critical science pedagogy, and intersectionality in the science classroom.

Shruti Krishnamoorthy is a Ph.D. student at New York University, interested in high school science education. She completed an undergraduate and master's degree in medical biophysics at the University of Western Ontario and spent three years teaching high school science in the Toronto area. Having grown up in the Middle East, India and Canada, her current research interests include students' identity and agency in the classroom, and using critical, post-colonial and feminist theories to look at science curriculum and pedagogy in South India and New York City.

Catherine Milne is Professor in Science Education at New York University. Her research interests include urban science education, socio-cultural elements of teaching and learning science, the role of the history of science in learning science, the development and use of multimedia for teaching and learning science, and models of teacher education. She is co-Editor-in-Chief for the journal *Cultural Studies of Science Education* and co-editor of two book series for Springer Publishers and Sense Publishers.

Kathryn Scantlebury is a Professor in the Department of Chemistry and Biochemistry at the University of Delaware and a visiting research professor at Uppsala University's Center for Gender Research in Uppsala, Sweden. Her research interests focus on feminist and gender issues in various aspects of science education, including preservice teacher education, teachers' professional development, urban education, and career paths in science. Scantlebury is a Fellow of the American Association for the Advancement of Science, co-editor of *Gender and Education* and co-editor of two book series for Brill | Sense.

ALBERTO BELLOCCHI, KATHY MILLS, REBECCA OLSON,
ROGER PATULNY AND JORDAN MCKENZIE

14. EMOTION WORK AT THE FRONTLINE OF STEM TEACHING

ABSTRACT

STEM education disciplines are facing a dilemma internationally. There is high demand for qualified high quality teachers in science, technology, and mathematics subjects in schools and a trend towards high attrition rates of teachers in the early years. Teacher attrition has been associated with, in part, the high demand for emotion work required on a daily basis in classrooms. Despite this fact, there is dearth of research on teachers' emotion management in specific disciplinary fields such as science and mathematics education. In this chapter, we address the need for research on teacher emotion work in science and mathematics education through sociological analysis of the lived experiences of two authors. Using narrative vignettes of classroom experiences, we consider what the sociology of emotions can contribute to understanding the emotion work required in science and mathematics classrooms. Following an analysis of the two vignettes, we offer suggestions for research, policy, and practice for addressing STEM teachers' emotional work through teacher education courses and professional learning.

Keywords: emotion work, emotion management, STEM education, emotionalism, sociology of emotion

BURNING OUT AND TURNING AWAY

> You start to *feel worthless*. Really *crushed*. And you want to tell them to stick it but you don't because you're *scared* of not having a job. I'm *fearful* all the time. I feel *sick in my stomach*. I'm not a *suicidal* person but I could see how someone could be driven to that. (Dalton, 2014, emphasis added)

Attrition of teachers within the first five years in the profession is an internationally recognized problem (e.g., Ingersoll, Merrill, & Stuckey, 2014) that often gains media attention, as illustrated from the above newspaper quote. Emotions such as fear, worthlessness and being sick in the stomach are not uncommon and some teachers have begun collecting undesirable experiences in web-logs[1]. Public disclosures about traumatic circumstances of teachers' work reveal the highly emotional nature

 DOI:10.1163/9789004389663_014

of the job. They may also represent a call for help from the outside or a call to action within the teaching community. It is unsurprising then that such experiences lead to burnout and ultimately to teachers leaving their jobs. Although there are ample contrary cases of productive and pleasant teaching experiences, our focus in this chapter is on understanding the emotional nature of the job, how this may interrelate with teacher attrition, and how we can attend to the problem through teacher education.

Attrition is not an issue specific to science education; however, there is high demand for science, technology, engineering and mathematics (STEM) teachers globally (Goldhaber, Krieg, Theobald, & Brown, 2016), and it will be challenging for governments to meet these demands if teachers are burning out and turning away in their early years of the profession. Managing emotions like those described in the above quote constitutes the emotion work or labour (Hochschild, 1983) performed by teachers during daily encounters. A concept developed by sociologist Arlie Hochschild (1983), emotional work or emotional management refers to the effort expended to control our own emotions and the emotions of others, when complying with organisationally defined feeling rules, or normative social expectations. In this chapter, we consider the issue of emotion work by reflecting on two narrative vignettes provided by Alberto and Kathy about their past high school and primary school science and mathematics teaching experiences respectively. Our discussion focuses on the following question: How can perspectives from the sociology of emotion inform understandings about emotional work in teaching?

We conclude with recommendations to address the problem of teacher attrition, with a focus on the preparation of preservice teachers for the forgotten but critical dimension of emotional work.

METHODOLOGICAL CONSIDERATIONS

Investigating teachers' emotional work through personal narratives is consistent with the idiom of *emotionalism* (Gubrium & Holstein, 1994) whereby researchers seek to recount emotional experiences through raw detail. In reporting unfettered experiences, our accounts as former teachers can address the way in which we came to terms with "the nature and conditions of [our] existence" and how we make "sense of the wider, more differentiated, and more complex … contexts of [our] experience[s]" (Brockmeier & Harré, 1997, p. 264). We distance ourselves from ontological fallacies that would receive our narratives as representations of truth or some historical (fixed) reality that was experienced by the authors (cf. Brockmeier & Harré, 1997). Rather, the narrative vignettes presented in this chapter, constitute Kathy and Alberto's (re-) constructions of lived teaching experiences that establish an empathetic connection and solidarity with colleagues who are in the profession or who may have abandoned it. The importance of this chapter, then, does not come from a pretence of isomorphism between the stories and the teachers' past experiences, but rather from what the stories and our interpretation of them awakens in our readers as possibilities for shared understanding and for informed action through "the vicarious testing of life possibilities" (Clandinin & Connelly, 2000, p. 42).

A TIMELY CHANGE FOR SCIENCE EDUCATION RESEARCH

In the face of STEM teacher shortages and the parallel problem of high teacher attrition, the increasing attention to emotion in STEM education research is timely (Bellocchi, Quigley, & Otrel-Cass, 2017). A focus on *emotion management* in science education began with the pioneering work of Michalinos Zembylas (e.g., 2004). In his ethnographic study of an elementary science teacher, Zembylas (2004) demonstrates how the performance of emotional work is an important aspect of teaching. Some noteworthy outcomes from that study included the identification of positive benefits from the teacher's emotional work including increased satisfaction, self-esteem, and psychological well-being. Zembylas' work has since been followed by a cascade of studies focusing on emotional regulation of high school students (e.g., Tomas, Rigano, & Ritchie, 2016), and preservice science teachers (e.g., Bellocchi, Mills, & Ritchie, 2015). Aside from these recent studies, past educational research on emotions has been informed predominantly by psychological perspectives (cf. Pekrun & Linnenbrink-Garcia, 2015). For this reason, we illustrate the relevance of *sociology of emotions* for expanding our understandings about the lives and work of teachers and the issues facing policy-makers in relation to attrition.

We frame our chapter around two narrative vignettes provided by Kathy and Alberto focusing on past school science and mathematics teaching experiences. As each vignette is presented, each of the chapter authors then discusses key emotional issues that address our guiding question. Before outlining our individual backgrounds, a brief biography of our collaboration is required. Alberto and Kathy are work colleagues who recently commenced collaborative research on emotion and online learning in science teacher education (see Bellocchi, Mills, & Ritchie, 2015) and in supporting Australian indigenous children in learning about emotion through multi-modal expression using video and images (Mills, Bellocchi, Patulny, & Dooley, 2017).

More recently, we have begun a new collaboration with Rebecca, Jordan and Roger drawing on their sociological backgrounds to understand emotional work in teaching and learning. In the chapter, Rebecca draws on her knowledge of emotion in health-care and her studies of inter-professional education to explore the narratives through sociological perspectives. Jordan draws on his grounding in social theory to provide perspectives on identity and well-being, and Roger offers a macro-sociological perspective connecting the emotion work evident in the narratives to issues of gender and emotion in education. As we each reflect on the narrative vignettes, we collectively develop a broader story in this chapter; one that informs our understandings of STEM teachers' emotional work, and one that considers how these understandings can inform STEM teacher education.

TALES AND INTERPRETATIONS OF EMOTIONAL WORK IN STEM TEACHING

Vignette 1 Emotionality and Primary Science Teaching

Kathy: As I reflect on my years as a primary teacher, I relive a wave of strong emotions. I taught maths and science along with a range of school subjects to children from ages 5 to 11 years. Research into teacher's emotions has only recently been recognised as incredibly vital (Mills & Ritchie, 2014).

During the first few years of my career, I felt that my teaching was driven by a genuine sense of purpose. I was contributing in many valuable ways to the life outcomes of my students, their families, and the wider community. Some might have described me as the idealist, the visionary, the influencer, the performer, the curriculum expert, the organizer, the scientist, the artist, and the wise sage. I relished the opportunity to lead inspirational sessions for staff at the start of the day. I enjoyed teaching the students about the physical and natural world through observation and hypothesis testing – to see them grapple with new concepts – and to open their minds to previously inaccessible ideas. Despite being the top of my year at college, I was earning below-award wages at an independent school away from my home city, but to me it didn't matter because I felt that I was making a difference. At night, I would often stay back late until I was satisfied with my curriculum preparations.

Colleagues were very supportive of each other at my first teaching post. If one of us was having a bad day, we would open the room to make a double teaching space and play dance music with the students to provide some comic relief. I once gathered my Year 1 students to sneak under the classroom windows of the next room to surprise them with an impromptu puppet show during that teacher's science lesson. I relished our science excursions which encouraged parent involvement, like "overnight camp" to observe the night sky through telescope, sleeping bags spread across the classroom floor. These early teaching experiences made me feel emotionally energized, with a deep happiness and satisfaction. No matter what difficulties I confronted in life, I found it easy to make the classroom a secure, supportive environment for my students – a place of belonging and engagement, a place of productivity and hope for the future.

Over time, I worked in both regional and urban schools. The emotional costs of teaching gradually seemed to become heavier. I became well respected by the school leadership and gained more curriculum leadership roles, but also gained a few enemies among my teaching colleagues. I became buffeted in the fray between teachers and the bureaucratic requirements of external curriculum policies. Staff did not want to change their curriculum to meet the new requirements. I too was resistant to certain school-based regulatory requirements, particularly mandates to use pedagogies that were antithetical to research evidence. For example, one school-based program required staff to read their lessons word-for-word from a teacher manual in front of the students. This was to compensate for all the mythical "bad" teachers who didn't plan proper lessons. At one time, a senior staff member required that all written work from the early year's students must be copied from the blackboard for notation accuracy, which limited thinking, learning, and creativity. I saw that centralised and school-based administrative decisions were often leading to the deprofessionalisation of the teaching workforce. Increased standardized testing and accountability, and the government regulation of schooling were exasperating both the students and myself alike.

Occasionally, there were stresses from parents who didn't accept the grades in report cards, but whose children needed learning support. Often school administrators placed demands on teachers for better student outcomes, but provided low levels of support in terms of teaching hours, class sizes, provision of teacher aids, specialist teachers and learning support and staff development. Workloads were high, especially during reporting periods. I watched as many of my female colleagues left teaching even within the first year after graduation, some never taught, and many never returned. Some left from the pain of managing difficult classrooms, others hurt by gossip and false accusations that had spread throughout the parent community. Somehow, I survived. I was resilient.

When referring to my teaching career to non-teaching friends, they would laugh at my use of the term "career". "Teaching is not a career, it's just a job", they would quip. At the time, I was primary Head of Curriculum and class teacher, mentor of pre-service teachers on field experience, completing two postgraduate degrees, and lecturing in the evenings at a teacher's college. Once a week, I coordinated a team of volunteers to run an afterschool club for underprivileged children. In social discussions about occupations outside of work, it was clear that some of my friends in other professions considered themselves to have an elevated status to teachers. I would simply laugh, "Aha, but if you love what you do, you never have to work a single day of your life!"

Disaffected students were a normal part of classroom life, and these students and their parents needed more time, counselling, and attention from teachers. I no longer had the energy to offer free tutoring after school for students who were struggling with troublesome maths and science concepts, as I had enthusiastically done in the beginning. The learning and behavioural problems of students seemed to be becoming greater, the learning gaps wider, and the inequities irreparable. I sometimes felt powerless to change the system. I was beginning to feel like a cog in a wheel, moving through the motions. I recall sorting out mundane disputes between students in a detached manner. In sum, around the end of my seventh year in the frontline battles of primary classrooms, I was essentially sapped of emotional resources. On my last day of teaching, I packed up my car filled with teaching resources I had invested my wages in – an aquarium, plants, and models of the solar system. I took one last look at my primary classroom and decided it was time to move on.

Rebecca: Important parallels can be drawn across Kathy's vignette and the scholarship on emotions in the health professions. The first is that emotions are relevant to STEM teaching and medicine. Although this may seem obvious, given the focus of this chapter, it serves as an important reminder. Disciplines and thought communities allow us to recognise some aspects of our work and push us to ignore others (Zerubavel, 1997). For years, emotions have been regarded as irrelevant or even detrimental to medicine (McNaughton, 2013). The same might be said of teaching, and perhaps science – where emotions are often treated as a threat to neutral observation – teaching in particular. Emotions are, however, integral to teaching, being a professional, and indeed to being human; Kathy's vignette is a poignant reminder of this fact.

A second parallel can be drawn regarding emotional work. If emotions are fundamental to teaching, then teachers could be said to perform emotional work. Teaching and most notably nursing (James, 1992), along with all other work involving interactions between humans, involves emotional work. This might include suppressing one's own feelings of frustration and lifting the spirits of a patient, engendering excitement about astronomy amongst year three students or helping to change the mood of a colleague who is having a bad day. In Kathy's vignette, we see her early experiences defined by an emotionally supportive organisational culture that encouraged teachers to use music, puppet shows and creativity to sustain an engaging learning atmosphere.

Roger: Whereas Kathy relates her experiences in a professional and gender-neutral tone, they implicitly reveal several gendered aspects to her work, and to the emotional work she describes. Women are more commonly represented in care industries, and care industries are more often undervalued (Husso & Hirvonen, 2012), a sentiment that sometimes runs through the teaching profession. An example of this can be seen in Kathy's recollections of her friend's quip that "Teaching is not a career, it's just a job."

Women are also more likely to engage in emotional work. Lois (2003) distinguishes men and women's emotional work, describing *tight* emotion work aimed at quickly directing and controlling the emotions of others (such as by police and fire fighters) as predominately male work, and *loose* emotion work aimed at listening and sympathising (such as by counsellors and nurses) as predominately female. Theodosius (2008) notes that women are more likely to engage in emotional work by warrant of their greater involvement in care and teaching work. It is important, though, to acknowledge Bolton and Boyds' (2003) critique of Hochschild, that emotional work is not always a form of alienation; sometimes it is an expression of the best in us as workers and colleagues. Kathy describes a supportive, playful environment in her early career, and recounts a wonderful story about performing a sneak puppet show for a fellow teachers' classroom. This is a good example of how emotional work can improve interpersonal wellbeing when enacted for purposes described by Bolton and Boyd (2003) as *prescriptive* (out of professional pride/commitment) or *philanthropic* (gift exchange to colleagues).

However, Kathy recounts how her emotional load became heavier as she gained more leadership roles. This is in keeping with research that identifies difficulties in reconciling organizational needs with the care needs of the jobs faced by many women, as noted by Husso and Hirvonen (2012). However, this could also have been part of a movement into a more competitive, masculine culture in the higher level, administrative side of the school. Men tend to dominate the senior management roles in most organisations, and studies have found that women are more likely to engage in upward management of *bullying* male bosses in managerial and legal professions (Lively, 2000). Kathy might have experienced a movement into a less mutually supportive, more masculine, competitive, and seemingly more conservative environment, in terms of enforcing outdated pedagogical practices.

Another gendered facet of her story was how she put in extra time to teach struggling children with special needs after class. This is again reflected in the literature on women performing much unrecognised labour in care contexts, and the personal dissonance this tension between organisational and human goals creates in teachers (King, 2012). But it is perhaps best revealed by its absence—that by the end of her time teaching, when she was burnt out, Kathy only performed what was officially expected of her in her teaching role, rather than performing above and beyond the role as a matter of course – as seems to be the expectation faced by many women in care work.

Jordan: Kathy's vignette is an enlightening reminder that emotions are always more than responses to stimuli or codes of social interaction. Although emotions certainly do fill these roles, they also contribute substantially to the development of a sense of self. Kathy's willingness to prioritise fulfilling and rewarding work over pay rates, convenient hours, and lighter workloads is supported by her pursuit of what Weber famously referred to as a *calling* in his canonical work on the Protestant Ethic. But a more contemporary reading based on the sociology of emotions literature would indicate that Kathy's emotional experiences in the workplace contributed to a powerful and motivating sense of self (Lasky, 2005), that became a source of fulfilment more pervasive than income or workload. This morale is reflected in the research literature on school teachers that indicates that if teachers can build a sense of self upon experiences of work that make a positive difference, they will be willing to make sacrifices to other aspects of their working life. Yet, if the institutional and bureaucratic pressures of the job challenge the authenticity of this charitable and generous sense of self, then teachers face a contradiction from the pressures of managing two contrasting selves. Arguably, this finding is reflected in Kathy's experiences. What is theoretically interesting here is the evidence that emotional experiences were not simply experiences for Kathy, but a key aspect of her Verstehen – or her way of engaging with and making sense of the world. It was through her emotional experiences that Kathy made sense of the bureaucratic school structure, the challenges of her students and the demands of their parents. Most importantly to her role as a teacher, it was the combination of positive and negative emotional experiences that supported the construction of an authentic self (Ruti, 2014) capable of contextualising the high and lows of teaching into a meaningful narrative that can connect with her initial desire to inspire students. Perhaps the challenges to authenticity posed by working within rationalised institutions are too much to overcome in the long term, leaving Kathy no choice but to sacrifice the sense of self that she worked so hard to accomplish or move on to another line of work.

Alberto: What we see in Kathy's story is how emotions such as empathy and care for the wellbeing of her students and colleagues, led her to create micro-social structures such as after-school learning support groups, and puppet shows with her year 1 students to lift the mood of her teaching colleague in an adjacent class. Not only is Kathy's personal emotional work evident from the vignette, but also examples like the puppet show illustrate the collegial work performed by teachers to support the emotional experiences of others. Whereas the notion of emotional work set out by Hochschild suggests a strong individual focus, Kathy's story indicates the possibility for collective emotion work to be just as significant in social groups as any labour that is conducted individually or in lack of physical presence of others. This could be a highly important factor in the retention of science teachers as EunJin Bang, Anne Kern, Julie Luft, and Gillian Roehrig (2007) found through their longitudinal study of science teaching. Although their focus was on secondary science, teachers in their study who left within the first year did so because of unfavourable school climate, feeling unsupported from colleagues, lack of administration support, and students' classroom conduct.

Vignette 2 Under Fire from F-Bombs and C-Bombs

Alberto: I was allocated 8th grade mathematics, some junior science classes, and an 11th grade chemistry class in my first year of teaching. I was not sure exactly what to expect because groups of students and classes can all have their unique personalities. It was in the first weeks of that year of teaching when I recall asking myself "What am I getting into." The mathematics class allocated to me was considered to be a low-achievment class based on previous academic performance.

During one lesson early in the year, we were in a small classroom located within a demountable building. The students had settled into the room and the lesson was underway. At one point, I faced the chalkboard to write some instructions for the lesson. As I did so, an argument developed behind me. I was not sure how it had started, but I will never forget how it ended. From what I recall of the incident, a boy sitting in the front row very close to where I was standing, had an altercation with a female student a few seats away in the row behind him. I think the girl had made a comment about his personal hygiene and the boy, taking offense to her comment, began to defend himself verbally.

As the girl threw an insult in his direction, he retorted with "F..k you b…h!" At this point, and for reasons that never became clear, another boy sitting in the very back row (i.e., approximately three rows behind the aggrieved boy), began throwing small pieces of eraser and verbal slurs at the first boy, while the girl retorted in kind with "f..k you c..t!" Now a three-way screaming match was exploding in a matter of seconds right behind me. If I had been asked hypothetically that morning, or any time before that day, "How do you deal with situations like this one," I am sure I would have rattled off what I had learnt in my teacher education course on classroom management. I am also sure, that my reply would have looked nothing like what I did in the moment of that lived experience. In the initial phase of the altercation, I asked the students to end their argument and refocus on their work.

At this stage, I was not expecting the full-blown verbal warfare that was about to explode as the f-bomb and c-bombs were yet to be dropped (colloquial Australian expressions for "f..k" and "c..t"). Once I heard these expressions, I spun around immediately, I could feel my anger swelling, making my body shake but this was shrouded in hesitance and fear. I pointed at the eraser-thrower at the back "You, stop that!" I turned and pointed at the c-bomber girl "You. In that corner [pointing to one part of the room]!" I then turned and pointed to the f-bomber boy in the front row, "You. OUTSIDE!" I am no shrinking violet, but hearing that kind of language use by 14-year-old students tapped a nerve. The school principal disliked seeing students on classroom verandas and this played on my mind, leaving me to feel nervous, as I addressed the girl in the classroom, leaving the f-bomber boy waiting outside. Despite my anger at the extent of profanities used, I found it in myself to ask her what had happened. After collecting her version of events, I went to the veranda, closed the door, and asked the boy to explain his version. I can't remember much more past this point, other than the fact that after teaching a few more classes that day, I went home and asked myself if this is what I had signed up for. I didn't feel like a "science-maths teacher" on that day. I felt like I was doing something else, and that I had to be something else. This experience was the first and most memorable that led me to question whether these types of experiences represented what I had signed up to do.

Somehow, I looked past that event and despite others like it, albeit not quite as intense as that one, I taught for eight years before leaving to take up a new career in academia. In fact, as each year went by teaching became easier to manage, I thoroughly enjoyed the job and my relationships with the students. I had become more experienced, students had begun to know me from previous years or through siblings whom I had taught, families knew me, and I had developed a good reputation within the school community. From that initial maths experience and related episodes when teaching science, I learnt to regulate my own reactions and feelings towards students and classroom experiences that were unfavourable. At that time, I had also become more aware of the powerful role that effective social bonds with students, their families, and colleagues can have in improving teaching and learning experiences.

I can now account for my past teaching experiences by differentiating three key stages of transformation regarding emotional events. The first phase was marked by unbridled reactions to events that unfolded in the classroom (e.g., the f- and c-bomb episode). Once I had come to realise the impact this was having on me mentally and physiologically, a new stage began. In this second stage, I suppressed any expressions of how I truly felt so that I would not act on my anger, for instance, through actions like shouting. Yet, after some time, I felt that I was cheating my students by not letting them see how I truly felt and realised that trying to suppress the expression of my feelings was not good for me. I also realised that covering up my emotional expressions did not convey the right signals for students to judge my feelings, and I understood that this would limit their own courses of action in response to what I displayed. At this point, I entered a third stage, where I chose to show my true feelings without reacting to them in the way that I had during stage 1. That is, I would let students see my anger, joy, or disappointment for example, through my facial expressions. However, with feelings like anger, I would avoid actions such as raising my voice. Instead, I would signal my feelings non-verbally (e.g., angry face), and then articulate how I felt ("I am annoyed at that"), before issuing a consequence for any unsuitable student behaviour. During this third stage, I also made deliberate efforts to share personal stories to establish social bonds with my classes as a pre-emptive measure. I wanted students to see me as a fellow human being rather than simply as their science or maths teacher or the automaton that they must have experienced in my second stage. I eventually returned to the ideas about managing classrooms from my university studies and started to construct more sophisticated strategies based on the interaction between my readings and classroom experiences in phase 3. For example, I have written previously about metaphors used to differentiate my approach to junior school classes (policeman metaphor) when compared to senior classes (Steve Irwin, entertainer metaphor) (Ritchie, Bellocchi, Pöltl, & Wearmouth, 2006), which evolved continually as I progressed through my three career stages.

Jordan and Roger: Alberto's reflections in this section capture Holmes' (2010) notion of 'Emotional Reflexivity' whereby the performance and understanding of emotional states can be reflexively interpreted and reinterpreted by both the performer and the audience. Rather than suggesting that individuals are necessarily reflexive about their emotions and risking the implication that all emotional states are intentional, Holmes argues that reflexive theories "need emotionalizing" (Holmes, 2010, p. 139). In so doing, Holmes points to the various ways that the reflexive process of interpreting social meanings, relationships and actions requires consideration of the role of emotions. We can see this clearly in Alberto's reflections on emotion management and the challenge of determining how honest he ought to be with his students regarding his own emotions.

Whereas traditionally the emotional roles of teachers were more fixed and structured by tradition, such that teachers were expected to present cold and unemotional personas that masked care or affection, Alberto faces the challenge of renegotiating how to engage with students in effective and sincere ways. However, this cannot be accomplished without some recognition of his own emotional experiences that naturally push or pull him in particular directions. He cannot rely on reason or emotion alone to do this, but must find a balance that enables him to incorporate the skills and knowledge acquired through his education, with the variety of emotional interactions taking places in the rather complex dynamic of a classroom.

Alberto's reflections highlight how the roles of reason and emotion are not only inseparable and interdependent, but also constantly challenged and renegotiated throughout his personal development as a teacher. This reflexive process is not attributable to rational action, but to a process of self-awareness and reflection that entangles experience, reason, and emotion. Alberto's recollections here hint at how masculinity can shape emotional responses to situations of conflict, and their resolution. He describes a situation between students that is gendered – an altercation between a girl and two boys – and his response could perhaps be interpreted as gendered. He recalls his own confusion as a precursor to the event – with little awareness of the tension in the room and the possibility of swift escalation to full aggression and even violence – and his own, reactionary, groundswell of anger and outrage at the language, emotions that strongly enervated his response.

Although his reactions were not inappropriate to the context – there is a role for discipline in teaching – his response was clearly assertive, authoritative, and perhaps typically masculine in its confusion as to the emotional dynamics underlying the situation, and in the degree of anger that came to characterise the response. In some contexts, such *masculine* approaches to handling emotionally charged situations in the classroom might not be so appropriate – or at least not optimal when compared to taking pre-emptive emotion work to avoid the situation occurring in the first place – in terms of the emotional toll it takes on teachers. Alberto recalls how the event made him reflect on what went wrong that day, and encouraged him to engage in greater efforts to befriend students and families to pre-emptively avoid such emotionally charged situations repeating themselves. This demonstrates a potential shift away from a *masculine* emotional culture of automatic authoritative responsiveness to classroom conflict, towards a more reflective – or gender-neutral – response.

Alberto's vignette again reveals the complexity of emotion management. Hochschild (1979) notes that emotion management is not just about locking down or repressing emotions, it also about evoking emotions in us or other people. This can be seen in Alberto's different forms of emotional work used on junior students (stern emotional control) and senior students (jokes and humour) (cf. Ritchie et al., 2006).

The former strategy does touch on other aspects of the masculine approach to emotionality, or more precisely to the often-cited need for men to appear rational, calm, and even stern rather than emotional in taking charge of situations (Louis, 2013). Jack Barbalet (1998) argues that emotions and rationality form complements rather than substitutes, though we traditionally place the latter above the former, and this is reflected in the development of Alberto's emotional skillset. In his self-described 'stage 1' period of teaching, Alberto reacted in a manner he felt was *over-emotional*, with feelings of anger, and actions such as shouting, that were unmediated by reflection or engagement in emotional work.

To curb these experiences, he shifted in stage 2 to a classic rational approach of suppressing emotions and engaging a tighter form of emotional control, which is somewhat akin to Hochschild's (1979) concept of *surface acting*. This 'surface' approach to *papering over* emotions does not convey an authentic emotional experience, is disappointing for students and debilitating for teachers (as Alberto recounts), and is worrisome given findings linking *surface acting* to increased levels of stress, mental health problems, and dissatisfaction with work (Näring, Briët, & Brouwers, 2007). Alberto's third stage was the real breakthrough, in that by this time he had learned to be much more aware of his own and others' emotions, and to neither to allow nor repress unfiltered emotions. Instead, he reflected on their appropriateness, and the appropriateness of their component actions, such as hand gestures, vocalizations and facial configurations for particular situations and classes.

CONTRADICTIONS AND COMMONALITIES IN THE NARRATIVES

Rebecca and Jordan: Both Kathy and Alberto's vignettes highlight the emotional work that teachers perform. Kathy describes the work she did to lift students' and colleagues' emotional states. Alberto describes the effort he put into reflecting on his anger, and regulating his reactions to maintain control over the classroom and facilitate learning. Alberto's account also hints at the limits of emotional work. After his unexpected efforts at disarming "c" and "f" bombs in the classroom, he writes, 'I didn't feel like a "science-maths teacher." I felt like I was doing something else and that I had to be something else.' Athough we may be able to manipulate our outward presentation to comply with role requirements, whether or not we can sustain this emotion work depends on whether the role we are acting, feeling, and being fits within our identity and our other roles. If it does not fit, role strain and burnout will likely eventuate (Lois, 2006).

Alberto's vignette also shows that emotions, and the emotional dimensions of teaching, are not solely experienced cognitively, but viscerally and socially – within interactions. He describes his embodied response to the confrontation in his classroom: anger so strong it prompted him to start shaking. This description reminds us to use theoretical tools that account for more than the conscious elements of our emotional experiences. Theories of emotions such as Hochschild's concept of emotional work leave something to be desired on this front. Although they acknowledge the biological aspects of emotion, the focus is primarily on the manipulation of emotional experiences within our conscious control.

Alberto's vignette highlights an example of how relationships of power are not always stable and predictable, but contested and malleable. This point is not uncommon in the literature of political sociology (Lukes, 1986), the presence of such a blatant disregard of Alberto's authority as the teacher in this scenario forced him to fulfil a role that is clearly outside of what aspiring teachers seek to play. The task of switching from the role of police officer (as he describes it) to the supportive mentor and friend requires an enormous amount of emotional energy and as Alberto suggests, this was not something that his tertiary education had fully prepared him. Although Hochschild describes scenarios where performance roles and their related statuses are challenged, the moment of indecision that Alberto describes seems to lie outside of her theory of emotional work. Perhaps it is in this moment that Alberto experiences a contradiction between deep and surface acting, or simply a tension between expectation and reality.

The temporal dimension of each account offers another point of analysis. Alberto provides snapshots of significant moments of emotional work, whereas Kathy describes a trajectory. She shifts from motivated visionary in her early years to detached and depleted automaton in her final years – propelled by increasing demands on her time, coupled with decreases in autonomy over her work. A similar phenomenon occurs in medicine. In their seminal study of medical education, Howard Becker, Blanche Geer, Everett Hughes, and Anselm Strauss (1961) found that first year students are often motivated by humanitarian ideals and eager to make a difference. By the time they complete their studies, many are disillusioned. The overwhelming volume of content and the short amount of time in which to learn it pushes many to replace their previous ideals with a singular goal: survival. This parallel depicts time as a resource necessary to professionals – teachers and health professionals alike – undertaking the preparatory, support and emotion work that allows them to take pride in their work. It also points to time as a necessary resource for enjoying aspects of one's work, such as a puppet show for the class next door, and processing less desired emotions. Whereas some emotions are fleeting, experienced in the moment, complex emotions take time to feel (Olson, 2015).

PREPARING STEM TEACHERS FOR THE EMOTIONAL WORK AHEAD

The teaching profession suffers from chronic annual turnover and attrition worldwide, with related ramifications for the STEM education disciplines (e.g., Bang et al., 2007). Research on the reasons why teachers leave the profession are aligned to the vignettes explored here: 75% of teachers leave because of low salary; 39% of teachers leave to pursue a better career; and 29% leave because of dissatisfaction with teaching conditions. The four main problems with working conditions cited by teachers are: student discipline problems; lack of support from the administration; poor student motivation; and lack of teacher influence of school and classroom decision-making (Ingersoll & Smith, 2003). Clearly, increasing the rewards and working conditions of STEM teachers, or teachers in general, to reduce stress is part of the solution, rather than blaming the victims and recommending that teachers simply become more resilient to oppressive workforce structures.

However, understanding and addressing the emotional work of teaching is another important dimension of reducing burnout among teachers. The emotional work of teachers includes behaviours such as *surface acting*, or displaying emotions that are not felt. It also involves *deep acting*, in which the teacher undertakes certain activities in order to feel a socially required emotion. Another potential dimension of emotional work is the suppression of emotions (Näring et al., 2007). What can be done to prepare STEM pre-service teachers for the emotional work of the job before they arrive at the classroom door?

We are conscious of avoiding simplistic arguments that would lead teachers into learning coping mechanisms for dealing with extreme situations. Some of the experiences reported by teachers in the media, such as our opening quote, are beyond anything that can be considered reasonable. In cases of violence, abuse, and forms of institutional bullying, different strategies from those we offer are needed to redress the often complex issues that lie beyond matters of teaching practice, self-awareness, and classroom management. Our aim here is to consider those strategies that can be taught in teacher education courses to prepare preservice teachers in dealing with low to medium intensity issues, such as those represented in Kathy and Alberto's vignettes.

Analysis of the vignettes has pointed to several ways in which scholarship from the sociology of emotions can inform practice and the preparation of pre-service teachers. First, the emotional dimensions of teaching should be acknowledged within school systems and within university curricula aimed at training pre-service teachers. If the high rates of burnout amongst teachers and parallel shortages of STEM teachers are to be addressed, emotions should be a primary focus of teacher education, policy, and research. Emotions can be effortless and easy, but they can also be complex, confusing, draining and require ongoing management. Thus, teaching involves emotional work.

University curricula and ongoing professional development could more explicitly include reflective lessons on emotional work. In doing so, it could be made clear that this form of labour is not just an individual skill, it is dynamic. The emotional work required of a teacher is responsive to students' needs and group dynamics, while also mirroring the organisational climate and expectations of the school and the profession. In Kathy's vignette, we also saw the potential for *collective emotional management* as a strategy of collegial support in helping others to deal with their teaching circumstances. Preparing pre-service teachers for this kind of supportive role is achievable in university courses.

Second, such curricula on emotional work in undergraduate and postgraduate teaching degrees could be effectively guided by Billett's (2011) socio-personal conceptualisation of work-integrated learning. This model highlights (1) the different experiences and kinds of learning that take place at universities and on placements; (2) the need for three forms of knowledge required for successful work performance: conceptual, procedural and dispositional; and (3) the active role of the learner in integrating knowledge, in all three forms, across university and placement experiences. In Billet's construct, conceptual knowledge is factual and theoretical; procedural knowledge refers to specific steps and strategies; and dispositional knowledge refers to the embodiment of relevant values.

Learning about emotional work, conceptually, in university would be the first step towards raising pre-service teachers' awareness of this facet of their (and other teachers') work while on placements. Encouragement of ongoing reflection on one's emotions and emotion work could help pre-service teachers to construct and embody emotional work practices that will work for their social and personal circumstances and, potentially, decrease experiences of role strain and burnout. Use of *emotion metaphors*, as developed by Zembylas (2004), could be one way of scaffolding teacher reflections. Identifying the use of metaphorical language in reflective accounts of teachers' emotional experiences can help in identifying the way in which they perceive these events and the emotional work involved in teaching. More recently, techniques such as breathing exercises have been used effectively with science teacher educators and this approach could be extended to preservice STEM teachers (cf. Tobin, King, Henderson, Bellocchi, & Ritchie, 2016). A famous quote by Henry Adams (1871, p. 261) argues, "A teacher affects eternity – you never can tell where this influence stops". The emotional cost of teaching is high, but it needs to be nurtured to support the lifelong difference made to future generations.

NOTE

1 One example, from which we recovered the news story is the following site: http://www.theteachersareblowingtheirwhistles.com/teachersstories.htm

REFERENCES

Adams, H. (1905). Failure. In H. Adams (Ed.), *The education of Henry Adams*. Retrieved from https://archive.org/details/educationofhenry00adamrich

Bang, E., Kern, A. L., Luft, J. A., & Roehrig, G. H. (2007). First-year secondary science teachers. *School Science and Mathematics, 107*, 258–261. Retrieved from https://doi.org/10.1111/j.1949-8594.2007.tb18287.x

Barbalet, J. M. (1998). *Emotion, social theory, and social structure: A macrosociological approach.* Cambridge: Cambridge University Press.

Becker, H. S., Geer, B., Hughes, E. C., & Strauss, A. L. (1961). *Boys in White: Student culture in medical school.* New Brunswick, NJ: Transaction Publishers.

Bellocchi, A., Mills, K. A., & Ritchie, S. M. (2016). Emotional experiences of preservice science teachers in online learning: The formation, disruption and maintenance of social bonds. *Cultural Studies of Science Education, 11*, 629–652. Retrieved from https://doi.org/10.1007/s11422-015-9673-9

Bellocchi, A., Quigley, C., & Otrel-Cass, K. (2017). Emotions, aesthetics, and wellbeing in science education: Theoretical foundations. In A. Bellocchi, C. Quigley, & K. Otrel-Cass (Eds.), *Exploring emotions, aesthetics, and wellbeing in science education research* (pp. 1–9). Dordrecht: Springer. Retrieved from https://doi.org/10.1007/978-3-319-43353-0_1

Billett, S. (2011). *Curriculum and pedagogic bases for effectively integrating practice-based experiences.* Brisbane: Griffith University, Australian Learning & Teaching Council.

Bolton, S., & Boyd, C. (2003). Trolley dolly or skilled emotion manager? Moving on from Hochschild's managed heart. *Work, Employment and Society, 17*, 289–308. Retrieved from https://doi.org/10.1177/0950017003017002004

Brockmeier, J., & Harré, R. (1997). Narrative: Problems and promises of an alternative paradigm. *Research on Language and Social Interaction, 30*, 263–283. Retrieved from https://doi.org/10.1207/s15327973rlsi3004_1

Clandinin, D. J., & Connelly, F. M. (2000). *Narrative inquiry: Experience and story in qualitative research.* San Francisco, CA: Jossey-Bass.

Dalton, T. (2014, July 19). Life on the frontline of teaching. *The Australian.* Retrieved from http://www.theaustralian.com.au/news/features/life-on-the-front-line-of-teaching/story-e6frg8h6-1226989800189

Goldhaber, D., Krieg, J., Theobald, R., & Brown, N. (2016). Refuelling the STEM and special education teacher pipelines. *The Phi Delta Kappan, 97*, 56–62. Retrieved from https://doi.org/10.1177/0031721715619921

Gubrium, J. F., & Holstein, J. A. (1997). *The new language of qualitative method.* New York, NY: Oxford University Press.

Hochschild, A. R. (1979). Emotion work, feeling rules, and social structure. *American Journal of Sociology, 85*, 551–575.

Hochschild, A. R. (1983). *The managed heart.* Berkeley, CA: University of California Press.

Holmes, M. (2010). The emotionalization of reflexivity. *Sociology, 44*, 139–154. Retrieved from https://doi.org/10.1177/0038038509351616

Husso, M., & Hirvonen, H. (2012). Gender agency and emotions in the field of care work. *Gender, Work and Organisation, 19*, 29–51. Retrieved from https://doi.org/10.1111/j.1468-0432.2011.00565.x

Ingersoll, R. M., Merrill, L., & Stuckey, D. (2014). *Seven trends: The transformation of the teaching force* (CPRE Report (#RR-80)). Philadelphia, PA: Consortium for Policy Research in Education, University of Pennsylvania. Retrieved from https://doi.org/10.1037/e579212012-001

Ingersoll, R. M., & Smith, T. M. (2003). The wrong solution to the teacher shortage. *Educational Leadership, 60*, 30–33.

James, N. (1992). Care=organisation + physical labour + emotional work. *Sociology of Health and Illness, 14*, 488–509. Retrieved from https://doi.org/10.1111/1467-9566.ep10493127

King, D. (2012). *'It's frustrating! Managing emotional dissonance in aged care work'* [Special Issue]. *Australian Journal of Social Issues, 47*, 51–70.

Lasky, S. (2005). A sociocultural approach to understanding teacher identity, agency and professional vulnerability in a context of secondary school reform. *Teaching and Teacher Education, 21*, 899–916. Retrieved from https://doi.org/10.1016/j.tate.2005.06.003

Lively, K. J. (2000). Reciprocal emotion management: Working together to maintain stratification in private law firms. *Work and Occupations, 27*, 32–63. Retrieved from https://doi.org/10.1177/0730888400027001003

Lois, J. (2003). *Heroic efforts: The emotional culture of search and rescue volunteers.* New York, NY: New York University Press.

Lois, J. (2006). Role strain, emotion management, and burnout: Homeschooling mothers' adjustment to the teacher role. *Symbolic Interaction, 29*, 507–530. Retrieved from https://doi.org/10.1525/si.2006.29.4.507

Lukes, S. (1986). *Power.* New York, NY: New York University Press.

McNaughton, N. (2013). Discourse(s) of emotion within medical education: The ever-present absence. *Medical Education, 47*, 71–79. Retrieved from https://doi.org/10.1111/j.1365-2923.2012.04329.x

Mills, K. A., Bellocchi, A., Patulny, R., & Dooley, J. (2017). Indigenous children's multimodal communication of emotions through visual imagery. *Australian Journal of Language and Literacy, 40*, 95–108.

Mills, K. A., & Ritchie, S. M. (2014, April). *E-motion diaries: Blogging emotional responses to online learning in higher education.* Paper presented at the annual meeting of the American Educational Research Association, Philadelphia, PA.

Näring, G., Briët, M., & Brouwers, A. (2007). Beyond demand-control: Emotional work and symptoms of burnout in teachers. *Work and Stress: An International Journal of Work, Health and Organisations, 20*, 303–315. Retrieved from https://doi.org/10.1080/02678370601065182

Olson, R. E. (2015). *Towards a sociology of cancer caregiving: Time to feel.* Surrey: Ashgate.

Pekrun, R., & Linnenbrink-Garcia, L. (2014). *International handbook of emotions in education.* New York, NY: Routledge. Retrieved from https://doi.org/10.4324/9780203148211

Ritchie, S. M., Bellocchi, A., Pöltl, H., & Wearmouth, M. (2006). Metaphors in transition: Beginning teachers' lived experience. In P. J. Aubusson, A. G., Harrison, & S. M. Ritchie (Eds.), *Metaphor and analogy in science education* (pp. 143–154). Dordrecht: Springer. Retrieved from https://doi.org/10.1007/1-4020-3830-5_12

Ruti, M. (2014). *The call of character.* New York, NY: Columbia University Press. Retrieved from https://doi.org/10.7312/columbia/9780231164085.003.0001

Theodosius, C. (2008). *Emotional labour in health care: The unmanaged heard of nursing.* New York, NY: Routledge.

Tobin, K., King, D. T., Henderson, S., Bellocchi, A., & Ritchie, S. M. (2016). Expression of emotions and physiological changes during teaching. *Cultural Studies of Science Education, 11*, 669–692. Retrieved from https://doi.org/10.1007/s11422-016-9778-9

Tomas, L., Rigano, D., & Ritchie, S. M. (2016). Students' regulation of their emotions in a science classroom. *Journal of Research in Science Teaching, 53*, 234–260.

Zembylas, M. (2004). Emotion metaphors and emotional labor in science teaching. *Science Education, 88*, 301–324. Retrieved from https://doi.org/10.1002/sce.10116

Zerubavel, E. (1997). *Social mindscapes: An invitation to cognitive sociology*. Cambridge, MA: Harvard University Press.

ABOUT THE AUTHORS

Alberto Bellocchi is an Associate Professor at the Queensland University of Technology and Principal Research Fellow of the Australian Research Council. His current three-year research fellowship focuses on the interactions between social bonds, emotions and learning. He conducts his research in high school and university teaching and learning contexts. Professor Bellocchi serves as a Lead Editor for *Cultural Studies of Science Education*, and is an editorial board member for the *Journal of Research in Science Teaching*. He is the lead co-editor of the international collection *Exploring Emotions Aesthetics and Wellbeing in Science Education Research* published in 2017 by Springer.

Kathy A. Mills is a Professor at the Learning Sciences Institute Australia, Australian Catholic University, and a Principal Research Fellow of the Australian Research Council. Her award-winning research across 90 high quality publications addresses multimodal communication in schooling, including pedagogies for teaching the language of emotion. Prof Mills serves in executive research leadership roles for the American Educational Research Association, is an Associate Editor of the *Australian Educational Researcher*, and serves on four journal editorial boards in the USA, New Zealand and Australia.

Rebecca E. Olson is a Senior Lecturer in Sociology at the University of Queensland. Her research intersects the sociology of health and illness and the sociology of emotion and affect. From 2013–2016, she was an Associate Editor of the Journal of Sociology and has published widely in the areas of inter-professional healthcare, cancer and informal care. Her recent monograph represents her most significant contribution to date: *Towards a Sociology of Cancer Caregiving: Time to Feel* (Ashgate, 2015).

Roger Patulny is a Senior Lecturer at the University of Wollongong, Australia. He is the co-founder and current co-convener of the Contemporary Emotions Research Network (CERN); co-founder and former convener of TASA-SEA; has been awarded several ARC grants (DP098810; LP140100033); and edited three special editions on emotions for AJSI and Emotion Review. His detailed profile and full publications can be found at: http://rpatulny.com

Jordan McKenzie is a lecturer in Sociology at the University of Wollongong, Australia. His research is informed by European social and critical theory, and these perspectives contribute to his current research in the sociology of emotion. In particular, Jordan's work critically engages with the current cultural fascination with happiness and the good life in order to better understand how emotional experience reflects modernization and social change. He is a co-covener for the Social Theory thematic group of The Australian Socological Association and the Contemporary Emotions Research Network (UOW).

MASAKATA OGAWA

15. SCIENCE EDUCATION RESEARCHER AS CONSULTANT-RESEARCHER

A Critical Reflection of the Nature of Science Education Research in Japan

ABSTRACT

Japanese science education researchers faced a crisis of losing their funding basis when the Ministry of Education announced a drastically new funding scheme plan in 2016, where the research field, science education, was omitted. As a science education researcher with over 30 years of professional experience, I could not help but to reflect on our research field in order to show or explain the *raison d'etre* of science education research to the public as well as the funding agency. The reflection was guided by a newly introduced model, in which 'science education research' and 'science education practice' are separate, and their mutual relationship is expressed as a function of 'consultant-researcher,' a new role model of science education researcher. The model provided insight into such normative questions, what is our research field and to whom our research should contribute.

Keywords: science education research, consultant-researcher, intervener, client, value orientations

INTRODUCTION

It came to the Japanese science education research community as a bolt from the blue; the Ministry of Education, Culture, Sports, Science and Technology (MEXT) and its funding agency, the Japan Society for the Promotion of Science (JSPS), released a new plan of application and review system of the Grants-in-Aid for scientific research (KAKENHI) in early May 2016, with the announcement inviting the submission of public comments. The most striking fact, we found, was that our research field, Science Education (Kagaku Kyoiku), disappeared from the list of research fields. While all of the disciplines and research areas (from humanities, social sciences, natural sciences, engineering and interdisciplinary areas) are currently divided into 320 research fields, the suggested plan will consist of 304 research fields. Most of the research fields could survive, but it appeared Kagaku Kyoiku, not. No reasons were disclosed. Science education research in Japan was indeed facing an imminent crisis. One of the top ten

 | DOI:10.1163/9789004389663_015

research fields in terms of the number of research proposals submitted every year was about to become extinct. Japanese colleagues were going to lose their research funding backgrounds. Thus, we submitted a significant number of public comments to the Ministry. Subsequently, the fixed version was announced on December 22, 2016, and we found that 'science education' survived as one of the research fields. It was the only one research field revived. The news was received as a wonderful Christmas gift by the Japanese science education research community, indeed.

As a science education researcher with over 30 years of professional experience and as one of the past presidents of the Japan Society for Science Education (JSSE), the academic association corresponding with the research field, *Kagaku Kyoiku*, I cannot help but to reflect on my research field in order to show or explain the *raison d'etre* of science education research, as a whole, to the public as well as the funding agency. This is, so to speak, research on science education research. What kinds of unique characteristics are shared with the research of science education? What is science education research for? Who does science education research serve? What kinds of roles do science education researchers play in our society or community? The chapter will outline my struggle with such questions.

Before starting, I will: (1) introduce the Japanese context of science education research to the readers and then, (2) proceed to develop a theoretical model explaining Japanese science education research and finally, (3) critical issues uncovered by the model, which have rarely been focused on, are explored among Japanese science education researchers.

CONTEXT: A RESEARCH FIELD, *KAGAKU KYOIKU*, IN JAPAN

Explaining the research field, *Kagaku Kyoiku*, in the Japanese Grants-in-Aid context may be a good starting point for my reflection, since most non-Japanese readers are unfamiliar with them, while even in an international setting, "there is no widely shared definition of science education" (McComas & Murie, 2014). Development of any academic research field, in general, is autonomous in its nature. Demarcation of whether or not a new research theme should be included in the research field concerned is determined by the research community itself through a kind of self-regulated academic mechanism (for example, review processes of submitted papers and conference abstracts). In this sense, the boundary of a research field is sometimes vague, ever changing or expanding, if the community is tolerant of accepting newly proposed themes. The research community of science education in Japan is one such community.

The origin of the community emerged in the late 1960s with a group of enthusiastic advocates of modernization of the school science program (called *Rika*), consisting of prominent scientists and mathematicians, not professionals in science teaching (science educators and science teacher educators). Since no specific research field on school science existed within the framework of the Grants-in-Aid system at that time, their strong lobby activities toward the Ministry of Education, in which they insisted that promotion of science and technology, in general, and science education, in particular, should be one of the top priority policies for Japan's national development in the near future, successfully worked out and the Ministry decided to start a nine-year (from 1967–1976) R&D project on *Kagaku Kyoiku* as a special arrangement within the Grants-in-Aid system. The model of the project was the US curriculum reform movement of the 1960s (DeBoer, 1991). Major themes of the project included development of: (1) new types of science teaching programs, (2) teaching methods utilizing technology-based educational apparatus and (3) curriculum materials, new experiments and new teaching instruments. After finishing the project in 1976, the Ministry decided to launch a new research field, *Kagaku Kyoiku*, within the Grants-in-Aid system, but it was allocated, not under a subcategory of *educational studies,* but *interdisciplinary studies*. At the same time, the core members of the project (led by scientists and mathematicians) joined together to establish a new academic association, the *Japan Society for Science Education* (JSSE), in 1977. The membership includes various academic disciplines: natural scientists, mathematicians, educational psychologists, educational technologists, researchers on science teaching and teacher training. At that time, there already existed several associations of science education research, including respective associations of physics education, chemistry education, biology education and earth science education. In addition, the *Society of Japan Science Teaching* (SJST), established in 1952, whose members consisted mainly of science methods professors of teacher training programs in national universities (most of them were newly established around 1951), science education supervisors of the respective local Boards of Education, science educators of their research institutes and practicing science teachers. Its research interests and concerns were on teaching and learning in *school science (Rika)* settings and science teacher training, exclusively. On the contrary, JSSE set its mission rather simply and extensively to invite researchers, not only members of such science education associations, but also other researchers from disciplines outside of the science education research associations. It notes, contributing to the progress and diffusion of "education in/about science" and "education by scientific and technological methods" (Japan Society for Science Education, 2016, November 30). The mission and establishment of the new research field, *Kagaku Kyoiku*, in the Grants-in-Aid system attracted many researchers, whose research interests were related to science education in its broadest sense, but not necessarily in the areas of school science teaching or learning, who also joined the JSSE.

One of the unique contexts of the Japanese school science education was, and has been that it is strictly under the national government's legal control of curricula and textbooks through the Course of Study system. Researchers interested in science education in school settings beyond the control of the Course of Study also joined in this new association. So, the membership of JSSE has been expanding and celebrates heterogeneity in its members' parent disciplines. Certain research collaborations and cooperation among JSSE members across the disciplines spontaneously emerged and became visible, through the occasions of annual conferences and research colloquia.

Now in 2016, many themes related to science education are investigated within the community. For example, in the keyword list of the research field, *Kagaku Kyoiku*, shown in the application procedures for the Grants-in-Aid (Japan Society for the Promotion of Science, 2016, p. 62), we found the following research themes: (1) science education in tertiary level (mathematics, physics, chemistry, biology, informatics, astronomy, earth and planets, and interdisciplinary), (2) science education in elementary and secondary school level (arithmetic and mathematics, *Rika* (school science) and informatics), (3) engineering education, (4) scientific literacy, (5) experiments and observations, (6) science curricula, (7) environmental education, (8) industrial and craft education, (9) science & society, and science & culture, (10) science teacher training, (11) science communication and (12) information literacy. Also, papers published in the JSSE's official research journal, the *Journal of Science Education in Japan* (JSEJ) and presentations at the JSSE annual conferences include research themes such as: (1) mathematics education, (2) engineering education, (3) technology education, (4) international aid and cooperation in science education, (5) ST-related human resource development, (6) collaboration and cooperation among schools, museums and community, (7) environmental education, (8) SSI or STS education and (9) risk or disaster education, in addition to the popular themes on teaching/learning school science, science teacher education and science curriculum material development. The research community identifies and accepts these research themes as *ours*. In this sense, the research field, *Kagaku Kyoiku*, in Japan is too broad compared with the research areas of *school science (Rika)* and science teacher training, which are also covered by other academic research communities shown above. The research community of *Kagaku Kyoiku* in Japan has enjoyed heterogeneity among its membership and shared the idea that it is this heterogeneity and open-mindedness that is one of the driving forces contributing to the ever-developing nature of the research field, smoothly coping with newly emerging societal issues related to science and technology.

NATURE OF SCIENCE EDUCATION RESEARCH: DEVELOPING A THEORETICAL MODEL

As shown above, research themes within the research field, *Kagaku Kyoiku*, in Japan vary widely. What are, then, the common nature across these research themes? No explicit criteria are shown or shared among the community in advance, but from time to time, new themes had been introduced by a certain group of community members or newcomers from other disciplines or research fields, and the community has shared and accepted them as their own. Thus, no systematic trial to examine a precise definition of *Kagaku Kyoiku* within the community has occurred, or rather, such a trial is intentionally avoided in order not to snip possible research buds off from the community. However, as long as our research is funded by a government agency (or public fund), we must try to explain what our research field is or what contributions our research field hopes to make. For this purpose, I must try to explore the commonalities shared across the diversity of research themes within the research field, *Kagaku Kyoiku*, in Japan.

In order to explore certain common natures of such a diversified research field, a kind of theoretical model is worth developing and useful as a heuristic tool. Heavy dependence on the status quo of the research field across various kinds of research themes sometimes misses the essentials. In this section, I will try to develop a theoretical model explaining the essentials of the research field, *Kagaku Kyoiku*, in Japan by focusing on the stakeholders (actors) who are involved in science education activities and their respective functions. Thus, the labels allocated to the stakeholders below, represent functions, not individuals or a group, carrying out the functions. Once the model is developed successfully, then, we can proceed to compare the model with reality and identify several issues, which we have not been aware of and which require more attention among Japanese science education researchers.

SCIENCE EDUCATION PRACTICE AND SCIENCE EDUCATION RESEARCH

At this point, I would distinguish two different science education activities from each other. They are, science education practice (SEP) and science education research (SER). The former, SEP, involves concrete, practical activities in *Kagaku Kyoiku*. Among them are; (1) science teaching/learning practice in which a science teacher teaches chemistry to tenth graders in a science lab room, (2) a group of scientists perform an outreach program, science café and (3) a group of science communicators invite the public to a science show. These activities are, in this chapter, named SEP. The stakeholders in the first example are the chemistry teacher and the students (learners). But behind the activities, there are *invisible (indirect)* stakeholders; the textbook writers, Course of Study (legal control of the MEXT), the regional Board of Education, the School Board, the school principal, the parents and so forth. In the case of the second example (science café), among the *invisible* stakeholders are; universities and research institutions the scientists are affiliated to, family of the participants, volunteer staff and the owner of the cafeteria, where the science café is performed. SEP activities occur during the interactions among the relevant stakeholders.

On the other hand, the latter, SER, involves research activities on such practical activities of SEP. For example, investigating: (1) how to improve the tenth graders' chemistry learning outcomes, (2) the public's need of the prospective topics for the science café or (3) what kinds of scenarios in the science show would attract the prospective audience; all of these research activities are SER. In this case, there exists two kinds of stakeholders, *visible (direct)* and *invisible (indirect)* stakeholders. The former is the SER researcher (SERer) or a team of SERers. The latter stakeholders are, for example, funding agencies, universities and/or research institutions of the SERer and clients requesting help, suggestions and recommendations (the regional Board of Education, science museums, PNOs and NGOs, a group of activists and so forth.).

Of course, there exist science education activities, in which SEP and SER are integrated into one. Action research programs, design research programs in science education and activist science education programs are among them and these have been very visible in science education research around the world. However, for my current considerations, these activities are regarded as exceptional cases, as the separation of SER from SEP seems to be a more primitive relationship (serving as a prototype) than the case of an integrated relationship. The separated relationship between the two, thus, should be the fundamental presupposition of my model construction.

COMMON NATURE OF THE SEP AND ITS ULTIMATE GOAL

Let me start from a reflection on my own view of the nature of *science communication*, one of the themes within SEP (Ogawa, 2013). The fundamental nature of the practice of *science communication* is reflected by considering the ideal goals of *science communication*, which are differentiated into the subfields: *Public Understanding of Science*, *Public Awareness of Science*, *Public Engagement with Science*, *Public Participation in Science* and so on, for the past 25 years. After examination from the two aspects, historical reflection and shared presuppositions, a common nature across science communication practices was extracted. It was "the intention to intervene in the relationship between science and society (or scientists and the public)" (p. 7). The key notions here are; (1) intention, (2) intervention and (3) the relationship between science and society. Ordinarily, the intention is held by the science communicators concerned (Ogawa (2013) used the terms, *driving actors* and the intervention, performed by the science communicators toward the general public, the *target actors*), in order to contribute to a *better* relationship between science and society.

Is it possible to share this nature with a practice like school science teaching/learning? In this case, the driving actors are science teachers and the target actors are students. A science teacher (driving actor) intervenes (teaches) chemistry (science) to tenth grade students (target actors) with the intention of making the students *better* future citizens (members of society). The relation is similarly found in the case of science teacher education. Science teacher educators (driving actors) intervene (teach) science method courses in order to make the prospective science teachers (target actors) *better* in-service science teachers, who will contribute to teaching future citizens (members of society). In most cases of the SEP themes, we can easily identify the respective driving actors and target actors, on whom certain kinds of intervention are performed. Thus, we may say that SEP shares a common nature, *driving actors' intention to intervene on target actors to make a better relationship between science and society (between science and human beings)*. However, in addition, we may not forget the possibility of the driving actors' other interventions for making a better relationship between science and society; interventions to; (1) science and (2) the relationship between science and society itself. Among the former interventions, included are, for example, recruiting a presenter and selecting the topic in the case of a science café, and selecting and developing teaching materials in the case of school science teaching/learning. Examples of the latter interventions are found in the activities of citizen sciences and activist approaches to science teaching/learning. Thus, three types of interventions can be included in the proposed model.

Here, the only two stakeholders, driving actors and target actors, are visible, but of course as mentioned earlier, more *invisible (indirect)* stakeholders exist around them. In the context of school science practices, principals, school boards, regional Boards of Education, parents, textbook writers/editors and textbook publishers are found as stakeholders. In the context of the science communication practices, scientific communities, media, policy makers, governmental bodies, interest groups, NPOs and NGOs are among them. While such indirect stakeholders are somewhat influential on the respective SEP activities of the direct stakeholders, the common nature of SEP remains unchanged, that is, *driving actors intend to intervene on their target actors in order to make a better relationship between science and society*. Thus, one of the ultimate goals of SEP can be identified as *to contribute to develop a better relationship between science (scientific enterprise) and society (people living there)*.

COMMON NATURE OF SER AND ITS ULTIMATE GOAL

Compared with the case of SEP, SER's ultimate goal seems to be rather simple. It should be just *contributing to make SEP better* through its research activities. SER activities should not stand alone without SEP activities. SER activities should, in principle, contribute or connect, at least, to be relevant to concrete SEP activities in order to make them *better*. Otherwise, research on the social phenomenon, science education, whether or not it be philosophical, epistemological, psychological, historical, sociological and anthropological, falls into the respective basic research disciplines of humanities and/or social sciences. The main purpose of such research remains in the activities to construct new (context-free) knowledge. Such research is theoretically excluded from the domain of SER.

One of the important points here is that the SER holds a role of *intervention* (intervention in the SEP activities). It does not simply remain ordinary academic research, which involves analyzing the target SEP activity through standardized disciplinary methodologies (psychological, sociological, anthropological, philosophical, etc.) and obtaining new knowledge or theoretical constructs on the target by the analysis, in order to *explain* or *interpret* the target activity. But, it serves as research, where the SER researcher (SERer) utilizes these data, information, knowledge and theories to *alter* the current situation of the target SEP activity toward a *better* one. In this sense, SER is research for *intervention.*

NATURE OF SERER: CONSULTANT-RESEARCHER

When considering the nature of the SERer, Randy Stoecker's (1999) argument on the nature of participatory research and the functional approaches that researchers adopt is helpful. He identified three functional approaches in participatory research; the initiator, the consultant and the collaborator. On the initiator of participatory research, he contended that while the research question should be generated by the community, not the researcher, many participatory action research projects would not occur without the initiative of someone with time, skill and commitment (Stoecker, 1999, p. 843). This type of researcher is named the initiator. In the participatory research model, community members should do the research themselves, but in many cases, academics find themselves consulting with community groups. This type of researcher is called the consultant. The point of this case is that the consulting is carried out at the community's request (p. 844). The last type of researcher named the collaborator, is a researcher with certain technical expertise who can work together with a community leader with a knowledge of community needs and perspectives (p. 844). Among the three approaches, he argued that the collaborator approach is an ideal, but "it may be less efficient than the consultant approach, asking community members to participate in ways that they are not interested in or do not have time for (p. 845)." Reflecting Stoecker's categorization in the context of SER, stakeholders within a SEP (including the driving actor and the target actor) are equivalent to the roles of the community members in the context of a participatory research project. Which type of academic (researcher) is appropriate to the role of the SERer within the SER? I prefer the consultant type as a standard for the proposed model, though I do not deny the other two types, as the ordinary SERers in Japan are neither a simple trigger of SEP activity, nor collaborators living together with the stakeholders of the SEP activity for a rather long time, though we know that there are certain activist SERers who have been working as collaborators in their own contexts.

One of the shared understandings of the necessary quality of a practicing teacher are the capacities to serve as a researcher. In the case of school science teaching/learning, a science teacher (driving actor) with quality research capacities can serve as a SERer. However, this is not the ordinary case in Japan. Statistically speaking, the about 4% of Japanese science teachers have master's degrees (lower-secondary schools) and 27% (upper-secondary schools) (MEXT, 2013). Also, the legal control (through the Course of Study system and others) over curricula has determined that practicing teachers need not serve substantially as researchers, though they are urged to reflect on their teaching as a practitioner. In most cases, a driving actor of SEP is not necessarily a SERer. The SERer, maybe a university professor, usually located outside of the SEP activity concerned. The SERer serves as a consultant as well as a researcher and we can name the SERer as *consultant-researcher*.

Here, I need to clarify the definitions of consulting, consultant and client for the discussion. Nora Jacobson, Dale Butterill, and Paula Goering (2005) broadly defined *consulting* as "a process of transferring expertise, knowledge, and/or skills from one party (the consultant) to another (the client) with the aim of providing help or solving problems" (p. 302). How then, should we conceptualize *consultant*? A consultant is, as Peter Block (2011) explained, "a person in a position to have some influence over an individual, a group or an organization, but has no direct power to make changes or

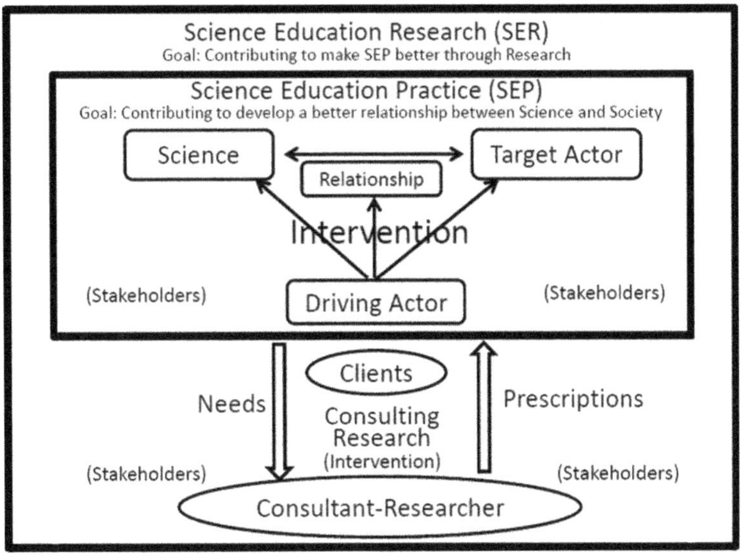

Figure 15.1. Model of science education research (SER). (The labels represent functions, not individuals or a group, carrying out the functions)

implement programs" (p. 2). In addition, Daniel Druckman (2000) identified seven consulting functions: advisers, technicians, applied theoreticians, study directors, bridge builders, facilitators and trainers (teachers). *Clients* are the recipients of all this advice from the consultant (Block, 2011). The client is sometimes a single individual, but other times, it may be a group, a department or a whole organization. The client is the person or persons whom the consultant wants to influence. In the discussion below, I will follow these definitions.

The natures of SEP and of SER and their relationship discussed here is summarized in Figure 15.1. Among the stakeholders of the SER activities are, for example, funding agencies, the SERer's institution and sometimes the general public (if the SEP activity is of a public nature) in addition to SEP activities.

In the next section, some of the issues which should be deliberated seriously will be pointed out in order to deepen the understanding of the nature of SER.

ISSUES REGARDING SER IN JAPAN: FROM THE THEORETICAL MODEL

SEP as a Public Policy

The ultimate goal of SEP in the model is identified as *contributing to the development of a better relationship between science and society* and it can be rephrased as *contributing to developing better ways of coexistence of human beings with enterprise and science*. The goal implies that each of the SEP activities in practice should, not simply intervene on the target actor(s) (and in some cases, science and/or relationship between science and the target actor(s), as well, shown in Figure 15.1 in order to solve their immediate problems or respond to their immediate need or context, but hold a careful vision of how this particular intervention contributes to *better* ways of coexistence of the target actor(s) with scientific enterprise beyond the immediate context. In this sense, SEP activities can be also regarded as intervention in society or the community itself.

This aspect leads me to the notion of *policy* in political sciences. Definitions of *policy* among political sciences contain certain kinds of commonalities, while a wide variety of definitions exist, as it includes many aspects and differential aspects are emphasized in each of the definitions (Guba, 1984; Levin, 1997). Some of the standard definitions, which are frequently referred to in the textbooks of policy studies are shown in Table 15.1.

Inspired by such definitions of policy, I proposed an idea of "science communication as policy," in which science communication is defined as "purposive intervention by a driving actor or a group of driving actors to alter the present state of the relationship between sciences and society toward their desired state" (Ogawa, 2013, p. 9). I argued that the driving actors need to: (1) express their respective value orientations before proposing any interventions, (2) identify and express their own view of the grand picture or goal, (3) identify the concrete target of their proposed interventions and (4) design concrete interventions reflecting their value orientations, goals and targets (Ogawa, 2013, p. 9). I also contended that the intervention is a *means* to achieve their own *goal.* The idea can be readily extended to the case of SEP. However, we need to be careful of value orientations among the stakeholders (including driving actors and target actors), which will be discussed later.

SEP possesses a nature of *policy*, then, how about the nature of *public policy*? Michael Kraft and Scott Furlong (2013) defined *public policy* as "what public officials

Table 15.1. Examples of definition of policy in political sciences

Authors	*Definitions*
Brewer & deLeon (1983)	A broad strategic statement of intent to accomplish aims (p. 30).
Clark & Kellert (1988)	A specific course of action designed to achieve a desired outcome (p. 7).
May (2003)	Policy intentions establish the goals and type of policy that are to be put into practice (p. 228).
Sapru (2004)	A purposive course of action taken or adopted by those in power in pursuit of certain goals or objectives (p. 5).
Anderson (2015)	A purposive course of action or inaction followed by an actor or set of actors in dealing with a problem or matter of concern (p. 7).
Verslius, van Keulen & Stephenson (2011)	A deliberate course of (in-)action selected from among available alternatives to achieve a certain outcomes (p. 11).

within government, and by extension the citizens they represent, choose to do or not to do about public problems," and continued the definition of public problems as "conditions the public widely perceives to be unacceptable and that therefore require intervention" (p. 3). These definitions generally fit the case of SEP, as the "relationship between science and society" or "coexistence of science and human beings" is one of the serious concerns for the general public (especially for the Japanese after the Fukushima disaster in 2011). In this sense, SEP can be regarded as a form of public policy as well as policy. The view that science education practice (in this particular chapter, SEP) holds the nature of public policy might be a new one for the science education community. While we realize school science (within the framework of schooling) has been under public policy, the current idea that every kind of SEP activity could fall into public policy may be beyond science educators' thoughts. But, as long as the ultimate goal of the intervention, SEP activity, is set as "contributing to develop a better relationship between science and society," it certainly serves the public good. If the SEP activities are public policy, that is, beyond the private arena, the driving actors (the interveners) should not simply pursue *their own* dreams, based upon *their own* value orientations. The same is true in the case of SER since the consultant-researchers serve as interveners of SEP activities. Their value orientations, in principle, should not be forced on the stakeholders of SEP activities.

CONSULTANT-RESEARCHER

As mentioned above, it is much more appropriate to regard the SERer as serving as a consultant and a researcher of the target SEP activity, at least, in the Japanese context. Thus, I shall call the SERer a *consultant-researcher*. The term is inspired by research on management accounting (van Helden, Aardema, ter Bogt, & Groot, 2010), which examined contributions to knowledge creation for practice in public sector management accounting by consultants and academics. The idea, *knowledge creation for practice* really fits the standpoint of SERers. G. Jan van Helden and colleagues (2010) critically analyzed and distinguished actual contributions to the practical knowledge creation by both researchers and consultants. For example, in the situation of knowledge creation, researchers are apt to adhere to their theories and research methods, but consultants rely heavily on their past experiences in comparable situations. Researchers' (academics') knowledge orientation is *fundamental* or *applied*, but that of consultants is exclusively *applied*. In terms of knowledge type created, researchers' knowledge is *explicit* while that of consultants is both *tacit* and *explicit*. Drivers of knowledge creation for researchers are *discipline-driven* and *problem-driven*, but for the case of consultants, *problem-driven* is predominant. Researchers are not so concerned whether or not their new knowledge is directly applied in practice, while consultants are seriously concerned that their new knowledge is directly applicable in practice.

Between the two, researchers and consultants, van Helden et al. (2010) identified a third category, *consultant-researcher*, who are *consultants with a part-time job in research* and active in *both worlds*. Some of their results indicated characteristics of consultant-researchers. In the case of the influence of practice on knowledge creation, they are described as follows:

> Consultant-researchers occupy an intermediate position between academics and consultants, in the sense that they want to comprehend practice. They use their research to reflect theoretically on their practical experiences, in order to "reveal true motives," "to unmask rhetorics (sic) used by practitioners," and "to generalize practical experience." (p. 87)

On the type of knowledge, they reported that:

> Consultant-researchers create knowledge which is similar to that of consultants, as it is primarily driven by problems stemming from practice and it has to be directly applicable in practice. However, disciplinary knowledge plays a substantial role in their work, and in this respect consultant-researchers have some similarities with researchers. Moreover, they argue that they use a broader body of knowledge than is strictly required for consultancy practice, and this includes evidence-based knowledge from academic research. (p. 88)

Thus, the idea, consultant-researcher seems to be promising when we are thinking of the SERer's roles and functions in SEP activities, though, in our case, it is the researcher that serves as consultant, which is different from the definition of van Helden et al. (2010) mentioned above. The SERer (as consultant-researcher), in response to the needs of *clients*, who are either among the internal stakeholders (including *driving actor* and *target actor*) of the SEP activity or its external stakeholders, starts with consulting and relevant academic research (analysis) of the target SEP activity and develops prescriptions of possible interventions based upon the clients' value orientations (not upon his/her own). Issues in the relationship between clients and their value orientations will be explored much deeper in the later section. In this sense, the SERers can serve as *consultant-researchers* in order to distinguish them from academic researchers in ordinary disciplines. They have a dual function, but the consultant function is more important to serve for the SERer than that of academic researcher. Even if they are concentrating a certain aspect of research on a SEP activity, the research should contribute to the consulting later on. Research, not aiming to contribute to SEP activity is meaningless in terms of SER. Such research should be academic research and not fall in the category of SER.

WHAT IS BETTER AND WHO IS BETTER? ISSUES ON VALUE

The theoretical model identifies the ultimate goal of SEP as "contributing to develop a better relationship between science and society" and that of SER as "contributing to make SEP better through research." This section proposes issues on the *better*, but before that, some explanations on *science* and *society* in this particular model are necessary. *Science* holds its broadest meaning and includes all of the components;

scientific knowledge, scientific way of knowing, scientific way of doing, doing science, the nature of science, scientific community, scientific enterprise, applications of science, etc. *Society* also consists of the following; society, community, individuals living there, public and private institutions, public services, democratic decision-making, etc. So, the *relationship between science and society* is said to include every aspect of the *relationship between the enterprise named science and human beings*. Or, it may be simply described as *ways of coexistence of science and human beings*. In this sense, a better relationship between science and society can be rephrased as *a better coexistence of human beings with the enterprise, science*.

The term, *better* is indeed value-laden. SERers in Japan have not deliberated the *what is better* and *who is better* issue, so seriously. There are some probable reasons; (1) Japanese society has consisted of a rather homogeneous population in that the National population census in 2015 shows the ratio of foreign residence is only 1.6% of total population, (2) people believe that the Japanese share certain kinds of value orientations and (3) the belief that science is value-free and worth learning for everybody has prevailed among SERers as well as SEPers. However, reality is different. Within the Japanese, there occurs a significant differentiation of their preference in terms of *better*. I identified one such example (Ogawa, 2014). I conducted a web survey of the Japanese public's preference of *the most preferable community-level decision-making styles on socio-scientific issues*. One of the results (Figure 15.2) indicated that their preferences were distributed evenly into the four distinctive styles of decision-making (except TYPE V), which is evidence of how Japanese people's values toward the ideals of democratic decision-making on socio-scientific issues are varied.

So, *value* matters even in Japan. An anthropologist, Clyde Kluckhohn (1951), argued that *value* is "a conception, explicit or implicit, distinctive of an individual or characteristic of a group, of the desirable which influences the selection from available modes, means, and ends of action" (p. 395). Through fieldwork in five

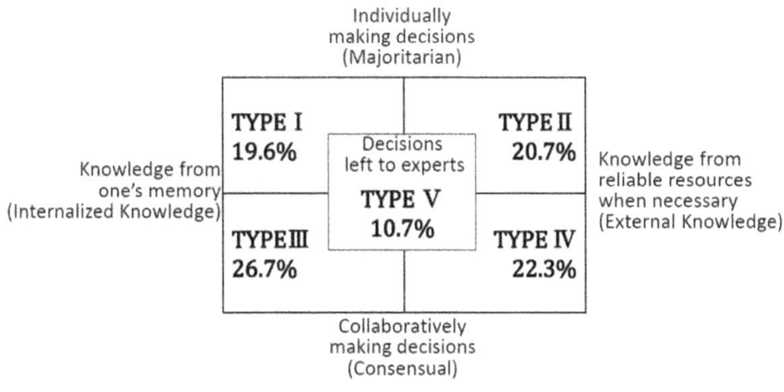

Figure 15.2. Japanese public's preferences of the most preferable community-level decision-making on socio-scientific issues

culturally distinct American southwestern communities, *value orientations* were defined as "complex but definitely patterned (rank ordered) principles resulting from the transactive interplay of analytically distinguishable elements of the evaluation process which give order and direction to the overflowing stream of human acts and thoughts" (Kluckhohn & Strodtbeck, 1961, p. 4). While the definition seems to be rather anachronistic, Forster Ndubisi (1991) rephrased it as "*value orientations* are ordered principles that are employed in selecting alternatives to action" (p. 56). This is more relevant to our present context. The Kluckhohn value orientations model (Kluckhohn & Strodtbeck, 1961) consisted of five orientations; (1) Perception of self and others (innate nature of humans), (2) Worldview (people's relationship to nature), (3) Temporal orientation (people's conception of time), (4) Forms of activity (the modality of human actions) and (5) Social relations (people's social relations to each other). For each of the value orientations, three to four postulated ranges of variations were identified. While a certain set of value orientations is supposed to be shared among the community members in the original work in the 1950s, in the contemporary era, community members' value orientations are reasonably expected to be differentiated and diverse, because of the influence of globalization and an information-driven society.

This implies that in the case of SEP and SER, value orientations of each individual stakeholder will have an influence. Since each individual stakeholder is supposed, in principle, to hold his/her unique set of value orientations toward the ideas of, for example, a *better* society, a *better* relationship between science and society, and a *better* coexistence of science and human beings, a field of SEP activity will be a kind of battlefield of different sets of value orientations of the stakeholders concerned. SEP is, as mentioned above, defined as "purposive intervention by a driving actor or a group of driving actors to alter the present state of the relationship between sciences and society toward desired state," then, the following questions will be raised; Whose purpose should the intervention (the SEP activity concerned) guide? Whose desired state of relationship between science and society, or whose desired state of coexistence of science and human beings should be the aim of the intervention? More concretely, whose value orientations should guide the SEP activity concerned, the driving actor, the target actor, other stakeholders including government, school boards, museums, scientific community or the SERer as intervener? Whose value orientations should be reflected to an image of a better (desired) relationship between science and society or better (desired) coexistence of science and human beings, the driving actor, the target driver, general public community, scientific community or government? Who should serve as the clients of the intervention, SER, the driving actor of SEP concerned, the target actor, community's interest group, local government, the local school or local institution concerned? Many questions will arise and possible answers will be highly diversified according to the contexts, as well as the value orientations of the individuals concerned. They are open questions for SERers.

However, curiously enough, Japanese SEPers and SERers seem to have shown indifference to the issue. In the context of school science, science teachers (SEPer, the driving actor) believe that they should teach (intervene) science classes to their students (the target actors) based upon the textbook and the Course of Study (the government's value orientations). They are not necessarily aware of their own value orientations, nor their students' value orientations to science, or relationship between science and their students. In the case of science communication, science communicators (the driving actors) are apt to develop and perform SEP activities based upon their own value orientations, without serious consideration of the target actors' (general public, attentive citizens or kids, for example) value orientations to the science at issue or the relationship between science and the target actors.

One of the most popular ways of doing is that the driving actors (science teachers or science communicators) unconsciously take their own position within science (that is, scientific community side). The science teachers and science communicators may usually share a similar set of value orientations to that of scientists (scientific community), but this does not necessarily mean that they must stay there as the driving actors. One of the traditional schemes showing the relationship among teacher, learner and subject matter, which has been popularly accepted by Japanese educators and teachers (Sakamoto, 1971), is called the *didactic triangle* (Figure 15.3), firstly introduced by Johann Herbart (1806) and is still popular, mainly in continental European traditions (Goodchild & Sriraman, 2012; Peterssen, 1983; Schoenfeld, 2012; Steinbring, 2005).

In this scheme, the teacher's position is independent from the subject matter (science), just as the case of the pupil. The teacher should be located outside of the subject matter. From outside, the teacher needs to deliberate on the educational values of the subject matter before deciding what and how the subject matter should and could be taught to the pupils concerned. The teacher's (intervener's) positioning against the subject matter (science) is crucial and he/she needs to decide the *sense of distance* to science. Not only SEPers, but also SERers in the theoretical model are required to

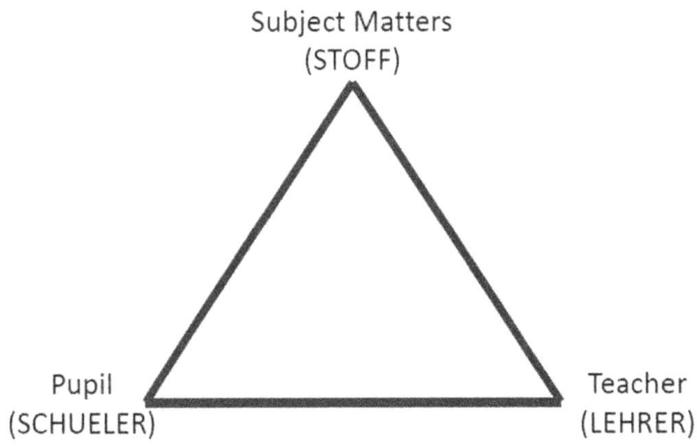

Figure 15.3. Didactic triangle

locate their own position before thinking of the intervention. Their positioning will be allocated after deliberation on various kinds of stakeholders' value orientations.

WHO SHOULD BE THE CLIENTS OF SERERS?

Clients of the SERers are also problematic. Since the objects of SERers' consulting are on the SEP activities, possible clients will be stakeholders of the SEP activities, that is, the driving actor, the target actors and some of the other stakeholders; for example, school principals, school boards, regional Board of Education and parents, in the case of school science teaching/learning.

Among the explicit stakeholders are driving actor(s) and target actor(s). In the case of school science teaching/learning, the driving actor is a science teacher and the target actors are students. If the science teacher requests a consultant-researcher for a prescription to improve his/her intervention (teaching) to help the target actors (students) develop a "better" relationship with science, the teacher is the client of the consultant-researcher. In this case, the teacher's needs will emerge from his/her own problems and value orientations (ideas of a "better" relationship between his/her students and science), and the consultant-researcher should analyze the current situation between the students and science from an outsider's (consultant's) point of view and propose possible alternatives of intervention design, based upon the teacher's (not the consulting-researcher's) value orientations. Once the teacher selects or determines an intervention among the alternatives to implement, the consultant-researcher will keep watching the process and outcomes of the intervention, and try to assess and evaluate the intervention's effectiveness in order to report it to the client (the science teacher). For driving actors in other settings (for example, scientists of the science café and science communicators of the science show), a similar relationship between the client and the consultant-researcher is expected.

Then, how about the case of target actors as clients? Generally, it is difficult to assume, as the target actors usually cannot have a chance to know a possible channel to contact the consultant-researcher beyond the direct stakeholder (the driving actor). For example, Japanese students rarely express their opinions or requests on what kinds of content they prefer even to the science teacher, mainly because they know that the course content is predetermined by the Course of Study and the textbooks as mentioned above. The students of the chemistry class hesitate to ask the teacher or the principal about engaging in a learning activity on water pollution of the river near the school. So, the target actors in the science classes (students) do not seem to serve as clients of a third person, the consultant-researcher. Similarly, in the case of a science café or science show, the target actors are mostly passive actors.

Recently, new types of activities (activist movement), like citizen sciences and consensus conferences on certain socio-scientific issues, are visible in Japan. On such occasions, a public group serves as both target and driving actors. They do not hesitate to invite a third person as the consultant-researcher to intervene in their activities. These examples are promising and encouraging, but from the stakeholders' relationship in the model, they still remain special cases.

At this point, we must add another issue for further deliberation. How to cope with the unspoken voices of the target actors (who are not the client, but the persons concerned) in the consulting processes (the intervention)? Just as the example of the students in the science classes, the unspoken voices of the target actors are not unusual in this model. In this point, fruits of research and practice on advocacy to promote children's right to participation (Percy-Smith & Thomas, 2010) seem to be promising.

Other stakeholders than the driving actors and the target actors are also possible clients. Or, rather, it is such stakeholders that are actually serving as the clients of the consultant-researcher. In the school science context, the regional Board of Education and the school principal are major clients, as it is very popular in Japan for them to invite university professors as supervisors or advisers of certain kinds of lesson study workshops, R&D projects and lecturing in their school settings. The university professors on such occasions are serving as the consultant-researchers. In the case of science museums' outreach activities, similar systems of supervisor or adviser are working. Such supervisors and advisers are also regarded as consultant-researchers.

Finally, I would like to add another important issue. Who should consultant-researchers serve, the client, the driving actors or the target actors? I think this is a good question for SERers to reflect their standpoints and own value orientations toward a better relationship between science and society (human beings). There is no correct answer, of course. It is homework for SERers.

CONCLUDING REMARKS

The common nature of the Japanese research field, *Kagaku Kyoiku*, has been explored by a theoretical model, in which SER and SEP are separate and their mutual relationship is expressed as a function of the consultant-researcher. Separation of science education research and practice and identification of relevant functions (not individuals) of stakeholders (driving actor, target actor, outside stakeholders and consultant-researcher) in each endeavor gives us an opportunity to reflect on what is our research field and to whom our SER should contribute. The ultimate goal of SEP was set as *contributing to develop a better relationship between science and society* (*between science and human beings* or *coexistence of science and human beings*) in the model. However, supposing that each of the stakeholders holds respective value orientations toward science or a better relationship between science and society, the SEPer (driving actor) and SERer (consultant-researcher) need to be careful of the fact that differentiation in the interpretation of the ultimate goal within the stakeholders are probable, since the term, *better* is value-laden. In the Japanese context, the general public as well as science education practitioners and researchers are not so seriously aware of the differences of value orientations among the people. Remember the didactic triangle (Figure 15.3). Why are the three components set independently of each other? The standing position of science education practitioners and teachers should be independent from science. This is very different from the positioning of scientists. They are sitting in the component, science, but the science education practitioners and teachers are not allowed to sit there, in principle. Science is a target of intervention for them. If you want to sit in the component, science, the decision is derived from your own value orientations. If you want to make your target actors (students) sit in science, you must be careful whether or not the target actors are willing to do so guided by their value orientations. The decision should be left to the target actors' hands, not to your own hands. How should we cope with students, saying "I do like science learning very much, and am interested in science, but do not want to be a scientist" or public (the target actors), confessing "I know science is very important for our life, but it's not for me"? Science education researchers as well as science education practitioners must be aware of the particular set of value orientations under such a slogan, science for all. Should we serve the target actors or science? We should continue to reflect ourselves, since we need to explain what our research field, *Kagaku Kyoiku,* is to the public as well as the funding agency.

REFERENCES

Anderson, J. E. (2015). *Public policymaking: An introduction* (8th ed.). Stamford, CT: Cengage Learning.

Block, P. (2011). *Flawless consulting: A guide to getting your expertise used* (3rd ed.). San Francisco, CA: Jossey-Bass.

Brewer, G., & deLeon, P. (1983). *The foundations of policy analysis*. Homewood, IL: Dorsey Press.

Clark, T. W., & Kellert, S. R. (1988). Toward a policy paradigm of the wildlife sciences. *Renewable Resources Journal, 7*, 7–16.

DeBoer, G. E. (1991). *A history of ideas in science education: Implications for practice*. New York, NY: Teachers College Press.

Druckman, D. (2000). Frameworks, techniques, and theory: Contributions of research consulting in social science. *American Behavioral Scientist, 43*(10), 1635–1666. Retrieved from https://doi.org/10.1177/00027640021957944

Goodchild, S., & Sriraman, B. (2012). Revisiting the didactic triangle: From the particular to the general. *ZDM Mathematics Education, 44*, 581–585. Retrieved from https://doi.org/10.1007/s11858-012-0449-3

Guba, E. G. (1984). The effect of definitions of policy on the nature and outcomes of policy analysis. *Educational Leadership, 42*(2), 63–70.

Herbart, J. F. (1806). *Allgemeine paedagogik aus den zweck der erziehung abgeleitet*. Goettingen: Johann Friedrich Roewer.

Jacobson, N., Butterill, D., & Goering, P. (2005). Consulting as a strategy for knowledge transfer. *The Milbank Quarterly, 83*(2), 299–321. Retrieved from https://doi.org/10.1111/j.1468-0009.2005.00348.x

Japan Society for Promotion of Science. (2016, September). *Application procedures for grants-in-aid for scientific research-KAKENHI FY2017*. Retrieved from http://www.jsps.go.jp/j-grantsinaid/03_keikaku/data/h29/h29_koubo_fullpage_e.pdf

Japan Society for Science Education. (2016, November). *Welcome to Japan society for science education!* Retrieved from http://www.jsse.jp/jsse/modules/note8/

Kluckhohn, C. K. (1951). Values and value orientations in the theory of action. In T. Parsons & E. A. Shils (Eds.), *Toward a general theory of action* (pp. 388–433). Cambridge, MA: Harvard University Press.

Kluckhohn, F. R., & Strodtbeck, F. L. (1961). *Variations in value orientations*. Evanston, IL: Row & Peterson.

Kraft, M. E., & Furlong, S. R. (2013). *Public policy: Politics, analysis, and alternatives* (4th ed.). London: Sage Publications.

Levin, P. (1997). *Making social policy: The mechanisms of government and politics, and how to investigate them*. Philadelphia, PA: Open University Press.

May, P. J. (2003). Policy design and implementation. In B. G. Peters & J. Pierre (Eds.), *Handbook of public administration* (pp. 223–233). London: Sage Publications.

McComas, W. F., & Murie, K. (2014). Science education. In W. F. McComas (Ed.), *The language of science education* (p. 86). Boston, MA: Sense Publishers.

Ministry of Education, Culture, Sports, & Technology (MEXT). (2013). *Gakko Kyoin Tokei Chosa (Statistics of school teachers)*. Retrieved from http://www.e-stat.go.jp/SG1/estat/NewList.do?tid=000001016172

Ndubisi, F. (1991). Variations in value orientations: Implications for guiding community decision behavior in cross cultural settings. *Journal of Planning Education Research, 11*, 51–65. Retrieved from https://doi.org/10.1177/0739456x9101100107

Ogawa, M. (2013). Toward a 'design approach' to science communication. In J. K. Gilbert & S. Stocklmayer (Eds.), *Communication and engagement with science and technology: Issues and dilemmas* (pp. 3–18). New York, NY: Routledge.

Ogawa, M. (2014, December). *Public engagement with science for democratic participation? Japanese public's preferences of community-level decision making on socio-scientific issues*. Presented as an Invited Talk at the 2014 2nd International History, Philosophy of Science and Teaching Group, Asian Regional Conference, Taipei.

Percy-Smith, B., & Thomas, N. (2010). *A handbook of children and young people's participation: Perspectives from theory and practice*. London: Routledge.

Peterssen, W. H. (1983). *Lehrbuch Allgemeine Didaktik: 3., Ueberarbeitete und erweiterte Auflage*. Muenchen: Ehrenwirth.

Sakamoto, T. (1971). *Kyoiku Kogaku no Genri to Hoho (Educational technology: Principles and methodology)*. Tokyo: Meiji Tosho.

Sapru, R. K. (2004). *Public policy: Foundation, implementation, and evaluation* (2nd ed.). New Delhi: Sterling Publishers Private Limited.

Schoenfeld, A. H. (2012). Problematizing the didactic triangle. *ZDM Mathematics Education, 44*, 587–599. Retrieved from https://doi.org/10.1007/s11858-012-0395-0

Steinbring, H. (2005). Analyzing mathematical teaching-learning situations: The interplay of communicational and epistemological constraints. *Educational Studies in Mathematics, 59*, 313–324. Retrieved from https://doi.org/10.1007/0-387-30451-7_12

Stoecker, R. (1999). Are academics irrelevant? Roles for scholars in participatory research. *American Behavioral Scientist, 42*(5), 840–854. Retrieved from https://doi.org/10.1177/00027649921954561

van Helden, G. J., Aardema, H., ter Bogt, H. J., & Groot, T. L. C. M. (2010). Knowledge creation for practice in public sector management accounting by consultants and academics: Preliminary findings and directions for future research. *Management Accounting Research, 21*, 83–94. Retrieved from https://doi.org/10.1016/j.mar.2010.02.008

Verslius, E., van Keulen, M., & Stephenson, P. (2011). *Analyzing the European Union policy process*. Basingstoke: Palgrave Macmillan.

ABOUT THE AUTHOR

Masakata Ogawa is Professor of Science Education at Tokyo University of Science and Professor Emeritus of Kobe University. He served as President of the Japan Society for Science Education (2004–2008) and President of the East Asian Association for Science Education (2007–2009). His research interests lie around cultural studies and policy studies in science education and science communication.

SHAKHNOZA KAYUMOVA, LUCY AVRAAMIDOU
AND JENNIFER D. ADAMS

16. SCIENCE EDUCATION

Diversity, Equity and the Big Picture

ABSTRACT

In the awaking of the second millennium the world finds itself facing a series of globally interconnected and locally specific challenges: climate change, the quality and security of food, water and air pollution, political instability, refugee crisis, poverty, migration, racisms, sexisms, gender bias, and various other forms of sociopolitical and economic inequalities and oppressions. These challenges create new needs and raise expectations for science education, especially in addressing goals related to diversity, equity, and social change. In this chapter, we define theoretical constructs that feature centrally in efforts to address these goals in science education. Following on that, we offer as examples, brief descriptions of current research projects that we are engaged in, which are framed within these constructs and address goals and issues related to global challenges.

Keywords: diversity, equity, social change, social justice

INTRODUCTION

Significant and rapid development in science, engineering, and technology such as a production and proliferation of mobile and electronic devices, robotics, digital communication and information systems have enabled instant flow, and exchange of various forms of data, work, and capital. These developments and activities have forever transformed the nature and organization of *life*, including human, non-human, and more-than-human life. Despite all the advancements, in the awaking of the second millennium the world finds itself facing a series of globally interconnected and locally specific challenges: climate change, the quality and security of food, water and air pollution, political instability, refugee crisis, poverty, migration, racisms, sexisms, gender bias, and various other forms of sociopolitical and economic inequalities and oppressions. Many of these challenges are arguably intertwined with, if not the direct result of, scientific and technological activities, globalization processes, and associated neoliberal

 | DOI:10.1163/9789004389663_016

economies, values, ethics, and behaviors. As Daniel Faber and Deborah McCarthy (2003) cogently argue, "As a result, the issues of sustainable development and social/environmental justice have surfaced together as in no other period in world history…[both]a part of the same historical process" (p. 40). Science education, which cuts across the persistent divide of natural sciences and social sciences, is potentially uniquely positioned to support new generations' understanding and engagement and address the social, economic and environmental dimensions of these global challenges. Scientific knowledge is vital for effective action, however, a new vision of contemporary societies that are socially inclusive, just and equitable, ought to be at the heart of the account of our efforts in averting a continuous, growing global crisis. Hence, addressing goals related to diversity and equity in science education becomes of crucial importance in light of the big picture of a rapidly changing world.

Concurrently, different reform efforts and related policy documents have been published in various parts of the world – e.g., Next Generation Science Standards (NGSS Lead States, 2013) in the U.S. context, Responsible Science and Innovation in Europe (European Commission [EC], 2018). The question is: *how aligned are these recommendations for reform in science education with current global challenges*? An examination of these recommendations shows that there exists a discrepancy between contemporary global challenges and reform efforts, which focus on goals related to economical competition instead of goals related to students' well-being and social justice.

One such example can be found in a report published in 2012 by the European Federation of National Academies of Sciences and Humanities entitled, *A Renewal of Science Education in Europe*. The report places emphasis on the importance of 'inquiry-based education' as a means for supporting students' interest in science and advancing their scientific literacy. However, these recommendations proved to be disconnected from the nature of the various and diverse socio-political contexts of Europe, during times when a continuing economic turbulence has been resulting into new social realities (youth unemployment) population demographics (refugees, migrants) and inequalities.

Another example comes from a more recently published report by the European Commission called, *Science Education for Responsible Citizenship* (EC, 2015), which offers a 21st century vision for science for society within the broader European agenda. The report places emphasis on the process of aligning research and innovation to the values, needs and expectations of society, referred to as "responsible research and innovation". Four of the six main objectives of the report are summarized into the following, as an example:

- Science education should be an essential component of a learning continuum for all, from pre-school to active engaged citizenship.
- Science education should focus on competences with an emphasis on learning through science and linking science with other subjects and disciplines.

- The quality of teaching, from induction through pre-service preparation and inservice professional development, should be enhanced to improve the depth and quality of learning outcomes.
- Collaboration between formal, non-formal and informal educational providers, enterprise and civil society should be enhanced.

This document is framed around the challenge of increasing the number of students interested in scientific careers because Europe faces a shortfall of scientists. Unlike previously published reports, this one offers a more comprehensive vision of reform in science education. However, the reform recommendations fall short in two distinct ways: (a) they are atheoretical as they are not rooted in any learning theories, paradigms, or frameworks; and, (b) they do not reflect how global challenges have shaped this vision for science for society and therefore lack attention to the need for more inclusive, equitable, and just societies. The word 'theory' is not used once in the entire 88-pages long document with no references made to research on science teaching and learning. Moreover, no attention to issues related to diversity, in terms of ethnicity, language, gender, religion, is paid. In addition, a discussion of the role of the discourse of science or how students talk and write science is missing as well as the role of the local context and the involvement of families in children's education. Lastly, goals related to reducing inequality, promoting social change and social justice are completely absent.

These issues, among others related to scientific practices as well as teacher education, were discussed in an international week-long workshop entitled, "Science Education for a Changing World", which was funded by the Lorentz Center in the Netherlands and led by science educators Lucy Avraamidou, Lynn Bryan, Sibel Erduran, and Martin Goedhart. The workshop brought together a group of 25 researchers from 13 different countries (Canada, Chile, China, Cyprus, Korea, Lebanon, Luxembourg, Netherlands, Spain, Sweden, Turkey, UK, and USA). The driving question of the workshop was: *How can scientific knowledge be utilized through education to address global challenges?* Four working groups were formed around four main research areas:

- The role of science education and teacher preparation in society
- Nature of science and social justice
- Identity, equity, and gender
- Scientific practices and argumentation

Each of these groups worked to develop a set of recommendations for reform in science education, in light of the current geosociopolitical context and with the purpose of addressing global challenges. In what follows, we share only a snapshot of the recommendations put forward by the third group who explored issues related to equity, identity and gender. In addition to the authors, the group consisted of the following researchers, to whom we are immensely grateful for providing ideas and advice on specific sections of this chapter: Anna Danielsson, Johanna Larsson, Sara Wilmes and Sonya Martin.

ATTENTION TO EQUITY

Achieving equitable learning opportunities and outcomes for, and among, all students is one of the most pressing challenges facing science education. Conceptions of equity is one of the most ubiquitous, and yet undertheorized concepts in science education (Dimick, 2012). Most often, equity in science education is associated with notions of distribution, access, opportunity, empowerment, and broadening participation. Teachers and teacher educators committed to social justice issues examine inequities in structural, cultural, and curricular organization of schools, namely racist, classist, sexist, and gendered discourses and practices, as one of the main reproductive systems of social and economic disparities in the society, and work to dismantle them by "empowering" students socially, politically, and academically. Elsewhere, Shakhnoza Kayumova, Chad McGuire, and Susanne Cardello (2018), posit that equity and justice efforts limited to notions of broadening participation, achievement, and access, might be insufficient to transform systems of domination and to understand complexities inherent in globally complex and locally specific sociopolitical, economic, spatial, and environmental disparities. If anything, Kayumova, McGuire and Cardello argue, that empowerment must be about supporting and legitimizing children's, families', and diverse communities' existing knowledge bases, skills and power to contribute to decision making and their rights to respond, response-ability, to the issues and topics discussed about, and on behalf of them and their children.

Shakhnoza Kayumova and Deborah Tippins (2018) underline the importance of expanding the notions of equity and justice by examining the dominant epistemological and ontological assumptions in the cultural and curricular organization of science education (e.g., how nature-culture relations are taken up in the current science standards), research methods and methodologies employed and understanding and framing of the justice issues, which left unexamined, might risk perpetuating the very inequities that equity-oriented scholarship seeks to transform. For example, much of the research that is considered to be an equity oriented might become complicit, if not inadvertently participating, in perpetuation of deficit discourses about student populations (e.g., girls, cultural and linguistic minorities, Black and Brown children) and communities when they do not question and examine taken for granted assumptions and methodologies in studying these populations. In the context of the U.S., Shakhnoza Kayumova, Elif Karsli, Martha Allexsaht-Snider, and Cory Buxton (2015) have documented that language identities and repertoires of emergent bilingual and multilingual children become a part of myriad axes of marginalization intertwined with their social positioning (e.g. race, class, and gender) given that science classrooms operate on certain dominant norms of English language usage. This is also true for many European counties, which are seeing a new surge of immigrant and refugee populations (United Nations High Commissioner for Refugees, 2016).

Consequently, dominant English language based performativities of students operate as yet another academic normativity along which notions of cognition, intelligence, capability, and other forms of cognitive assumptions and deficit discourses become conceived and against which students from culturally and linguistically diverse, communities and their differential learning performances are judged (Kayumova & Ji, 2018).

Science education research in mainstream channels has been dominated by established binary-categories (e.g., boys vs. girls, native vs. immigrants) and compare/contrast models that both establish artificial commonalities and use differences as a way to construct, justify and reproduce systems of power in relation to science education. A paradigm shift is needed where researchers adopt multiple sociocultural and diverse theoretical lenses, rooted in sociocultural, critical, and radical approaches, epistemologies and ontologies (i.e., critical race theory, radical feminism, anticolonial theory), and/or consider bringing theoretical perspectives into conversation with one another in novel ways that when examining learners participation in, and relationship to science, as an alternative prism that allows us to look into students' lives and to address goals related to equity, diversity, and power differentials. These goals are directly linked to the following key questions, which we argue ought to be addressed by contemporary research in science education:

- How do we address science teaching, learning and research challenges in increasingly diverse contexts that are shaped by migration, multiculturalism and resulted diversity?
- What approaches can we use so that the social, political, historical and educative aspects are considered in our research? How do we capture the complexities of these issues?
- What new questions and approaches can we use so that we do (a) not reproduce the inequities in science teaching and learning contexts; and, (b) advance equity in science teaching and learning contexts? Further, how are we (re)defining equity in science?
- What are the ways that power is reproduced in science education research spaces? What could we do to elucidate, unpack and disrupt these power dynamics?

One way of understanding the different approaches to science education that center diversity is through "anti-oppressive education". Kevin Kumashiro (2000) has categorised four kinds of anti-oppressive education which focuses on the following: (a) education for the other; (b) education about the other; (c) education that is critical of privileging and othering; and, (d) education that changes students and society. As science education researchers and science teachers, we need to be knowledgeable about the plurality of such approaches and knowing when to use what and the affordances and constraints of the different approaches. We also emphasize the need to identify and employ notions of science as socially constructed, in research and in teaching, in ways that honor the social, cultural, and historical contexts that have influenced and that continue to contribute to the nature of science.

The science education community also needs to consider the diverse ways of knowing, being, describing the natural world among diverse socio-cultural groups, which might be different from the dominant ways in which hegemonic science education understands and explains it. For instance, aboriginal and indigenous communities' ways of relating to the nature and culture defy traditional nature-culture binaries by which classical science operates. Utilizing indigenous and aboriginal theories in understanding diverse students and their family's relations to dominant science perspectives, can help science education to be more inclusive of the cultures and groups traditionally left out of hegemonic science. Moreover, it will also help us to overcome a dichotomy of what appears to be a Western and indigenous knowledge systems, and support the equity efforts.

Specific constructs feature centrally in addressing these goals and essentially in adopting a science *in* society instead of a science *for* society paradigm. These are defined next.

Diversity

Diversity refers to the inclusion of different types of people. Science education often emphasizes ethnicity/race, gender and socioeconomic status/social class as important social constructs through which one can examine the issues of diversity. However, we must also consider the various ranges and constructs of diversity that also influence science teaching, learning and research such as dis/ability, linguistic, sexuality, gender identity, political, religious, geographical, age [etc.]. Our approaches to diversity need to take into consideration the potentialities of this also about using different theoretical lenses in order better to elucidate the ways in which inclusion and exclusion, of peoples and knowledges, happens within science education. For example, using critical whiteness studies to unpack power structures embedded in science education and other "norms" that are taken for granted in science-related contexts.

Equity

In science education, equity is often taken to mean broadening participation, achievement, and/or access. A reconceptualisation of equity is needed, that takes into account how science education is intrinsically linked to power, culture, epistemology, and identity. Doing well on achievement measures does not necessarily, by itself, imply equity. Similarly, broadening participation and access are not enough to transform systems of domination. To this end, we underline the importance of examining issues of power and equity within the structural, cultural, and curricular organization of science education, teaching and learning. We also propose considering and recognizing how as researchers our practices of knowledge production and dissemination are consequential to equity efforts, and if not afforded internal scrutiny and continuous reflexivity, might become complicit in the reproduction of the very inequities we seek to transform.

Intersectionality

Intersectionality is used to highlight the inseparability of categories of social differences such as race, gender, sexuality, ethnicity, and nationality (Crenshaw, 1989). Intersectionality calls for an emphasis on the systemic power dynamics that arise as multiple dimensions of social interactions across individual, institutional, cultural and societal spheres of influence and is used to address inequality and discrimination in relation to engagement with science. As a term, coined by Kimberlé Crenshaw in 1989 to counter the disembodiment of Black women from law, intersectionality captured the inadequacy of legal frameworks to address inequality and discrimination resulting from the ways race and gender traversed to shape the employment experiences of Black women. Intersectionality scholarship can assist in revealing key concerns; namely, the existence of a multiplicity of power-relations, of the interlacing of social dynamic categorizations based on gender, race, and ethnicity and other social constructs.

Identity

Using 'subjectivity' or other theoretical constructs might be better to avoid falling in the everyday meaning of the word identity, and emphasizing understanding identity construction through social participation and lived experience (Avraamidou, 2014, 2018). We can also consider identity as it relates to the ways in which students position themselves with relation to others: "Who one is and who one desires to be at any given moment is always under negotiation and is contingent upon the resources on has access to and the social, cultural, and historical context in which one seeks to author oneself against the expectations of others" (Holland, Lachiotte, Skinner, & Cain, 2001, p. 120). For instance, students' diverse ways of being and knowing are structured by their social and cultural experiences of race, class, gender, and sexuality. These social constructs play crucial role by which students are recognized or misrecognized as they perform dominant science practices. Identity can also be used as a lens to understanding science teachers' learning and development.

Creativity

Creativity is becoming a more popular construct in science especially in association with notions of innovation and scientific entrepreneurialism. However, narrow conceptions of creativity in science only serve to reproduce the inequities and hegemony that currently dominates science teaching, learning and research. Adams (this volume) calls for an expanded view of creativity that both challenges existing scientific epistemologies and centers addressing global challenges from a critical and socially just perspective. In science, there is a need to engage in, what Sylvia Wynter (2003) calls "epistemic disobedience" in order to "imagin[e] alternative choices and futures," especially for the most vulnerable people and places (susceptible to climate change, food insecurity, economic and/or political instability, etc.). By presenting a notion of scientific creativity that centers wellness, equity and social justice and emphasizes the creation of new knowledge, products and artifacts towards transforming the world for the greater good, we would have a STEM teaching, learning and research framework that would promote the collective well-being of the planet.

BOLD VISIONS, CURRENT PROJECTS

The adoption of such theoretical constructs calls for a re-examination of the methodological approaches that we use to respond to research questions related to equity, diversity, and inclusion. Below, we offer a brief list of such methods, which we use in our own research practices, following with examples of current projects:

- Use of life history and narrative approaches that recognize the contextual nature and continuity of experiences with science across time and contexts (Avraamidou, 2018).
- Methodological approaches that value the co-creation of knowledge through centering learners' lived experiences and voices, such as participatory action research and dialogic data collection methods (Strong, 2016).
- Use of methodological approaches that seek to disrupt boundaries as for example, the use of plurilingualism as an analytical lens to view communicative resources versus a lens of multi-or bilingualism.
- Use of methodological approaches that forefront research-to-practice so that we are learning about diversity and equity in learning environments as we also create learning environments that value diversity and learning.
- Use of methods and methodological approaches that will allow us to position our participants as critical research partners (Kayumova, 2018), by blurring the hierarchical boundaries of researcher-researched.

In what follows, we offer as examples, brief descriptions of current research projects that we are engaged in, which are framed within the constructs we defined above and use these methodological approaches.

Project 1: Beyond Gender – Intersectional Identities as a Lens to Examining Women's (Non-)Participation in Science, *Lucy Avraamidou*

Framed within the construct of science identity, this project presents a qualitative multiple-case study on the under-representation of women in scientific careers. Grounded within a combined theoretical framework of intersectionality, identity, and narrative, the project aims at examining the lived experiences and educational pathways of eight purposefully selected female scientists who work in various STEM-related fields in the Netherlands. In exploring the participants' life histories, I pay special attention to the obstacles and barriers they faced throughout their studies and careers. The design of the study is situated within literature that has used gender as a theoretical construct to examine the under-representation of women in science, which, however, as I argue, provides a limited and single-sided understanding of why women are underrepresented in the sciences. The analysis of the participants' life-histories in relation to science are expected to produce knowledge that that can be used as input for: (a) carrying out large-scale empirical research; (b) rethinking institutional change through intersectionality; and, (c) designing inclusive and equitable workplaces. From a theoretical perspective, the goal of this project is to put an argument forward about the use of relational identities and intersectionality as a lens to examining women's career trajectories in STEM. Such a lens provides a more comprehensive, diverse, and multidimensional frame while it pays attention to the ways in which culture, religion, ethnic status, class, family status, sexuality, and race, intersect with science identity. This deeper and comprehensive understanding of how different identities, social markers, and life-experiences might influence women's science career trajectories allows us to shift the focus away from a deficit model where gender alone is considered.

Project 2: ILETES-Informal Learning Environments and Teacher Education for STEM, *Jennifer D. Adams*

This research examines the relationship between teacher education and learning to teach vis-a-vis the racialized and politicized structures of urban schools. As a collective of researchers, we specifically address questions around the relationship between teachers' (ongoing) learning and understandings, (re)definitions and enactments of informal science education in formal classrooms and the reiterative relationship between practice and identities. Emerging from this project are understandings of Black and Latino teachers, as racialized bodies, in science classrooms and they ways that they either take up, resist and/or transform discourses around students of African and Latinx descent and STEM. Using a framework of racial storylines (Nasir, Snyder, Shaw, & Miraya Ross, 2012) and seeking events that charge identity (Saldanha, 2006), this project learns how messages about race present in schools and how these discourses influenced teaching enactments and identity. Within the different schooling contexts teachers recreated meanings of informal science education and enacted teaching in relation to how they viewed themselves vis-à-vis their students. Their goals for creating STEM learning experiences were shaped and enacted around STEM futures that they imagined for their students and to counter the prevailing deficit discourses around students of color and STEM.

Project 3: Power, knowledge and identity in science and technology classrooms: Teachers' enactments of disciplinary discourses as establishing inclusion and exclusion, *Anna Danielsson, Maria Berge, Malena Lidar, Åke Ingerman, Leif Östman and Heather Mendick*

> The aim of the project is to explore how knowledge and power are mutually constituted in meaning- making processes in physics and technology classrooms, in relation to the teachers' enactment of a disciplinary discourse. By analysing the interrelationship of knowledge and power, the project seeks to contribute to the knowledge about processes of inclusion and exclusion in physics and technology. The empirical data consists of video-recordings of physics and technology lessons in five different schools (lower and upper secondary school) and interviews with upper secondary school students. The project is multi-theoretical; the selected publication makes use of a combination of pragmatism and a Foucauldian power perspective, other sub-studies make use of positioning theory or a social semiotic perspective. While the project is mostly focused on the co-constitution of power and knowledge in classroom interactions, there is also a strand exploring students' identity constitution.

Project 4: Analyzing the nexus between advantaged social positioning and science identity development among ethno-linguistically diverse learners, *Shakhnoza Kayumova*

> The United States has a large population of children from diverse ethnic-racial and linguistic backgrounds. Over 5.4 million of these students are identified as English learners (ELs), with 4.4 million being Spanish-speaking (U.S. Census Bureau, 2012). Currently, the increasing EL student population (NCES, 2015); growing demands for science-, technology-, engineering-, and mathematics-related (STEM) jobs; mounting disparities in Els' science achievement outcomes, and consequently their substantial underrepresentation in the STEM workforce, constitute intersecting sources of inequality (Landivar, 2013). The aim of this longitudinal mixed-method study is to analyze how ethno-linguistically diverse learners' language identities affect their science identity development using positioning theory as a primary lens. Filling an important gap in the literature, the study brings new empirical evidence related to the intersection of language-based perceptions and science identity development. The project is supported by National Science Foundation Early Career Grant.

INSTEAD OF A CONCLUSION

Collectively, the examples of these projects are grounded in the assumption that in order to make meaningful and transformative changes in science education we need to consider the characteristics of an era of globalization. Such characteristics are the diverse culture and racial origins of students, and language barriers caused by dramatic social changes happening in the world, such as border crossing and migration. These social changes call for a re-visioning of science education that addresses equity and social justice, and which involves a conceptualization of science that goes beyond binary oppositions, beyond borders and boundaries, given that borders are limiting and considering that the boundaries of science are constantly shifting. In doing so, we ought to acknowledge that learning environments are nowadays hybrid and multi-contextual, and learning experiences are complex and multi-dimensional. As researchers interested in issues related to social justice and social change, we problematize various aspects of science education and we propose new ways of conceptualizing science education shaped at the intersection of personal, local and global realities. At the heart of the account of this proposition is an examination of the role of science education in addressing global challenges, in the context of current sociopolitical contexts as we question the taken-for-granted spaces, boundaries of science, and traditional discourses.

As our concluding remark, we would like to leave our readers with a quote from a feminist scientist Karen Barad: "Justice, which entails acknowledgment, recognition, and loving attention, is not a state that can be achieved once and for all" (2007, p. X). Any prescriptive account of justice can be essentialized and reproduce the very injustices. As Barad (2007) rightly argues:

> There is only the ongoing practice of being open and alive to each meeting, each intra-action, so that we might use our ability to respond, our responsibility, to help awaken, to breathe life into ever new possibilities for living justly. The world and its possibilities for becoming are remade in each meeting. (p. x)

REFERENCES

Avraamidou, L. (2014). Studying science teacher identity: Current insights and future research directions. *Studies in Science Education, 50*, 145–179. Retrieved from https://doi.org/10.1080/03057267.2014.937171

Avraamidou, L. (2018). Stories we live, identities we build: How are elementary science teachers' science identities shaped by their lived experiences. *Cultural Studies of Science Education*. Retrieved from https://doi.org/10.1007/s11422-017-9855-8

Barad, K. (2007). *Meeting the universe halfway: Quantum physics and the entanglement of matter and meaning*. Durham, NC: Duke University Press.

Crenshaw, K. (1989). Demarginalizing the intersection of race and sex: A Black feminist critique of antidiscrimination doctrine, feminist theory, and antiracist politics. *University of Chicago Legal Forum, 139*, 139–167.

Dimick, A. S. (2012). Student empowerment in an environmental science classroom: Toward a framework for social justice science education. *Science Education, 96*, 990–1012. Retrieved from https://doi.org/10.1002/sce.21035

European Commission. (2015). *Science education for responsible citizenship*. Brussels: Directorate-General for Research and Innovation, Science with and for Society. Retrieved from http://ec.europa.eu/research/swafs/pdf/pub_science_education/KI-NA-26-893-EN.pdf

European Commission. (2018). *Responsible research and innovation*. Retrieved from https://ec.europa.eu/programmes/horizon2020/en/h2020-section/responsible-research-innovation

Faber, D. R., & McCarthy, D. (2003). Globalization and the struggle for ecological democracy: Linking sustainability and environmental justice. In J. Agyeman, R. D. Bullard, & R. Evans (Eds.), *Just sustainabilities: Development in an unequal world*. Cambridge, MA: MIT Press.

Holland, D., Lachicotte, W., Skinner, D., & Cain, C. (1998). *Identity and agency in cultural words*. Cambridge, MA: Harvard University Press.

Kayumova, S., Karsli, E., Allexsaht-Snider, M., & Buxton, C. (2015). Latina mothers and daughters: Ways of knowing, being, and becoming in the context of bilingual family science workshops. *Anthropology & Education Quarterly, 46*, 260–276. Retrieved from https://doi.org/10.1111/aeq.12106

Kayumova, S., McGuire, C., & Cardello, S. (2018). From empowerment to response-ability: Rethinking socio-spatial, environmental justice, and nature-culture binaries in the context of STEM education. *Cultural Studies of Science Education*, 1–25. Retrieved from https://doi.org/10.1007/s11422-018-9861-5

Kayumova, S., & Tippins, D. (2018). Obsessed with accountability? Science teachers under the microscope. In L. Bryan & K. Tobin (Eds.), *Thirteen questions in science education* (pp. 91–101). New York, NY: Peter Lang.

Kumashiro, K. V. (2000). Toward a theory of anti-oppressive education. *Review of Educational Research, 70*(1), 25–53. Retrieved from https://doi.org/10.2307/1170593

Landivar, L. C. (2013). *Disparities in STEM employment by sex, race, and hispanic origin* (American Community Survey Report, ACS-24). Washington, DC: U.S. Census Bureau.

Nasir, N. S., Snyder, C. R., Shaw, N., & Ross, K. M. (2012). Racial storylines and implications for learning. *Human Development, 55*, 285–301. Retrieved from https://doi.org/10.1159/000345318

NGSS Lead States. (2013). *Next Generation Science Standards: For states, by states*. Washington, DC: The National Academies Press.

Saldanha, A. (2006). A geophilosophy to come. *Theory and Event, 9*(4), 1–8. Retrieved from https://doi.org/10.1353/tae.2007.0012

Strong, L. (2016). The intersection of identity, culture and science engagement. *Cultural Studies of Science Education, 11*, 379–385. Retrieved from https://doi.org/10.1007/s11422-015-9680-x

United Nations High Commissioner for Refugees. (2016). *Missing out: Refugee education in crisis*. Retrieved from http://uis.unesco.org/sites/default/files/documents/missing-out-refugee-education-in-crisis_unhcr_2016-en.pdf

Wynter, S. (2003). Unsettling the coloniality of being/power/truth/freedom: Towards the human, after man, its overrepresentation: An argument. *CR: The New Centennial Review, 3*(3), 257–337. Retrieved from https://doi.org/10.1353/ncr.2004.0015

ABOUT THE AUTHORS

Lucy Avraamidou is a Rosalind Franklin fellow and Associate Professor of Science Education at the Institute for Science Education and Communication at the University of Groningen in the Netherlands. Her research is associated with theoretical and empirical explorations of teachers' learning and development with the use of interpretive approaches in the area of science identity and factors that impact its development.

Shakhnoza Kayumova is currently an Assistant Professor in the Department of STEM Education and Teacher Development at the University of Massachusetts Dartmouth and research scientist at Kaput Center for STEM Education Research and Innovation. Her interdisciplinary work focuses on the issues related to equity, diversity, and language issues in science education.

Jennifer D. Adams is an associate professor and Canada Research Chair in Creativity and STEM at The University of Calgary where she holds a dual appointment in the Department of Chemistry and Werklund School of Education. Her research focuses on the intersection between creativity and STEM teaching and learning in postsecondary contexts. She has expertise in STEM teaching and learning in informal science contexts, place-based and environmental education and critical perspectives in science.

INDEX

D

www.ingramcontent.com/pod-product-compliance
Lightning Source LLC
Chambersburg PA
CBHW072242240926
51846CB00012B/147
9789004389649